STRUCTURES OF PERMANENT MAGNETS

STRUCTURES OF PERMANENT MAGNETS
GENERATION OF UNIFORM FIELDS

Manlio G. Abele

A WILEY-INTERSCIENCE PUBLICATION

JOHN WILEY & SONS, INC.

New York • Chichester • Brisbane • Toronto • Singapore

This text is printed on acid-free paper.

Copyright © 1993 by John Wiley & Sons, Inc.

All rights reserved. Published simultaneously in Canada.

Reproduction or translation of any part of this work beyond that permitted by Section 107 or 108 of the 1976 United States Copyright Act without the permission of the copyright owner is unlawful. Requests for permission or further information should be addressed to the Permissions Department, John Wiley & Sons, Inc., 605 Third Avenue, New York, NY 10158-0012.

Library of Congress Cataloging in Publication Data:
Abele, Manlio G., 1920–
    Structures of permanent magnets: generation of uniform fields/ Manlio G. Abele.
        p.   cm.
    Includes bibliographical references and index.
    ISBN 0-471-59112-2 (alk. paper)
  1. Permanent magnets. 2. Permanent magnets—Industrial applications. 3. Magnetic materials. I. Title.
QC757.9.A24 1993
621.34—dc20   92-35449

Printed in the United States of America

10 9 8 7 6 5 4 3 2 1

To Audrey

CONTENTS

Preface   xi

1. Magnetostatic Fields   1

         Introduction   1
- 1.1   Basic Equations   1
- 1.2   Differential Operators in Orthogonal Curvilinear Systems   7
- 1.3   Equipotential Surfaces   12
- 1.4   Solution of the Field Equations   18
- 1.5   Multipoles   25
- 1.6   Energy of a Magnetostatic Field   31
- 1.7   Magnetic Materials   33
        References   38

2. Basic Geometries of Permanent Magnets   39

         Introduction   39
- 2.1   One-Dimensional Magnet   39
- 2.2   Uniformly Magnetized Cylindrical Magnet   42
- 2.3   Hollow Cylinder of Magnetic Material   46
- 2.4   Example of Field Confinement   52
- 2.5   Spherical Magnet   55
- 2.6   Magnetic Material with Nonzero Susceptibility   62
        References   66

3. Generation of a Uniform Field in Prismatic Cavities   67

         Introduction   67
- 3.1   Existence of a Uniform Field Solution   68
- 3.2   Structures of Uniformly Magnetized Prisms   76
- 3.3   Yoked Two-Dimensional Prismatic Structures   89
- 3.4   Yokeless Two-Dimensional Prismatic Structures   96
- 3.5   Invariance Properties of Yokeless Structures   101
        References   106

4. Yokeless Magnets — 107

Introduction 107
4.1 Two-Dimensional, Single Layer Yokeless Magnets 107
4.2 Magnetic Structures with Similar Polygonal Contours of External and Internal Boundaries 116
4.3 Figure of Merit of Yokeless Magnets 125
4.4 Figure of Merit of Multilayered Magnets 134
4.5 Three-Dimensional Yokeless Magnets 138
References 144

5. Yoked Magnets — 145

Introduction 145
5.1 Two-Dimensional Yoked Magnetic Structures 145
5.2 Square Cross-Sectional Cavity 154
5.3 Position of $\Phi = 0$ Line within the Magnet Cavity 163
5.4 Closed Curvilinear Boundary of the Magnet Cavity 171
5.5 Generation of a Unidirectional Flux of $\vec{B}$ 177
5.6 Three-Dimensional Yoked Magnets 182
References 188

6. Hybrid Magnets — 189

Introduction 189
6.1 Comparison of Yoked and Yokeless Structures 189
6.2 Figures of Merit of Hybrid Structures 200
6.3 Insertion of Thin Layers of High Magnetic Permeability Materials in a Magnetic Structure 208
6.4 Three-Dimensional Magnets 216
6.5 Magnets with Rectangular Prismatic Cavities 223
6.6 Closure of the Cavity in Traditional Permanent Magnets 231
References 241

7. Materials with Linear Magnetic Characteristics — 242

Introduction 242
7.1 Field Computation in Magnetic Structures with Small, Nonzero Susceptibility 242
7.2 Singularities in Multilayer Magnets 249
7.3 Yoked Magnets 254
7.4 Field Distortion in a Two-Dimensional Square Cross-Sectional Magnet 257
7.5 Field Computation in the Presence of Ferromagnetic Media with Linear Magnetic Characteristics 265
7.6 Field Inside a Ferromagnetic Medium 269
References 274

8. Open Magnetic Structures   275

Introduction   275
8.1   Cylindrical Magnet of Finite Length   275
8.2   Open Yokeless Spherical Magnet   287
8.3   Prismatic Yokeless Magnets of Finite Length   294
8.4   Square Cross-Sectional Yokeless Magnet   298
8.5   Image Method of Field Computation in Yoked Magnets   309
8.6   Boundary Integral Equations Computational Method   314
References   322

9. Design Considerations   323

Introduction   323
9.1   Comparison of Designs with Different Magnetic Materials   323
9.2   Compensation of Field Distortion Caused by Magnet Opening   336
9.3   Compensation of Field Distortion with Ferromagnetic Materials   345
9.4   Compensation of Field Distortion Caused by Nonzero Magnetic Susceptibility   351
9.5   Compensation of Magnetization Tolerances   355
9.6   Concluding Remarks   359
References   360

Appendix I.   Units   361

Appendix II.   Legendre Polynomials of the First Kind   362

Appendix III.   Complete Elliptic Integrals   363

Appendix IV.   Coefficients of the Spherical Harmonics for the Yokeless Square Cross-Sectional Magnet   364

Index   369

PREFACE

Magnetic fields that do not vary with time can be generated either by steady electric currents or by permanent magnets. The selection of the appropriate source is normally dictated by the specific application. Prior to the discovery of new high energy product magnetic materials, the use of permanent magnets had been limited to small electric machines and electronic devices. Large magnets for heavy industrial applications are normally electromagnets, and extremely strong magnetic fields are generated with superconductive magnets, where the design is not limited by losses in the electric conductors and by saturation of the ferromagnetic materials.

An attractive feature of permanent magnets is obviously the absence of external sources of electric power. However, high energy product magnetic materials are expensive, and the advantages of a permanent magnet technology are lost unless efficient designs of large and powerful magnets with a minimum amount of magnetic material are developed.

The designer of a permanent magnet has two basic ingredients to generate the magnetic field in a given region: permanently magnetized materials that provide the energy of the magnetic field and ferromagnetic materials that make it possible to concentrate the field within the region of interest. Ferromagnetic materials are characterized by strongly nonlinear magnetic characteristics, which automatically relegate the design problem to a category of nonlinear problems whose solutions require the use of numerical methods. In many instances the magnet design involves a trial and error procedure requiring a number of iterations before a satisfactory solution is achieved. The reduction of the design problem to the level of a numerical integration of the field equations has been one of the reasons for the lack of progress in the development of a general magnet design theory.

In a conventional design approach the desired configuration of the magnetic field is achieved using the ferromagnetic components and, in particular, the pole pieces. The basic equations of a magnetostatic field are used to design the shape of these components and then the magnet is analyzed on the basis of an equivalent magnetic circuit of lumped elements characterized by their magnetic reluctance. The lumped elements correspond to a largely arbitrary classification of the components of the magnet as air gaps, pole pieces, and yokes. In the magnetic circuit model, the magnetic material assumes the role of a power supply in an electric circuit, and the analysis of the circuit is used to optimize the design by adjusting the values of the magnetic reluctances of its elements to maximize the energy product of the magnetic material.

The conventional or lumped elements design approach fails when a magnet must be considered as a continuous distribution of magnetized material. Such a situation arises in a design aimed at taking full advantage of the properties of modern high energy product rare earth materials. The main reason for the inadequacy of conventional methods to generate efficient designs with these materials stems from the striking differences of their properties compared to the older magnetic alloys, like Alnico.

The developments of rare earth permanent magnets have impacted the magnet technology by introducing several important factors in the applications as well as in the theoretical aspects of the design. First, their high energy product has made it possible to generate strong fields with magnetic structures much smaller than conventional magnets. Second, the property of the rare earth magnets of being practically impervious to strong external demagnetizing forces has opened new fields of applications of permanent magnets. From a theoretical point of view, the quasi-linear demagnetization characteristics of the rare earth magnets have made it possible to analyze two-dimensional and three-dimensional magnetic structures with exact mathematical models. The understanding of the properties of the exact solution of the field equations has led to the development of magnetic structures so strikingly different from conventional magnets that names like "magic ring" and "magic sphere" were coined to identify them.

The integration of the field equations in assigned geometries of magnetic materials provides the solution to what is called a *direct problem*. The designer is faced with the *inverse problem* of determining the geometry of magnetic materials capable of generating the assigned field. The major difficulty in achieving this goal is theoretical. A magnetostatic field is described by elliptic equations whose solutions are indeterminate, i.e., the field in an assigned region can be generated by an infinity of magnetic structures. A theory is needed to provide the designer with a quantitative way of selecting the particular solution of the inverse problem suitable for a design requirement.

A first step in the development of a design theory is the methodology presented in this book, which summarizes the work of the author on the theory of magnetic structures with linear characteristics. Structures of magnetized materials of zero magnetic susceptibility in the presence of ideal media of infinite permeability are derived with exact mathematical solutions for arbitrarily assigned geometries of the magnet cavity. The solutions are extended to the type of magnetic characteristics exhibited by rare earth permanent magnets, thereby providing the designer with an explicit design methodology for generating complex magnetic structures.

The subject of this book is basic magnet design, rather than engineering and fabrication problems. The book addresses the category of magnetic structures designed to generate a uniform field. Though this may appear to be a rather stringent limitation, the generation of a uniform field covers the vast majority of practical applications. An application which poses a severe challenge to a designer is the development of magnets for nuclear magnetic resonance (NMR) imaging in medicine, not only because of the high degree of field uniformity, but also because of field strength and magnet dimensions. Field uniformities better than 10^{-4} are required in volumes of linear dimensions dictated by the size of a human body cross-section. An important aspect of the optimization of structures for NMR applications is the development of magnets yielding values of the intensity of the magnetic field of the order of the remanence of the magnetic material.

The material is presented in three parts. First, a review of the fundamental concepts of magnetostatics and the analysis of solutions of magnetostatic problems in simple geometries are presented to introduce some of the basic properties of permanent magnets. The second part formulates the design of closed magnetic structures and defines the basic categories of these magnets. The third part analyzes the two major aspects of practical applications: opening of the magnetic structures and demagnetization characteristics of real magnetic material.

The reader is assumed to possess the basic knowledge of magnetostatics provided by engineering or physics courses in electricity and magnetism. For the benefit of readers who may be less familiar with the mathematical formulation of magnetostatic problems, the first chapter presents the derivation of the fundamental equations which are used throughout the text.

Chapter 2 analyzes the properties of the solutions of the field equations in simple geometries. The formulation of the inverse problem is the subject of Chapter 3, which presents the theory of structures composed of uniformly magnetized polyhedrons and introduces the basic design criteria. Chapters 4, 5, and 6 are devoted to a discussion of the basic categories of permanent magnets: yokeless structures where the field confinement is achieved without the use of high magnetic permeability materials and yoked as well as hybrid structures.

Whenever possible, the problems are formulated and solved in graphic form with vector diagrams. This approach eliminates cumbersome mathematical developments, and it conveys in a direct and visual way the design principles and the properties of the magnetic structures. Moreover, the graphic solution clearly conveys the insight of the sequence of steps of the design procedures, especially when a complex magnetic structure is involved.

A key aspect of the design approach introduced in Chapter 3 is the indeterminacy of the solution of the design of a permanent magnet. In principle, the uniqueness of the solution of a design problem can never be achieved, no matter how many criteria and constraints are imposed by the designer. However, as shown in this book, a logical sequence of criteria can be established that uniquely determines the basic categories of permanent magnets and, within each category, achieves a unique solution of the essential geometrical and magnetic characteristics. The design can then be developed in such a way that the conceptual problem of indeterminacy is relegated to details that play a minor role in the magnet configuration.

The effects of magnetic characteristics such as nonzero susceptibility of magnetic materials and finite permeability of ferromagnetic materials are discussed in Chapter 7. Chapter 8 is devoted to the analysis of the field distortion caused by the opening of a magnetic structure which is required in a practical magnet to gain access to the cavity of the magnet. Finally, Chapter 9 shows how the design criteria apply to the generation of a uniform field in a given region by means of magnetic materials of vastly different values of the energy product like ferrites and rare earth alloys. The examples developed in Chapter 9 are used to illustrate techniques designed to compensate the effects of practical constraints and real characteristics of available materials. In particular, the development and the limitations of shimming techniques are briefly discussed in this chapter.

The author is deeply indebted to his friend and colleague Dr. Henry Rusinek for his invaluable assistance in the writing of the book and in the preparation of the manuscript. Dr. Rusinek has developed the computational programs and performed the numerical calculations of the geometries and properties of the examples of magnetic structures.

Some of the material presented in the book is based on the work performed at the Department of Radiology of New York University. It gives this author great pleasure to acknowledge the support and interest of Professor and Chairman Norman E. Chase in the research on permanent magnet technology. The research activity has been partially funded by Esaote Biomedica of Italy as part of a research and development program on NMR imaging. The author wishes to thank Drs. Andrea Oberti and Franco Bertora for their cooperation and for providing the results of their computational and experimental work developed at Esaote Biomedica.

The author wishes to express his appreciation and gratitude to many colleagues, both in this country and abroad, who have participated in the work with discussions and support. Drs. Herbert Leupold and Ernest Potenziani II of the U.S. Army Electronics Technology and Devices Laboratory in Fort Monmouth, N.J. provided the experimental demonstration of the properties of field confinement and linear superposition of the field generated by yokeless multilayered magnets. Dr. Tsuyoshi Miyamoto of Sumitomo Special Metals of Japan shared the results of his work and provided material and an experimental model of a yokeless magnet.

Mr. Paul Kimura of Sumitomo Corporation of America deserves special thanks for his assistance and support. Finally the author wishes to thank Mr. Hsiao Dee Lieu and Ms. Ronit Cohen for their excellent work in the preparation and typing of the manuscript.

STRUCTURES OF
PERMANENT MAGNETS

CHAPTER 1

Magnetostatic Fields

INTRODUCTION

This chapter is devoted to the derivation of the general formulas which constitute the basis of the linear theory of static magnetic fields. The properties of these fields are reviewed with the emphasis placed on the introduction of concepts and equations needed for the solutions of the problems analyzed in the following chapters.

1.1 BASIC EQUATIONS

The magnetic properties of a medium are characterized by a phenomenon called polarization, which occurs when a magnetic dipole moment is induced in a medium in the presence of an external magnetic field. In the linear theory of magnetostatics, the polarization of the medium is described by an equation which relates two vector quantities $\vec{B}$ and $\vec{H}$ called the magnetic induction and the intensity of the magnetic field, respectively:

$$\vec{B} = \mu_0 \vec{H} + \vec{J}. \tag{1.1.1}$$

$\vec{J}$ is the magnetic polarization density which is equal to the magnetic dipole moment per unit volume of the medium. In free space and in nonmagnetic materials $\vec{J} = 0$. The value of the constant μ_0 depends upon the system of units used to measure $\vec{B}$ and $\vec{H}$. In the rationalized MKS system of units $\vec{B}$ and $\vec{H}$ are measured in Tesla (webers/square meter) and ampere-turns/meter, respectively, and the value of μ_0 is

$$\mu_0 = 4\pi \cdot 10^{-7} \text{ henry/meter} \tag{1.1.2}$$

(App. I). In general $\vec{J}$ may be considered as the sum of two terms

$$\vec{J} = \vec{J}_0 + \vec{J}_m . \tag{1.1.3}$$

Vector $\vec{J}_0$ is independent of the intensity of the magnetic field and represents the permanent polarization that results from a magnetization process with an external magnetic field. In Eq. 1.1.1 $\vec{J}_0$ is the value of $\vec{B}$ in the absence of $\vec{H}$. The value $\vec{B}_r = \vec{J}_0$ is usually referred to as the remanence of the magnetic medium. Throughout the text of the book, magnitude and orientation of the remanence will be identified by symbol $\vec{J}$ with an appropriate subindex.

Vector $\vec{J}_m$ is the polarization induced by the magnetic field. In an isotropic medium, within the limits of the linear theory, $\vec{J}_m$ is proportional to the magnetic field:

$$\vec{J}_m = \mu_0 \chi_m \vec{H} \ . \tag{1.1.4}$$

The dimensionless quantity χ_m is called the magnetic susceptibility. Equation 1.1.1 can be written again in the form

$$\vec{B} = \mu \vec{H} + \vec{J}_0 \ , \tag{1.1.5}$$

where

$$\mu = \mu_0 (1 + \chi_m) \tag{1.1.6}$$

is the permeability of the medium. Frequently the dimensionless quantity

$$\kappa_m = \frac{\mu}{\mu_0} = 1 + \chi_m \tag{1.1.7}$$

is used to specify the permeability.

In an anisotropic medium, the relationship between $\vec{J}_m$ and $\vec{H}$ depends upon the orientation of $\vec{H}$, and the components of $\vec{J}_m$ are related to the components of $\vec{H}$ by the system of equations

$$J_{m,i} = \mu_0 \sum_{j=1}^{3} \chi_{m,i,j} H_j \qquad (i = 1, 2, 3) \ , \tag{1.1.8}$$

where coefficients $\chi_{m,i,j}$ are the components of a symmetric tensor χ_m.

Isotropic paramagnetic media are characterized by positive values of the magnetic susceptibility. A medium with a negative value of χ_m is called diamagnetic. Usually in both paramagnetic and diamagnetic media, χ_m is small compared to unity.

In a magnetostatic field, vectors $\vec{B}$ and $\vec{H}$ satisfy the two fundamental equations

$$\nabla \times \vec{H} = \vec{j} \tag{1.1.9}$$

and

$$\nabla \cdot \vec{B} = 0 \ , \tag{1.1.10}$$

where $\vec{j}$ is a time independent electric current density which satisfies the equation of continuity

$$\nabla \cdot \vec{j} = 0 \ . \tag{1.1.11}$$

Equation 1.1.10 postulates the absence of true magnetic charges. As a consequence of the solenoidal property of the magnetic induction, the lines of force of $\vec{B}$ close either upon themselves or at infinity.

By virtue of Eq. 1.1.10, vector $\vec{B}$ can be written as the curl of a vector $\vec{A}$

$$\vec{B} = \nabla \times \vec{A} \ . \tag{1.1.12}$$

$\vec{A}$ is called the vector potential, and by virtue of Eqs. 1.1.1 and 1.1.9, it satisfies the equation

$$\nabla^2 \vec{A} = -\mu_0 \vec{J} - \nabla \times \vec{J}. \tag{1.1.13}$$

Thus the polarization of the medium is equivalent to an electric current density

$$\vec{j}' = \frac{1}{\mu_0} \nabla \times \vec{J}. \tag{1.1.14}$$

Assume now that the electric current density $\vec{j}$ is zero everywhere. The curl of $\vec{H}$ in Eq. 1.1.9 vanishes and one can write

$$\vec{H} = -\nabla \Phi, \tag{1.1.15}$$

where Φ is called the scalar magnetostatic potential. In a uniform medium, by virtue of Eq. 1.1.10, function Φ satisfies Poisson's equation

$$\nabla^2 \Phi = \frac{1}{\mu} \nabla \cdot \vec{J}_0. \tag{1.1.16}$$

By analogy with the Poisson equation in electrostatics, the permanent polarization of the medium is equivalent to a volume density υ of magnetic charge

$$\upsilon = -\nabla \cdot \vec{J}_0. \tag{1.1.17}$$

In the particular case of a solenoidal remanence, one has

$$\nabla \cdot \vec{J}_0 = 0, \tag{1.1.18}$$

and Eq. 1.1.16 reduces to Laplace's equation

$$\nabla^2 \Phi = 0. \tag{1.1.19}$$

A solution of Eq. 1.1.19 is called a harmonic function. To review some of its basic properties, assume that Φ is finite and continuous with its first and second derivatives within the volume V of integration of Eq. 1.1.19. Consider a sphere S of radius ρ within volume V. An important property of harmonic functions is that the average value $\langle\Phi\rangle$ of the potential over the sphere is equal to the value of Φ at the center P of the sphere, i.e.,

$$\Phi(P) = \frac{1}{4\pi\rho^2} \int_{S=4\pi\rho^2} \Phi(S)\, dS, \tag{1.1.20}$$

where $\Phi(S)$ is the value of Φ over the surface of the sphere. Equation 1.1.20 can be readily proved by computing the derivative of the right hand side with respect to ρ. One has

$$\frac{d}{d\rho}\langle\Phi\rangle = \frac{1}{4\pi\rho^2} \int_S (\nabla \Phi) \cdot \vec{n}\, dS, \tag{1.1.21}$$

where $\vec{n}$ is a unit vector perpendicular to S and oriented toward the region outside of the sphere. By virtue of the divergence theorem

$$\frac{d}{d\rho}\langle\Phi\rangle = \frac{1}{4\pi\rho^2}\int_{V_s} \nabla^2\Phi\, dV = 0, \qquad (1.1.22)$$

where V_s is the volume of the sphere. Thus the average value of Φ over the sphere is a constant, independent of the radius of the sphere. As a consequence, for $\rho \to 0$, $\langle\Phi\rangle$ must be equal to the value of Φ at the center P of the sphere [1-3].

As a consequence of Eq. 1.1.20 a harmonic function can have neither a maximum nor a minimum within the volume of integration of Eq. 1.1.19. It also follows that if Φ is constant over a closed surface S, Φ has the same constant value at each point of the volume V enclosed by S. In particular if Φ is zero over S, it is also identically equal to zero within S. This property can be proved by applying the divergence theorem to the vector $\Phi\nabla\Phi$. One has

$$\int_V \nabla\cdot(\Phi\nabla\Phi)\, dV = \int_S \Phi(\nabla\Phi)\cdot\vec{n}\, dS. \qquad (1.1.23)$$

Because of the identity

$$\nabla\cdot(\Phi\nabla\Phi) = \Phi\nabla^2\Phi + (\nabla\Phi)^2, \qquad (1.1.24)$$

and by virtue of Laplace's equation, one has

$$\int_V (\nabla\Phi)^2\, dV = \int_S \Phi(\nabla\Phi)\cdot\vec{n}\, dS. \qquad (1.1.25)$$

Thus if Φ is zero over S, one has

$$\nabla\Phi = 0 \qquad (1.1.26)$$

within S; i.e., Φ is constant and equal to zero within S.

Based on this result, one can prove that the solution of Laplace's equation within S is uniquely determined by its value over the closed surface S. If one assumes the existence of two solutions Φ_1, Φ_2 with the same value over S, the difference

$$\Phi = \Phi_1 - \Phi_2 \qquad (1.1.27)$$

would be a solution of Laplace's equation, and by virtue of Eq. 1.1.25, it would be zero everywhere. Thus $\Phi_1 = \Phi_2$.

Finally, assume that the solution of Laplace's equation satisfies the condition

$$\vec{n}\cdot(\nabla\Phi) = 0 \qquad (1.1.28)$$

at each point of the closed surface S. If two solutions Φ_1, Φ_2 exist with the same value

$$\vec{n}\cdot(\nabla\Phi_1) = \vec{n}\cdot(\nabla\Phi_2) \qquad (1.1.29)$$

over surface S, inside S one has

$$\Phi_1 - \Phi_2 = C\ , \qquad (1.1.30)$$

where C is an arbitrary constant. Thus the value of Φ is uniquely determined at each point of V, except for an arbitrary additive constant.

In general, a magnetostatic problem involves a structure of media with assigned magnetic properties and assigned geometries of their interfaces. The problem is formulated by assigning the distribution of electric currents and remanences in each medium, and the uniqueness of the solution results from the boundary conditions at the interfaces between media.

The equations of the boundary conditions can be developed by considering a surface S which separates two media of different values of μ. Assume in each of the two media a distribution of electric current densities $\vec{j}_1$, $\vec{j}_2$ and a distribution of remanences $\vec{J}_{0,1}$, $\vec{J}_{0,2}$. Also assume that a surface distribution of electric current $\vec{\tau}$ flows on surface S. Vector $\vec{\tau}$ is oriented parallel to S and its dimension is a current density per unit length.

Assume first that surface S is replaced by a thin layer of thickness δl where χ_m and $\vec{J}_0$ change continuously from the values $\chi_{m,1}$, $\vec{J}_{0,1}$ in the first medium to the values $\chi_{m,2}$, $\vec{J}_{0,2}$ in the second medium. Also assume that the surface current density $\vec{\tau}$ is replaced by a distribution of current density $\vec{j}$ inside the layer such that

$$\lim_{\delta l \to 0} \vec{j}\,\delta l = \vec{\tau}. \qquad (1.1.31)$$

Let δV be the infinitesimal volume of the layer enclosed by the right circular cylinder of height δl shown in Fig. 1.1.1(a). δS is the area of the base of the cylinder. The integration of Eq. 1.1.10 over volume δV yields

$$\int_{\delta V} \nabla\cdot\vec{B}\,dV = \int \vec{B}\cdot\vec{n}\,dS = 0\ , \qquad (1.1.32)$$

where the surface integral is extended over the surface of the cylinder and $\vec{n}$ is a unit vector perpendicular to its surface and oriented outwards with respect to δV. In the limit $\delta l \to 0$, Eq. 1.1.32 yields

$$(\vec{B}_1\cdot\vec{n}_1 + \vec{B}_2\cdot\vec{n}_2)\,\delta S = 0\ , \qquad (1.1.33)$$

where $\vec{B}_1$ and $\vec{B}_2$ are the values of the magnetic induction at two points P_1 and P_2 of the two media infinitely close to each other. One has

$$\vec{n}_2 = -\vec{n}_1\ . \qquad (1.1.34)$$

6 MAGNETOSTATIC FIELDS

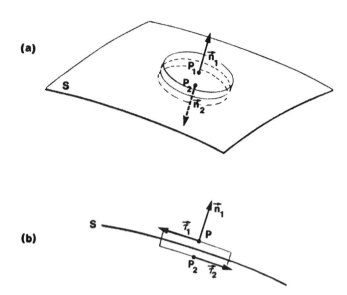

Fig. 1.1.1. Closed surface (a) and closed path (b) across the interface S between two media for the computation of the boundary conditions.

Hence the magnetic induction satisfies the boundary condition

$$(\vec{B}_1 - \vec{B}_2) \cdot \vec{n}_1 = 0 \; ; \tag{1.1.35}$$

i.e., the normal component of $\vec{B}$ is continuous across S. Equation 1.1.35 can be written in the form

$$(\mu_1 \vec{H}_1 - \mu_2 \vec{H}_2) \cdot \vec{n}_1 = - (\vec{J}_{0,1} - \vec{J}_{0,2}) \cdot \vec{n}_1 \; , \tag{1.1.36}$$

where $\vec{H}_1, \vec{H}_2$ are the values of the intensity at P_1 and P_2, and μ_1, μ_2 are the permeabilities of the two media.

Consider now the rectangular path shown in Fig. 1.1.1(b) with two sides of length δs on the surface of the layer of thickness δl. The area of the rectangle is

$$\delta S_r = \delta s \; \delta l \; . \tag{1.1.37}$$

By virtue of Stokes' theorem, the integration of Eq. 1.1.9 over the rectangular path of Fig. 1.1.1(b) yields

$$\int \vec{H} \cdot \vec{\tau} ds = \vec{j} \cdot \vec{n}_r \, \delta S_r \; , \tag{1.1.38}$$

where unit vector $\vec{\tau}$ is tangent to the rectangular path and $\vec{n}_r$ is a unit vector perpendicular to δS_r. The relative orientation of $\vec{\tau}$ and $\vec{n}_r$ is such that an observer oriented along $\vec{n}_r$ sees the circulation of $\vec{\tau}$ as counterclockwise. By virtue of Eq. 1.1.31, in the limit $\delta l \to 0$, Eq. 1.1.38 yields

$$\vec{H}_1 \cdot \vec{\tau}_1 + \vec{H}_2 \cdot \vec{\tau}_2 = \vec{\tau} \cdot \vec{n}_r \ . \tag{1.1.39}$$

One has

$$-\vec{\tau}_2 = \vec{\tau}_1 = \vec{n}_r \times \vec{n}_1 \ . \tag{1.1.40}$$

Thus the intensity of the magnetic field satisfies the second boundary condition

$$\vec{n}_1 \times (\vec{H}_1 - \vec{H}_2) = \vec{\tau} \ ; \tag{1.1.41}$$

i.e., the discontinuity of the tangential component of the intensity $\vec{H}$ is equal to the surface current density on surface S.

1.2 DIFFERENTIAL OPERATORS IN ORTHOGONAL CURVILINEAR SYSTEMS

The explicit form of the equations of the differential operators introduced in Section 1 can be developed in a generalized system of coordinates by assuming that the position of a point P is defined by a set of coordinates u_1, u_2, u_3 which are single valued functions of the cartesian coordinates x, y, z of P.

Each equation

$$u_i(x, y, z) = constant \qquad (i = 1, 2, 3) \tag{1.2.1}$$

represents a coordinate surface, and the intersection of two coordinate surfaces of different values of i defines a coordinate line. Thus, the coordinate line u_1 is the intersection of two coordinate surfaces defined by $i = 2$, $i = 3$, etc.

Assume that a second point P' has cartesian coordinates

$$x + dx \ , \qquad y + dy \ , \qquad z + dz \ . \tag{1.2.2}$$

The distance ds between P and P' is given by

$$ds^2 = \sum_{i=1}^{3} [h_i^2 du_i^2 + 2f_i du_j du_k] \qquad (j \neq k \neq i) \ , \tag{1.2.3}$$

where

$$h_i^2 = \left[\frac{\partial x}{\partial u_i}\right]^2 + \left[\frac{\partial y}{\partial u_i}\right]^2 + \left[\frac{\partial z}{\partial u_i}\right]^2$$

$$f_i = \frac{\partial x}{\partial u_j} \frac{\partial x}{\partial u_k} + \frac{\partial y}{\partial u_j} \frac{\partial y}{\partial u_k} + \frac{\partial z}{\partial u_j} \frac{\partial z}{\partial u_k} \ . \tag{1.2.4}$$

8 MAGNETOSTATIC FIELDS

The quantity

$$\frac{f_i}{h_j h_k} \tag{1.2.5}$$

is the cosine of the angle between coordinate lines u_j and u_k at point P. If the three quantities f_i ($i = 1, 2, 3$) are zero everywhere, the coordinate lines are perpendicular to each other at each point P, and the system of coordinates u_1, u_2, u_3 is called an orthogonal curvilinear system. In such a system the element of length ds is

$$ds = \left[\sum_{i=1}^{3} h_i^2 \, du_i^2 \right]^{1/2}. \tag{1.2.6}$$

An element of surface dS_i perpendicular to coordinate line u_i is

$$dS_i = h_j h_k \, du_j \, du_k \qquad (j \neq k \neq i), \tag{1.2.7}$$

and an element of volume dV is

$$dV = h_1 h_2 h_3 \, du_1 \, du_2 \, du_3. \tag{1.2.8}$$

The components of the gradient of a function Φ along coordinate lines u_i are

$$(\nabla \Phi)_i = \frac{1}{h_i} \frac{\partial \Phi}{\partial u_i} \qquad (i = 1, 2, 3). \tag{1.2.9}$$

Thus $\nabla \Phi$ is given by the equation

$$\nabla \Phi = \frac{1}{h_1} \frac{\partial \Phi}{\partial u_1} \vec{u}_1 + \frac{1}{h_2} \frac{\partial \Phi}{\partial u_2} \vec{u}_2 + \frac{1}{h_3} \frac{\partial \Phi}{\partial u_3} \vec{u}_3, \tag{1.2.10}$$

where $\vec{u}_1, \vec{u}_2, \vec{u}_3$ are the unit vectors tangent to the coordinate lines u_1, u_2, u_3. To derive the equation of the divergence of a vector $\vec{F}$, apply the divergence theorem to the element of volume dV shown in Fig. 1.2.1. One has

$$\nabla \cdot \vec{F} \, dV = \sum_l \vec{F} \cdot \vec{n}_l \, dS_l, \tag{1.2.11}$$

where the sum is extended over the six faces of the element of volume and $\vec{n}_l$ is a unit vector perpendicular to face dS_l and oriented outwards with respect to dV. The outgoing flux of $\vec{F}$ across the two faces perpendicular to coordinate line u_i is

$$\frac{\partial}{\partial u_i} (h_j h_k F_i) \, du_1 \, du_2 \, du_3 \qquad (j \neq k \neq i). \tag{1.2.12}$$

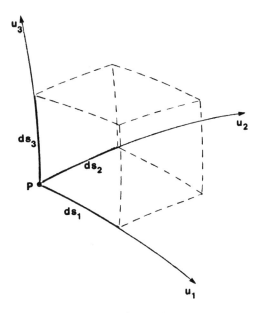

Fig. 1.2.1. Element of volume in an orthogonal curvilinear system.

Thus, the equation of the divergence of $\vec{F}$ is

$$\nabla \cdot \vec{F} = \frac{1}{h_1 h_2 h_3} \left[\frac{\partial}{\partial u_1}(h_2 h_3 F_1) + \frac{\partial}{\partial u_2}(h_3 h_1 F_2) + \frac{\partial}{\partial u_3}(h_1 h_2 F_3) \right] \quad (1.2.13)$$

and by virtue of Eqs. 1.2.10 and 1.2.13, the Laplacian of a scalar function Φ is

$$\nabla^2 \Phi = \frac{1}{h_1 h_2 h_3} \left[\frac{\partial}{\partial u_1} \left[\frac{h_2 h_3}{h_1} \frac{\partial \Phi}{\partial u_1} \right] + \frac{\partial}{\partial u_2} \left[\frac{h_3 h_1}{h_2} \frac{\partial \Phi}{\partial u_2} \right] + \frac{\partial}{\partial u_3} \left[\frac{h_1 h_2}{h_3} \frac{\partial \Phi}{\partial u_3} \right] \right]$$

$$(1.2.14)$$

The equation of the curl of a vector $\vec{F}$ is derived by applying Stokes' theorem to elements of surfaces dS_i. One has

$$(\nabla \times \vec{F})_i \, h_j h_k \, du_j \, du_k = \sum_l \vec{F} \cdot \vec{\tau}_l \, ds_l \quad (i = 1, 2, 3), \quad (1.2.15)$$

where the sum is extended over the four sides of dS_i and $\vec{\tau}_l$ is a unit vector tangent to side ds_l and oriented in a counterclockwise direction with respect to an observer oriented along the coordinate line u_i, as shown in Fig. 1.2.2.

For $i = 1$, the right hand side of Eq. 1.2.15 is equal to

$$\left[\frac{\partial}{\partial u_2}(h_3 F_3) - \frac{\partial}{\partial u_3}(h_2 F_2) \right] du_2 du_3 \quad (1.2.16)$$

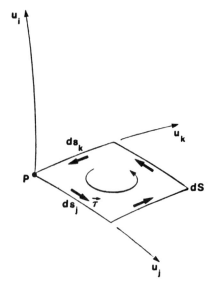

Fig. 1.2.2. Element of surface in an orthogonal curvilinear system.

and by rotation of the subindices, one obtains the other two values of the right hand side of Eq. 1.2.15 for $i=2$ and $i=3$. Thus the equation of the curl of $\vec{F}$ can be written in the form

$$\nabla \times \vec{F} = \begin{vmatrix} \dfrac{\vec{u}_1}{h_2 h_3} & \dfrac{\vec{u}_2}{h_3 h_1} & \dfrac{\vec{u}_3}{h_1 h_2} \\ \dfrac{\partial}{\partial u_1} & \dfrac{\partial}{\partial u_2} & \dfrac{\partial}{\partial u_3} \\ h_1 F_1 & h_2 F_2 & h_3 F_3 \end{vmatrix}. \qquad (1.2.17)$$

Two examples of orthogonal curvilinear systems are presented in Fig. 1.2.3. Figure 1.2.3(a) shows a system of cylindrical polar coordinates r, ψ, z where r is the distance of point P from the axis z of the system of cartesian coordinates x, y, z and ψ is the angle between plane $y = 0$ and the plane formed by point P and the z axis. One has

$$u_1 = r, \qquad u_2 = \psi, \qquad u_3 = z$$
$$x = r \cos \psi, \qquad y = r \sin \psi, \qquad z = z \qquad (1.2.18)$$

and functions h_i defined by Eq. 1.2.4 are

$$h_1 = 1, \quad h_2 = r, \quad h_3 = 1. \qquad (1.2.19)$$

Figure 1.2.3(b) shows a system of spherical polar coordinates ρ, θ, ψ where ρ is the distance of point P from the origin O, θ is the angle of the line (OP) relative to the z

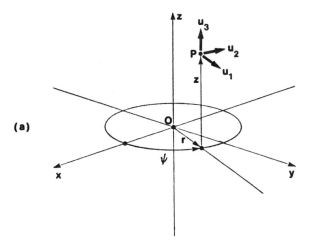

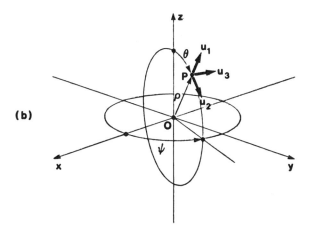

Fig. 1.2.3. Cylindrical polar coordinates (a) and spherical polar coordinates (b).

axis, and ψ is the angle between plane $y = 0$ and the plane formed by point P and the z axis. One has

$$u_1 = \rho, \qquad u_2 = \theta, \qquad u_3 = \psi$$
$$x = \rho \sin\theta \cos\psi, \qquad y = \rho \sin\theta \sin\psi, \qquad z = \rho \cos\theta \qquad (1.2.20)$$

and functions h_i are

$$h_1 = 1, \quad h_2 = \rho, \quad h_3 = \rho \sin\theta. \qquad (1.2.21)$$

1.3 EQUIPOTENTIAL SURFACES

In the absence of electric currents, in a region where the scalar potential Φ is regular, a magnetostatic field can be mapped as a family of equipotential surfaces, where each surface is identified by a particular value of Φ.

Assume for instance a two-dimensional problem in a frame of cartesian coordinates x, y with the origin at a point P_0. In a small region about P_0 the scalar potential can be expanded in the series

$$\Phi = \Phi_0 + \left[\frac{\partial \Phi}{\partial x}\right]_0 x + \left[\frac{\partial \Phi}{\partial y}\right]_0 y + \frac{1}{2}\left[\frac{\partial^2 \Phi}{\partial x^2}\right]_0 x^2$$

$$+ \frac{1}{2}\left[\frac{\partial^2 \Phi}{\partial y^2}\right]_0 y^2 + \left[\frac{\partial^2 \Phi}{\partial x \partial y}\right]_0 xy + \cdots , \quad (1.3.1)$$

where Φ_0 is the value of Φ at $x = y = 0$. From Eq. 1.3.1 one obtains the x and y components of intensity $\vec{H}$

$$H_x = -(\nabla \Phi)_x = -\left[\frac{\partial \Phi}{\partial x}\right]_0 - \left[\frac{\partial^2 \Phi}{\partial x^2}\right]_0 x - \left[\frac{\partial^2 \Phi}{\partial x \partial y}\right]_0 y - \cdots$$

$$H_y = -(\nabla \Phi)_y = -\left[\frac{\partial \Phi}{\partial y}\right]_0 - \left[\frac{\partial^2 \Phi}{\partial y^2}\right]_0 y - \left[\frac{\partial^2 \Phi}{\partial x \partial y}\right]_0 x - \cdots . \quad (1.3.2)$$

An equipotential line satisfies the differential equation

$$\frac{dy}{dx} = -\frac{H_x}{H_y}. \quad (1.3.3)$$

Assume that intensity $\vec{H}$ is zero at point P_0. Thus,

$$\left[\frac{\partial \Phi}{\partial x}\right]_0 = \left[\frac{\partial \Phi}{\partial y}\right]_0 = 0, \quad (1.3.4)$$

and by virtue of Eq. 1.3.2, in a small region about P_0 Eq. 1.3.3 of the equipotential lines becomes

$$\frac{dy}{dx} = -\frac{\alpha x + \beta y}{\beta x + \gamma y}, \quad (1.3.5)$$

where

$$\alpha = \left[\frac{\partial^2 \Phi}{\partial x^2}\right]_0, \quad \beta = \left[\frac{\partial^2 \Phi}{\partial x \partial y}\right]_0, \quad \gamma = \left[\frac{\partial^2 \Phi}{\partial y^2}\right]_0. \quad (1.3.6)$$

Since Φ is a solution of Laplace's equation,

$$\gamma = -\alpha , \qquad (1.3.7)$$

and Eq. 1.3.5 can be rewritten in the form

$$\frac{dy}{dx} = -\frac{x + ay}{ax - y} , \qquad (1.3.8)$$

where

$$a = \frac{\left[\frac{\partial^2 \Phi}{\partial x \partial y}\right]_0}{\left[\frac{\partial^2 \Phi}{\partial x^2}\right]_0} . \qquad (1.3.9)$$

Both numerator and denominator on the right hand side of Eq. 1.3.8 vanish at $x = y = 0$, and as a consequence, the orientation of the equipotential line $\Phi = \Phi_0$ at point P is indeterminate. Equation 1.3.8 has the solution

$$x^2 - y^2 + 2axy = C , \qquad (1.3.10)$$

where C is an arbitrary constant. The equipotential lines are hyperbolas whose asymptotes are given by Eq. 1.3.10 for the particular value $C = 0$. The two asymptotes are the lines

$$\frac{y}{x} = a \pm (a^2 + 1)^{1/2} . \qquad (1.3.11)$$

Point P_0 is common to the two equipotential lines 1.3.11, i.e., $C = 0$ corresponds to the value $\Phi = \Phi_0$ of the scalar potential. Positive and negative values of the constant of integration in Eq. 1.3.10 identify the equipotential lines whose value of Φ is larger or smaller than Φ_0. A plotting of the family of lines 1.3.10 is shown in Fig. 1.3.1. One observes that the value of C is equivalent to the elevation in a mapping of contour lines. For this reason, point P_0 in Fig. 1.3.1 is called a saddle point.

The lines of force of $\nabla\Phi$ satisfy the differential equation

$$\frac{dy}{dx} = \frac{H_y}{H_x} \qquad (1.3.12)$$

which, by virtue of Eq. 1.3.7, becomes

$$\frac{dy}{dx} = \frac{ax - y}{x + ay} . \qquad (1.3.13)$$

The solution of Eq. 1.3.13 is the family of hyperbolas

$$x^2 - y^2 - \frac{2}{a}xy = C' , \qquad (1.3.14)$$

14 MAGNETOSTATIC FIELDS

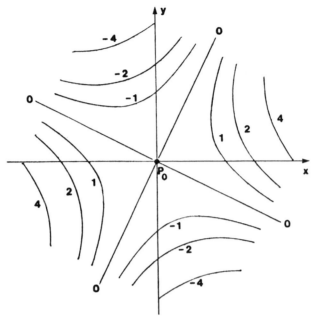

Fig. 1.3.1. Saddle point.

where C' is again an arbitrary constant. Hyperbolas 1.3.14 and 1.3.10 are perpendicular to each other.

Let us write the equations of the families of hyperbolas in polar coordinates r, ψ such that

$$x = r\cos\psi, \quad y = r\sin\psi. \tag{1.3.15}$$

Equation 1.3.10 becomes

$$r^2\cos 2(\psi - \psi_2) = \frac{C}{\left[1 + a^2\right]^{1/2}}, \tag{1.3.16}$$

where angle ψ_2 is given by

$$\tan 2\psi_2 = a. \tag{1.3.17}$$

Angle ψ_2 is the angle between the major axis of hyperbolas 1.3.16 and the x axis. Equation 1.3.16 is a particular solution of Laplace's equation written in cylindrical polar coordinates r, ψ, z. If Φ is independent of the variable z, Laplace's equation reduces to

$$\frac{\partial^2 \Phi}{\partial r^2} + \frac{1}{r}\frac{\partial \Phi}{\partial r} + \frac{1}{r^2}\frac{\partial^2 \Phi}{\partial \psi^2} = 0. \tag{1.3.18}$$

A general solution of Eq. 1.3.18, regular at point P_0, is

$$\Phi(r,\psi) = r^n \sin n(\psi - \psi_n) + \Phi_0 , \qquad (1.3.19)$$

where n is a positive integer and ψ_n is an arbitrary angle. Point P_0 is common to n equipotential lines $\Phi = \Phi_0$:

$$\psi - \psi_n = \frac{m\pi}{n} \qquad (m = 0, 1, 2, \ldots, n-1) . \qquad (1.3.20)$$

At point P_0, all partial derivatives of Φ with respect to r and ψ whose order is equal to or lower than $n - 1$ are equal to zero.

Consider now a particular case of a three-dimensional problem where the scalar potential Φ satisfies the symmetry condition

$$\Phi(x, y, z) = \Phi(-x, -y, z) , \qquad (1.3.21)$$

where x, y, z are the cartesian coordinates with the origin at P_0. In a small region about P_0, Φ can be expanded in the series

$$\Phi = \Phi_0 + \left[\frac{\partial \Phi}{\partial z}\right]_0 z + \frac{1}{2}\left[\frac{\partial^2 \Phi}{\partial x^2}\right]_0 x^2 + \frac{1}{2}\left[\frac{\partial^2 \Phi}{\partial y^2}\right]_0 y^2$$

$$+ \frac{1}{2}\left[\frac{\partial^2 \Phi}{\partial z^2}\right]_0 z^2 + \left[\frac{\partial^2 \Phi}{\partial x \partial y}\right]_0 xy + \cdots , \qquad (1.3.22)$$

where the terms of the third or higher order are neglected. Again Φ_0 is the value of the scalar potential at P_0 and

$$\vec{H}_0 = -(\nabla \Phi)_0 = -\left[\frac{\partial \Phi}{\partial z}\right]_0 \vec{z} \qquad (1.3.23)$$

is the intensity of the magnetic field at point P_0. By neglecting the higher order terms in Eq. 1.3.22, in order for Φ to be a solution of Laplace's equation, one must have

$$\left[\frac{\partial^2 \Phi}{\partial x^2}\right]_0 + \left[\frac{\partial^2 \Phi}{\partial y^2}\right]_0 + \left[\frac{\partial^2 \Phi}{\partial z^2}\right]_0 = 0 . \qquad (1.3.24)$$

Assume that

$$\left[\frac{\partial^2 \Phi}{\partial y^2}\right]_0 = C \left[\frac{\partial^2 \Phi}{\partial x^2}\right]_0 , \qquad (1.3.25)$$

where C is an arbitrary positive constant. Equation 1.3.22 becomes

$$\Phi = \Phi_0 + \left[\frac{\partial \Phi}{\partial z}\right]_0 z + \frac{1}{2}\left[\frac{\partial^2 \Phi}{\partial x^2}\right]_0 \left[x^2 + Cy^2 - (1+C)z^2 + 2axy\right] + \cdots, \quad (1.3.26)$$

where constant a is given by Eq. 1.3.9. In Eq. 1.3.26, the factor

$$x^2 + Cy^2 - (1+C)z^2 + 2axy = constant \quad (1.3.27)$$

describes a family of hyperboloids. In particular, if one assumes that the constant is equal to zero, Eq. 1.3.27 describes a cone with vertex at P_0, with the z axis as its axis of symmetry and elliptical cross-section in planes perpendicular to z.

Major and minor axes of the ellipses are oriented at angles ψ_a, ψ_b relative to the x axis,

$$\tan 2\psi_a = \frac{2a}{1-C}, \quad \psi_b = \psi_a \pm \frac{\pi}{2}. \quad (1.3.28)$$

A negative value of the constant on the right hand side of Eq. 1.3.27 corresponds to the family of hyperboloids of two sheets in the region inside the cone. Conversely, a positive value of the constant corresponds to the hyperboloids of one sheet outside of the cone.

In the special case of symmetry of revolution about the z axis one has

$$C = 1, \quad a = 0 \quad (1.3.29)$$

and Eq. 1.3.27 reduces to the family of hyperboloids

$$r^2 - 2z^2 = constant, \quad (1.3.30)$$

where

$$r = (x^2 + y^2)^{1/2} \quad (1.3.31)$$

is the distance from the axis z. A zero value of the constant in Eq. 1.3.30 yields the equation of the right circular cone

$$r = \sqrt{2}|z|. \quad (1.3.32)$$

The geometry of the equipotential surfaces described by Eq. 1.3.30 is shown in Fig. 1.3.2.

The analysis of the properties of points common to a number of equipotential surfaces of two-dimensional magnetic fields can be extended to three-dimensional prob-

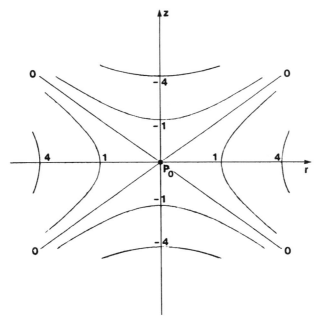

Fig. 1.3.2. Equipotential surfaces with symmetry of revolution about axis z.

lems by writing Laplace's equation in a frame of spherical polar coordinates ρ, θ, ψ with origin at P_0. Functions h_i, defined by Eqs. 1.2.4, for $u_1 = \rho$, $u_2 = \theta$, $u_3 = \psi$, are given by Eqs. 1.2.21, and Laplace's equation becomes

$$\frac{1}{\rho^2}\frac{\partial}{\partial \rho}\left[\rho^2\frac{\partial \Phi}{\partial \rho}\right] + \frac{1}{\rho^2 \sin\theta}\frac{\partial}{\partial \theta}\left[\sin\theta\frac{\partial \Phi}{\partial \theta}\right] + \frac{1}{\rho^2 \sin^2\theta}\frac{\partial^2 \Phi}{\partial \psi^2} = 0, \quad (1.3.33)$$

whose general solution, regular at P_0, is

$$\Phi(\rho, \theta, \psi) = \rho^l P_l^j(\xi) \sin j(\psi - \psi_i) \quad (j \leq l), \quad (1.3.34)$$

where

$$\xi = \cos\theta. \quad (1.3.35)$$

P_l^j is the associated Legendre function of the first kind

$$P_l^j(\xi) = (1 - \xi^2)^{\frac{j}{2}} \frac{d^j}{d\xi^j} P_l(\xi), \quad (1.3.36)$$

and P_l is the Legendre polynomial of the first kind (App. II)

$$P_l(\xi) = \frac{1}{2^l l!} \frac{d^l}{d\xi^l} (\xi^2 - 1)^l . \qquad (1.3.37)$$

Point P_0 is common to the $\Phi = 0$ equipotential planes

$$\psi - \psi_i = \frac{m\pi}{j} \qquad (m = 0, 1, \ldots, j-1) \qquad (1.3.38)$$

and to the $\Phi = 0$ equipotential cones

$$P_l^j(\xi) = 0 \qquad (1.3.39)$$

whose axis coincides with the $\theta = 0$ axis.

1.4 SOLUTION OF THE FIELD EQUATIONS

Assume that the induced polarization $\vec{J}_m$ in Eq. 1.1.3 is negligible compared to $\vec{J}_0$. The magnetized medium is then equivalent to an assigned distribution of remanence $\vec{J}_0$ in free space ($\chi_m = 0$). Assuming a given distribution of electric current density $\vec{j}$ and remanences $\vec{J}_0$, let us seek a particular solution of Eqs. 1.1.13 and 1.1.15 in an unbounded medium of zero magnetic susceptibility.

To derive the solution of Eqs. 1.1.13 and 1.1.16, consider first two scalar functions F and G within a volume V enclosed by a surface S. Because of the identity

$$\nabla \cdot (G \nabla F) = (\nabla G) \cdot (\nabla F) + G \nabla^2 F \qquad (1.4.1)$$

and by virtue of the divergence theorem, one has

$$\int_V \nabla \cdot (G \nabla F) \, dV = \int_V \left[(\nabla G) \cdot (\nabla F) + G \nabla^2 F \right] dV = \int_S G \frac{\partial F}{\partial n} dS , \qquad (1.4.2)$$

where n is the direction perpendicular to S oriented toward the region outside S. Equation 1.4.2 is called Green's first identity. Similarly one can write

$$\int_V \nabla \cdot (F \nabla G) \, dV = \int_V \left[(\nabla F) \cdot (\nabla G) + F \nabla^2 G \right] dV = \int_S F \frac{\partial G}{\partial n} dS . \qquad (1.4.3)$$

Thus, from Eqs. 1.4.2 and 1.4.3, one has

$$\int_V \left[G \nabla^2 F - F \nabla^2 G\right] dV = \int_S \left[G \frac{\partial F}{\partial n} - F \frac{\partial G}{\partial n}\right] dS , \qquad (1.4.4)$$

which is called Green's second identity.

Assume now that F is a solution of the inhomogeneous differential equation

$$\nabla^2 F = \Psi(x,y,z) , \qquad (1.4.5)$$

where Ψ is a known function of position, independent of F, and assume that G is a solution of the homogeneous equation

$$\nabla^2 G = 0 . \qquad (1.4.6)$$

Assume that the boundary S of volume V is composed of the two closed surfaces S_0 and S_1 as shown in Fig. 1.4.1. Surface S_0, which is contained inside S_1, is a small sphere of radius ρ_0 and center at point P. A solution of the homogeneous equation 1.4.6 is

$$G = \frac{1}{\rho} , \qquad (1.4.7)$$

where $\rho \geq \rho_0$ is the distance of a point of volume V from P. On the small sphere of radius ρ_0 one has

$$\frac{\partial}{\partial n} = -\frac{\partial}{\partial \rho} . \qquad (1.4.8)$$

Thus, by virtue of Eqs. 1.4.5 through 1.4.8, Eq. 1.4.4 transforms to

$$\int_V \frac{\Psi}{\rho} dV = -\int_{S_0} \left[\frac{1}{\rho} \frac{\partial F}{\partial \rho} + \frac{F}{\rho^2}\right] dS + \int_{S_1} \left[\frac{1}{\rho} \frac{\partial F}{\partial n} + \frac{F}{\rho^2} \frac{\partial \rho}{\partial n}\right] dS . \qquad (1.4.9)$$

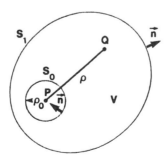

Fig. 1.4.1. Domain of integration of the field equations.

In the limit of an infinite distance of P from S_1, and assuming that F vanishes at infinity at least as ρ^{-1}, the second integral on the right hand side of Eq. 1.4.9 vanishes. Hence, in the limit $\rho_0 \to 0$, Eq. 1.4.9 yields

$$F(P) = -\frac{1}{4\pi}\int_V \frac{\Psi}{\rho} dV \ . \qquad (1.4.10)$$

Equation 1.4.10 is Green's solution of Eq. 1.4.5. Consequently, in free space ($\chi_m = 0$), the solution of Poisson's equation 1.1.16 is

$$\Phi(P) = -\frac{1}{4\pi\mu_0}\int_V \frac{\nabla \cdot \vec{J}_0}{\rho} dV \ . \qquad (1.4.11)$$

The assumption of an infinite distance of P from S_1 means that S_1 can be assumed to contain all magnetized materials; i.e., each point of S_1 is a point of free space where remanence $\vec{J}_0$ is identically equal to zero.

By virtue of the identity

$$\nabla \cdot \left[\frac{\vec{J}_0}{\rho}\right] = \frac{1}{\rho}\nabla \cdot \vec{J}_0 + \nabla\left[\frac{1}{\rho}\right] \cdot \vec{J}_0 \qquad (1.4.12)$$

and by applying the divergence theorem to the volume integral of the left hand side term of Eq. 1.4.12, Eq. 1.4.11 transforms to

$$\Phi(P) = \frac{1}{4\pi\mu_0}\int_V \nabla\left[\frac{1}{\rho}\right] \cdot \vec{J}_0 \, dV \ , \qquad (1.4.13)$$

where the reader is reminded that the gradient is taken at a point Q of volume V in Fig. 1.4.1.

In general, the remanence is not continuous throughout the volume of integration of Eqs. 1.4.11 and 1.4.13. Assume that the magnetized material is contained in a volume V_m limited by a surface S_m. A discontinuity of $\vec{J}_0$ across S_m may result in a singularity of $\nabla \cdot \vec{J}_0$. In order to compute the contribution of this singularity to the value of the integral in Eq. 1.4.11, let us replace S_m with a thin layer of thickness δl where the remanence changes continuously from zero outside the layer to the value $\vec{J}_0$ inside the material. Thus, if $\vec{J}_0$ is a continuous function of position inside V_m, Eq.

1.4.11 can be written in the form

$$\Phi(P) = -\frac{1}{4\pi\mu_0}\int_{V_m}\frac{\nabla\cdot\vec{J}_0}{\rho}dV - \frac{1}{4\pi\mu_0}\lim_{\delta l \to 0}\int_{\delta V}\frac{\nabla\cdot\vec{J}_0}{\rho}dV , \qquad (1.4.14)$$

where δV is the volume of the layer. By virtue of the divergence theorem, Eq. 1.4.14 transforms to

$$\Phi(P) = -\frac{1}{4\pi\mu_0}\int_{V_m}\frac{\nabla\cdot\vec{J}_0}{\rho}dV + \frac{1}{4\pi\mu_0}\int_{S_m}\frac{\vec{J}_0\cdot\vec{n}}{\rho}dS , \qquad (1.4.15)$$

where $\vec{n}$ is a unit vector perpendicular to S_m and oriented toward the free space surrounding V_m. By analogy with Coulomb's equation in electrostatics, the scalar potential of the magnetic field generated by the remanence $\vec{J}_0$ can be considered as the potential generated by a volume density υ of magnetic charges given by Eq. 1.1.17 and a surface density of magnetic charges

$$\sigma = \vec{J}_0\cdot\vec{n} . \qquad (1.4.16)$$

Thus each element of volume dV of the magnetized material contributes to the scalar potential $\Phi(P)$ as an equivalent magnetic charge

$$dm = -\nabla\cdot\vec{J}_0\, dV \qquad (1.4.17)$$

and each element of surface dS of the magnetized material contributes as an equivalent magnetic charge

$$dm = \vec{J}_0\cdot\vec{n}\, dS . \qquad (1.4.18)$$

Consider now the field generated by a distribution of electric current $\vec{j}$ in free space. In the absence of magnetized materials, by virtue of Eq. 1.4.10, the solution of Eq. 1.1.13 is

$$\vec{A}(P) = \frac{\mu_0}{4\pi}\int_V\frac{\vec{j}}{\rho}dV \qquad (1.4.19)$$

and because of Eq. 1.1.12, the intensity of the magnetic field generated by $\vec{j}$ at point P is

$$\vec{H}(P) = \frac{1}{4\pi}\int_V \nabla\times\left[\frac{\vec{j}}{\rho}\right]dV . \qquad (1.4.20)$$

22 MAGNETOSTATIC FIELDS

One has

$$\nabla \times \left[\frac{\vec{j}}{\rho}\right] = \nabla \left[\frac{1}{\rho}\right] \times \vec{j} + \frac{1}{\rho} \nabla \times \vec{j} . \qquad (1.4.21)$$

The curl of $(\vec{j}/\rho)$ in Eq. 1.4.20 is computed at point P. Consequently the second term of the right hand side of Eq. 1.4.21 vanishes, because $\vec{j}$ is independent of the coordinates of P, and Eq. 1.4.20 transforms to

$$\vec{H}(P) = \frac{1}{4\pi} \int_V \frac{\vec{\rho} \times \vec{j}}{\rho^2} dV , \qquad (1.4.22)$$

where $\vec{\rho}$ is the unit vector oriented in the direction from P to the point Q of the element of volume dV. Assume that the electric current flows in a closed line s, which is equivalent to a closed loop formed by a conductor of zero cross-section. Then in Eq. 1.4.22 one can write

$$\vec{j} dV = I \vec{s} ds , \qquad (1.4.23)$$

where I is the total electric current and $\vec{s}$ is a unit vector tangent to the element of length ds of the closed line. Equation 1.4.22 reduces to

$$\vec{H}(P) = \frac{I}{4\pi} \int_s \frac{\vec{\rho} \times \vec{s}}{\rho^2} ds . \qquad (1.4.24)$$

Thus the intensity $\vec{H}(P)$ can be interpreted as the integral of intensities $d\vec{H}$ generated by current I flowing in each element ds of the loop

$$d\vec{H} = \frac{I \, ds}{4\pi} \frac{\vec{\rho} \times \vec{s}}{\rho^2} . \qquad (1.4.25)$$

Equation 1.4.25 is the Biot-Savart law. In a cartesian frame of reference x, y, z, the x component of vector 1.4.24 is

$$H_x = \frac{I}{4\pi} \int_s \vec{\Gamma} \cdot \vec{s} \, ds , \qquad (1.4.26)$$

where vector $\vec{\Gamma}$ is

$$\vec{\Gamma} = \vec{x} \times \nabla \left[\frac{1}{\rho}\right] , \qquad (1.4.27)$$

$\vec{x}$ is the unit vector oriented along the x axis, and the components of $\vec{\Gamma}$ are

$$\Gamma_x = 0, \quad \Gamma_y = -\frac{\partial}{\partial z}\left[\frac{1}{\rho}\right], \quad \Gamma_z = \frac{\partial}{\partial y}\left[\frac{1}{\rho}\right]. \tag{1.4.28}$$

By virtue of Stokes' theorem, Eq. 1.4.26 transforms to

$$H_x = \frac{I}{4\pi}\int_S (\nabla \times \vec{\Gamma}) \cdot \vec{n}\, dS, \tag{1.4.29}$$

where S is an arbitrary surface whose boundary is the closed line s, and $\vec{n}$ is a unit vector perpendicular to the element of surface dS. An observer oriented along the unit vector $\vec{n}$ sees a counterclockwise direction of circulation of unit vector $\vec{s}$.

The components of $\nabla \times \vec{\Gamma}$ at a point of S of coordinates ξ, η, ζ are

$$(\nabla \times \vec{\Gamma})_\xi = -\frac{\partial^2}{\partial \eta^2}\left[\frac{1}{\rho}\right] - \frac{\partial^2}{\partial \zeta^2}\left[\frac{1}{\rho}\right]$$

$$(\nabla \times \vec{\Gamma})_\eta = \frac{\partial^2}{\partial \xi \partial \eta}\left[\frac{1}{\rho}\right] \tag{1.4.30}$$

$$(\nabla \times \vec{\Gamma})_\zeta = \frac{\partial^2}{\partial \zeta \partial \xi}\left[\frac{1}{\rho}\right].$$

As previously stated, ρ^{-1} is a solution of Laplace's equation. Thus, by virtue of Eqs. 1.4.30, one can write

$$\nabla \times \vec{\Gamma} = -\frac{\partial}{\partial x}\nabla\left[\frac{1}{\rho}\right], \tag{1.4.31}$$

where the gradient is computed at the point of coordinates ξ, η, ζ. By virtue of Eq. 1.4.31, Eq. 1.4.29 becomes

$$H_x = -\frac{I}{4\pi}\frac{\partial}{\partial x}\int_S \nabla\left[\frac{1}{\rho}\right]\cdot \vec{n}\, dS. \tag{1.4.32}$$

By exchanging variable x in Eq. 1.4.32 with variables y and z, one obtains the other two components of vector $\vec{H}$, which can then be written as

$$\vec{H}(P) = -\nabla \Phi, \tag{1.4.33}$$

24 MAGNETOSTATIC FIELDS

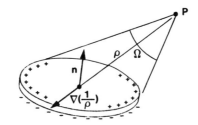

Fig. 1.4.2. Equivalence of currents and remanences.

where

$$\Phi(P) = \frac{I}{4\pi} \int_S \nabla \left[\frac{1}{\rho}\right] \cdot \vec{n} \, dS \ . \tag{1.4.34}$$

An equivalence can be established between the scalar potential 1.4.13 generated by a distribution of remanence J_0, and the scalar function Φ defined by Eq. 1.4.34, by replacing the fictitious surface S in Eq. 1.4.34 with a thin layer of thickness $\delta l \to 0$, as shown in Fig. 1.4.2. Assume that the volume

$$\delta V = \delta S \, \delta l \tag{1.4.35}$$

of the layer is magnetized with a remanence $\vec{J}_0$ oriented parallel to unit vector $\vec{n}$, such that

$$\lim_{\delta l \to 0} \vec{J}_0 \, \delta l = \vec{p}_0 \ , \tag{1.4.36}$$

where $\vec{p}_0$ is a vector called the magnetic dipole moment per unit area. By definition a dipole consists of two charges $\pm m$ separated by a distance d as shown in Fig. 1.4.3, and the dipole moment is the product $p = md$. By virtue of Eq. 1.4.16, $\pm J_0 dS$ are the two equivalent magnetic charges separated by a distance δl. Thus each element of volume δV of the layer is a dipole whose magnetic dipole moment is

$$d\vec{p} = \vec{p}_0 \, dS \ . \tag{1.4.37}$$

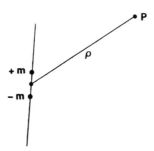

Fig. 1.4.3. Magnetic dipole.

Let us apply Eq. 1.4.13 to the element of volume δV. One has

$$\Phi(P) = \frac{1}{4\pi\mu_0} \lim_{\delta l \to 0} \int_{\delta V} \nabla\left[\frac{1}{\rho}\right] \cdot \vec{J}_0 \, dV \qquad (1.4.38)$$

and, by virtue of Eq. 1.4.36, Eq. 1.4.38 transforms to

$$\Phi(P) = \frac{p_0}{4\pi\mu_0} \int_S \nabla\left[\frac{1}{\rho}\right] \cdot \vec{n} \, dS = -\frac{p_0}{4\pi\mu_0} \Omega , \qquad (1.4.39)$$

where p_0 is the magnitude of the dipole moment $\vec{p}_0$ defined by Eq. 1.4.36, and Ω is the solid angle from point P subtended by surface S, as indicated in Fig. 1.4.2. The angle Ω is negative if the observer located at point P sees the positively charged side of the fictitious layer. The value of the scalar potential given by Eq. 1.4.39 becomes equal to the value of $\Phi(P)$ given by Eq. 1.4.34 if

$$p_0 = \mu_0 I , \qquad (1.4.40)$$

which establishes the equivalence of magnetized materials and electric currents.

1.5 MULTIPOLES

As shown in Section 4, the scalar potential generated by the remanence of the magnetic material can be computed as the potential generated by equivalent magnetic charges distributed on the surfaces of, as well as inside, the magnetized material. The equivalence between magnetic charge and remanence leads to the concept of magnetic multipoles, which is of particular importance in the study of the properties of complex magnetic structures.

To define the magnetic multipoles, consider first a point magnetic charge m_i located at a point P_i of coordinates ρ_i, θ_i, ψ_i in the frame of spherical polar coordinates ρ, θ, ψ shown in Fig. 1.5.1. Point charge m_i can be replaced by a surface charge density $\sigma_i(\theta, \psi)$ on the sphere of radius $\rho = \rho_i$, such that for $\theta \neq \theta_i$, $\psi \neq \psi_i$

$$\sigma_i = 0 , \qquad (1.5.1)$$

and the singularity of σ_i at $\theta = \theta_i$, $\psi = \psi_i$ satisfies the condition

$$m_i = \rho_i^2 \int_0^{2\pi} d\psi \int_0^{\pi} \sigma_i \sin\theta \, d\theta . \qquad (1.5.2)$$

26 MAGNETOSTATIC FIELDS

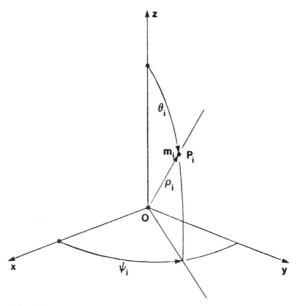

Fig. 1.5.1. Point charge m_i in a spherical frame of reference.

Surface charge density σ_i can be expanded in the series of spherical harmonics

$$\sigma_i = \sum_{l=0}^{\infty} \left\{ g_0^{(l)} P_l(\cos\theta) + \sum_{j=1}^{l} [g_{j,1}^{(l)} \cos(j\psi) + g_{j,2}^{(l)} \sin(j\psi)] P_l^j(\cos\theta) \right\}. \quad (1.5.3)$$

By virtue of the properties of P_l, P_l^j and Eqs. 1.5.1, 1.5.2, coefficients $g_0^{(l)}$, $g_{j,1}^{(l)}$, $g_{j,2}^{(l)}$ in Eq. 1.5.3 are equal to

$$g_0^{(l)} = \frac{2l+1}{4\pi\rho_i^2} \int_0^{2\pi} d\psi \int_{-1}^{+1} \rho_i^2 \sigma_i P_l(\xi) d\xi = \frac{2l+1}{4\pi\rho_i^2} P_l(\xi_i) m_i \quad (1.5.4)$$

and

$$g_{j,1}^{(l)} = \frac{2l+1}{2\pi\rho_i^2} \frac{(l-j)!}{(l+j)!} P_l^j(\xi_i) \cos(j\psi_i) m_i \quad (1.5.5)$$

$$g_{j,2}^{(l)} = \frac{2l+1}{2\pi\rho_i^2} \frac{(l-j)!}{(l+j)!} P_l^j(\xi_i) \sin(j\psi_i) m_i, \quad (1.5.6)$$

where

$$\xi_i = \cos\theta_i. \tag{1.5.7}$$

With the introduction of the surface charge distribution σ_i, the scalar potential generated by m_i can be computed by solving Laplace's equation in the two regions $\rho < \rho_i$ and $\rho > \rho_i$ and by imposing the boundary conditions at $\rho = \rho_i$.

In region $\rho < \rho_i$, the solution of Laplace's equation which is regular at $\rho = 0$ is

$$\Phi_{i,1} = \frac{1}{4\pi\mu_0} \sum_{l=0}^{\infty} \rho^l \left\{ q_{i,0}^{(l)} P_l(\xi) + \sum_{j=1}^{l} [q_{i,j,1}^{(l)} \cos(j\psi) + q_{i,j,2}^{(l)} \sin(j\psi)] P_l^j(\xi) \right\} \tag{1.5.8}$$

and in region $\rho > \rho_i$, the solution regular at $\rho = \infty$ is

$$\Phi_{i,2} = \frac{1}{4\pi\mu_0} \sum_{l=0}^{\infty} \frac{1}{\rho^{l+1}} \left\{ p_{i,0}^{(l)} P_l(\xi) + \sum_{j=1}^{l} [p_{i,j,1}^{(l)} \cos(j\psi) + p_{i,j,2}^{(l)} \sin(j\psi)] P_l^j(\xi) \right\}. \tag{1.5.9}$$

The values of the constants of integration $p_{i,0}^{(l)}, p_{i,j,1}^{(l)}, p_{i,j,2}^{(l)}, q_{i,0}^{(l)}, q_{i,j,1}^{(l)}, q_{i,j,2}^{(l)}$ are provided by the boundary conditions on surface $\rho = \rho_i$

$$\begin{aligned} \Phi_{i,2}(\rho_i) - \Phi_{i,1}(\rho_i) &= 0 \\ H_{i,\rho,2}(\rho_i) - H_{i,\rho,1}(\rho_i) &= \frac{\sigma}{\mu_0}, \end{aligned} \tag{1.5.10}$$

where $H_{i,\rho,1}$ and $H_{i,\rho,2}$ are the radial components of the intensity in the regions $\rho < \rho_i$, $\rho > \rho_i$, respectively. Equations 1.5.10 are identically satisfied for all values of θ and ψ. By virtue of Eqs. 1.5.3, 1.5.8 and 1.5.9, Eqs. 1.5.10 yield

$$p_{i,0}^{(l)} = m_i \rho_i^l P_l(\xi_i) \tag{1.5.11}$$

$$p_{i,j,1}^{(l)} = 2m_i \rho_i^l \frac{(l-j)!}{(l+j)!} P_l^j(\xi_i) \cos(j\psi_i) \tag{1.5.12}$$

$$p_{i,j,2}^{(l)} = 2m_i \rho_i^l \frac{(l-j)!}{(l+j)!} P_l^j(\xi_i) \sin(j\psi_i) \tag{1.5.13}$$

28 MAGNETOSTATIC FIELDS

and

$$q_{i,0}^{(l)} = \frac{p_{i,0}^{(l)}}{\rho_i^{2l+1}}, \quad q_{i,j,1}^{(l)} = \frac{p_{i,j,1}^{(l)}}{\rho_i^{2l+1}}, \quad q_{i,j,2}^{(l)} = \frac{p_{i,j,2}^{(l)}}{\rho_i^{2l+1}}. \tag{1.5.14}$$

Expansion 1.5.9 of the scalar potential $\Phi_{i,2}$ contains one term for $l = 0$. One has

$$P_0(\xi_i) = 1, \quad p_{i,0}^{(0)} = m_i; \tag{1.5.15}$$

i.e., term $l = 0$ in Eq. 1.5.9 is the scalar potential of a magnetic charge m_i located at the origin of the frame of reference ρ, θ, ψ.

Expansion 1.5.9 contains three terms for $l = 1$. One has

$$P_1(\xi_i) = \cos\theta_i, \quad P_1^1(\xi_i) = \sin\theta_i \tag{1.5.16}$$

and Eqs. 1.5.11, 1.5.12, 1.5.13 reduce to the three nonzero coefficients

$$p_{i,0}^{(1)} = m_i z_i, \quad p_{i,1,1}^{(1)} = m_i x_i, \quad p_{i,1,2}^{(1)} = m_i y_i, \tag{1.5.17}$$

where

$$x_i = \rho_i \sin\theta_i \cos\psi_i, \quad y_i = \rho_i \sin\theta_i \sin\psi_i, \quad z_i = \rho_i \cos\theta_i. \tag{1.5.18}$$

Coefficients 1.5.17 are the components of a vector

$$\vec{p}_i^{(1)} = m_i \vec{\rho}_i, \tag{1.5.19}$$

where $\vec{\rho}_i$ is a vector of magnitude ρ_i oriented in the direction from the origin O of the frame of reference to point P_i. Vector $\vec{p}_i^{(1)}$ is the dipole moment of m_i relative to origin O. Terms $l = 1$ in Eq. 1.5.9 are

$$\frac{1}{4\pi\mu_0} \frac{1}{\rho^3} \left[p_{i,1,1}^{(1)} x + p_{i,1,2}^{(1)} y + p_{i,0}^{(1)} z \right] = -\frac{1}{4\pi\mu_0} \nabla\left[\frac{1}{\rho}\right] \cdot \vec{p}_i^{(1)}, \tag{1.5.20}$$

where x, y, z are the cartesian coordinates of point P. Terms $l = 1$ in Eq. 1.5.9 are the potential of a dipole of moment $\vec{p}_i^{(1)}$ located at the origin O.

Expansion 1.5.9 contains five terms for $l = 2$. One has

$$P_2(\xi) = \frac{1}{2}(3\cos^2\theta - 1), \quad P_2^1(\xi) = 3\sin\theta\cos\theta, \quad P_2^2(\xi) = 3\sin^2\theta, \tag{1.5.21}$$

and the nonzero coefficients in Eqs 1.5.11, 1.5.12, 1.5.13 are

$$p_{i,0}^{(2)} = \frac{1}{2} m_i \rho_i^2 (3\cos^2 \theta_i - 1)$$

$$p_{i,1,1}^{(2)} = \frac{1}{2} m_i \rho_i^2 \sin 2\theta_i \cos \psi_i$$

$$p_{i,1,2}^{(2)} = \frac{1}{2} m_i \rho_i^2 \sin 2\theta_i \sin \psi_i \qquad (1.5.22)$$

$$p_{i,2,1}^{(2)} = \frac{1}{4} m_i \rho_i^2 \sin^2 \theta_i \cos 2\psi_i$$

$$p_{i,2,2}^{(2)} = \frac{1}{4} m_i \rho_i^2 \sin^2 \theta_i \sin 2\psi_i \; .$$

Coefficients 1.5.22 are the components of the quadrupole moment of m_i relative to the origin O, and terms $l = 2$ in Eq. 1.5.9 are the potential of a quadrupole located at O, which decreases with the radial distance ρ as ρ^{-3}.

For $l = 3$, expansion 1.5.9 contains seven terms. One has

$$P_3(\xi) = \frac{1}{2} (5\cos^2 \theta - 3) \cos \theta \; ,$$

$$P_3^1(\xi) = \frac{3}{2} \sin \theta (5\cos^2 \theta - 1) \; , \qquad (1.5.23)$$

$$P_3^2(\xi) = 15 \sin^2 \theta \cos \theta \; , \quad P_3^3(\xi) = 15 \sin^3 \theta \; ,$$

and the nonzero ($l = 3$) coefficients in Eqs. 1.5.11, 1.5.12, 1.5.13 are

$$p_{i,0}^{(3)} = \frac{1}{2} m_i \rho_i^3 (5\cos^2 \theta_i - 3) \cos \theta_i$$

$$p_{i,1,1}^{(3)} = \frac{1}{4} m_i \rho_i^3 (5\cos^2 \theta_i - 1) \sin \theta_i \cos \psi_i$$

$$p_{i,1,2}^{(3)} = \frac{1}{4} m_i \rho_i^3 (5\cos^2 \theta_i - 1) \sin \theta_i \sin \psi_i$$

$$p_{i,2,1}^{(3)} = \frac{1}{4} m_i \rho_i^3 \sin^2 \theta_i \cos \theta_i \cos 2\psi_i \qquad (1.5.24)$$

$$p_{i,2,2}^{(3)} = \frac{1}{4} m_i \rho_i^3 \sin^2 \theta_i \cos \theta_i \sin 2\psi_i$$

$$p_{i,3,1}^{(3)} = \frac{1}{24} m_i \rho_i^3 \sin^3 \theta_i \cos 3\psi_i$$

$$p_{i,3,2}^{(3)} = \frac{1}{24} m_i \rho_i^3 \sin^3 \theta_i \sin 3\psi_i \; .$$

30 MAGNETOSTATIC FIELDS

Coefficients 1.5.24 are the components of the octupole moment of m_i relative to O, and terms $l = 3$ in Eq. 1.5.9 are the potential of an octupole located at O, which decreases with ρ as ρ^{-4}.

In general, the terms of Eq. 1.5.9 for a given value of l are the scalar potential of a multipole of order l, whose moment is defined by $2l + 1$ coefficients given by Eqs. 1.5.11, 1.5.12, 1.5.13. The potential of the multipole of order l outside of the sphere of radius ρ_i decreases with ρ as $\rho^{-(l+1)}$.

Let us analyze now the scalar potential $\Phi_{i,1}$ given by Eq. 1.5.8 within the region $\rho < \rho_i$. Expansion 1.5.8 contains one term for $l = 0$, which corresponds to coefficient

$$q_{i,0}^{(0)} = \frac{m_i}{\rho_i} . \tag{1.5.25}$$

Thus term $l = 0$ in Eq. 1.5.8 is a uniform scalar potential equal to the value of $\Phi_{i,2}$ for $\rho = \rho_i$.

The three nonzero coefficients in Eqs. 1.5.14 for $l = 1$ result in the potential

$$\frac{1}{4\pi\mu_0} \frac{m_i}{\rho_i^3} (x_i x + y_i y + z_i z) = \frac{1}{4\pi\mu_0 \rho_i^3} \vec{\rho} \cdot \vec{p}_i^{(1)} , \tag{1.5.26}$$

where $\vec{\rho}$ is a vector of magnitude ρ oriented in the direction from the origin O to point P. Thus terms $l = 1$ in Eq. 1.5.8 correspond to a uniform magnetic field whose intensity is

$$\vec{H} = -\frac{\vec{p}_i^{(1)}}{4\pi\mu_0 \rho_i^3} . \tag{1.5.27}$$

Expansion 1.5.8 contains five terms for $l = 2$, which correspond to the five coefficients 1.5.22. Assume that the frame of spherical polar coordinates ρ, θ, ψ is rotated by angle θ_i about an axis which passes through O and is perpendicular to the plane $\psi = \psi_i$. In this new frame of reference ρ, θ', ψ', axis $\theta' = 0$ coincides with the line which passes through O and P_i, and the angular coordinate θ' of a point P is related to the θ, ψ coordinates of P by the equation

$$\cos \theta' = \cos \theta_i \cos \theta + \sin \theta_i \sin \theta \cos (\psi - \psi_i) . \tag{1.5.28}$$

In the new frame of reference ρ, θ', ψ', the five coefficients 1.5.14, for $l = 2$, reduce to one,

$$q_{i,0}^{(2)} = \frac{m_i}{\rho_i^3} , \tag{1.5.29}$$

and the term $l = 2$ in Eq. 1.5.8 is the potential

$$\frac{1}{4\pi\mu_0} m_i \frac{\rho^2}{\rho_i^3} P_2(\cos \theta') = \frac{1}{8\pi\mu_0} \frac{m_i}{\rho_i^3} (2z^2 - r^2) , \tag{1.5.30}$$

where

$$z = \rho \cos\theta', \quad r = \rho \sin\theta', \quad (1.5.31)$$

and function 1.5.30 coincides with the particular case 1.3.29 of the scalar potential given by Eq. 1.3.26, whose equipotential surfaces are the hyperboloids 1.3.30.

1.6 ENERGY OF A MAGNETOSTATIC FIELD

Energy has to be spent to generate a magnetostatic field. Depending upon the source of the field, the energy is spent either to generate the distribution of electric current or to magnetize the magnetic material.

Assume that the intensity of the magnetic field has a nonzero value in a volume V enclosed by surface S. Outside S, the scalar potential Φ has an arbitrary constant value which can be assumed to be zero. Furthermore, let us assume an ideal medium of zero magnetic susceptibility inside S.

By definition, the scalar potential at each point P inside S is the work done against the magnetic field to bring a unit magnetic charge from a point P_0 outside S to P. Thus the energy U spent to bring a charge m to point P is

$$U = m \int_{P_0}^{P} d\Phi = m\Phi, \quad (1.6.1)$$

where Φ is the scalar potential at P. Assume that Φ is the potential generated by a charge m_1 located at a point P_1. By virtue of Eq. 1.4.15, the work done against the field generated by m_1 to bring a charge m_2 to a point P_2 is

$$\frac{1}{4\pi\mu_0} \frac{m_1 m_2}{\rho_{1,2}} = \frac{1}{2} [\Phi_{1,2} m_2 + \Phi_{2,1} m_1], \quad (1.6.2)$$

where $\rho_{1,2}$ is the distance between the two charges. $\Phi_{1,2}$ is the potential of charge m_1 at the point of charge m_2 and $\Phi_{2,1}$ is the potential of charge m_2 at the point of charge m_1. If one assumes that a third charge m_3 is brought to a point P_3 at a distance $\rho_{1,3}$ from m_1 and $\rho_{2,3}$ from m_2, the total work necessary to locate the three charges at points P_1, P_2, P_3 is

$$U = \frac{1}{2} \sum_{i=1}^{3} \Phi_i m_i, \quad (1.6.3)$$

where Φ_i is the scalar potential at the point of charge m_i generated by the other two charges. Equation 1.6.3 can be extended to the general distribution of volume density and surface density of charges equivalent to the distribution of the remanence within V

and one has

$$U = \frac{1}{2}\int_V \Phi \upsilon \, dV + \frac{1}{2}\int_S \Phi \sigma \, dS \ , \qquad (1.6.4)$$

where volume charge density υ and surface charge density σ are given by Eqs. 1.1.17 and 1.4.16, respectively. Because of Eq. 1.1.10, one has

$$\upsilon = \mu_0 \nabla \cdot \vec{H} \qquad (1.6.5)$$

and

$$\frac{1}{2}\int_V \Phi \upsilon \, dV = \frac{\mu_0}{2}\int_V H^2 \, dV + \frac{\mu_0}{2}\int_S \Phi \vec{H} \cdot \vec{n} \, dS \ . \qquad (1.6.6)$$

In the second term on the right hand side of Eq. 1.6.6, $\vec{H}$ is the intensity of the magnetic field at a point inside S at an infinitesimal distance from S. By definition the scalar potential is zero on surface S. Thus, the second term on the right hand side of Eq. 1.6.6 is zero, and one has

$$U = \frac{\mu_0}{2}\int_V H^2 \, dV \ . \qquad (1.6.7)$$

The total energy U can be interpreted as an energy stored within the volume V and distributed with an energy density

$$W = \frac{\mu_0}{2} H^2 \ . \qquad (1.6.8)$$

Assume that intensity $\vec{H}$ is generated by a distribution of remanence $\vec{J}_0$ of the magnetized material contained inside volume V. If no flux of the magnetic induction $\vec{B}$ is generated anywhere within V, outside the magnetized material the intensity $\vec{H}$ is zero, and inside the material the intensity $\vec{H}$ is given by

$$\mu_0 \vec{H} = -\vec{J}_0 \qquad (1.6.9)$$

and Eq. 1.6.7 yields

$$U_0 = \frac{1}{2\mu_0}\int_V J_0^2 \, dV \ . \qquad (1.6.10)$$

U_0 is the energy stored in the magnetic material which can be interpreted as distributed with a volume density

$$W_0 = \frac{1}{2\mu_0} J_0^2 . \tag{1.6.11}$$

W_0 is the energy source that the designer of a permanent magnet can tap to generate a magnetic field outside of the magnetic material.

1.7 MAGNETIC MATERIALS

Let us conclude this introductory chapter with a brief summary of basic properties of magnetic materials. Section 1 has classified the media according to the sign of the magnetic susceptibility χ_m. Positive values of χ_m identify paramagnetic media and negative values of χ_m identify diamagnetic media.

In a classical model of diamagnetism, a time dependent magnetic field applied to the medium interacts with the orbiting electrons of the atoms according to the laws of classical electrodynamics. The interaction results in an induced electric current which opposes the change in time of the intensity $\vec{H}$ of the applied magnetic field. Thus, if the amplitude of $\vec{H}$ increases, the induced current generates a magnetic polarization $\vec{J}_m$ oriented in the direction opposite to $\vec{H}$. Diamagnetic atoms do not exhibit a permanent magnetic dipole moment.

By contrast, paramagnetism results from a permanent magnetic dipole moment, which characterizes atoms with unbalanced electron spin and orbital moments. In the absence of an external magnetic field, the permanent magnetic dipoles are oriented at random. However, when a magnetic field is applied to the medium, the dipoles tend to orient themselves along the applied field, and this results in a finite polarization $\vec{J}_m$ oriented as $\vec{H}$. For a sufficiently small intensity $\vec{H}$, $\vec{J}_m$ is proportional to $\vec{H}$, resulting in a positive value of parameter χ_m.

It is worthwhile pointing out that in magnetic media classified as paramagnetic, the interaction between field and permanent dipoles is strong enough to compensate for the diamagnetic effect which is always present.

Certain media may exhibit a spontaneous magnetization even in the absence of an external magnetic field. When the interaction forces between neighboring dipoles are sufficiently strong, each dipole tends to orient itself in a direction parallel to its neighbor. This alignment extends over a small volume which is called a "domain." Each domain exhibits a magnetic dipole moment, and the dipoles associated with the individual domains may be oriented at random relative to each other, in which case the medium exhibits no magnetization on a macroscopic scale. In the presence of an external magnetic field, the individual domains tend to align themselves along the applied field, and as the intensity of the magnetic field increases, a saturation effect occurs, which results in an upper limit of the polarization of the medium.

When the external magnetic field is removed, the individual domains do not return to their original orientation. This effect results in the well known hysteresis cycle. Media that exhibit this property are classified as ferromagnetic. The ferromagnetic effect is temperature dependent, and the magnetic dipole moments of the individual domains vanish at a critical temperature called the "Curie Temperature."

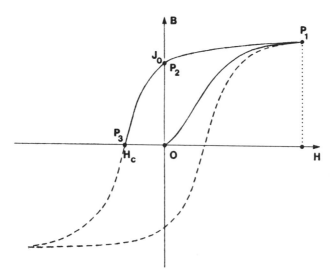

Fig. 1.7.1. Hysteresis cycle.

A permanent magnetization can also be attained in media where the interaction forces between neighboring dipoles result in an antiparallel orientation of neighboring dipoles. Although neighbor dipole moments cancel each other, certain crystal lattices exhibit a parallel orientation of a number of dipoles. This property, called ferrimagnetism, occurs in ferrites. The interaction between elementary dipoles in a ferrimagnetic medium differs from the interaction between dipoles in a ferromagnetic medium. However, the interaction between magnetic domains and an external magnetic field is the same in both categories of magnetic media.

A typical hysteresis cycle of a ferromagnetic medium is shown in Fig. 1.7.1. Line OP_1 is the curve of first magnetization, and the ordinate of point P_1 is the value of the magnetic induction induced by the maximum intensity H_1 of the magnetizing field. For a sufficiently large value of H_1, the polarization of the medium reaches saturation. As the intensity of the field within the medium is reduced to zero, the magnetic induction follows lines (P_1P_2) and, in the linear model of Eq. 1.1.1, the value of the magnetic induction at $H = 0$ is the permanent polarization J_0.

As the sign of the external field is reversed, the magnetic induction decreases following line (P_2P_3) in the second quadrant of Fig. 1.7.1. Line (P_2P_3) is the demagnetization characteristic and the abscissa of point P_3 is the value of the intensity H_c of the magnetic field required to cancel the induction in the material. H_c is called the coercive force. The broken line in Fig. 1.7.1 completes the hysteresis cycle.

The basic properties of the demagnetization curve of a permanent magnet are described in Fig. 1.7.2. The first quadrant shows the energy product $|BH|$ of the magnetic material, defined as the product of the magnitudes of the magnetic induction and the intensity at each point of the demagnetization curve. The energy product, plotted in the first quadrant versus the magnetic induction, is zero at $B = 0$ and $B = J_0$. It attains a maximum value $\overline{W}$ at the point where the demagnetization curve is tangent to the hyperbola

$$|BH| = \overline{W}, \qquad (1.7.1)$$

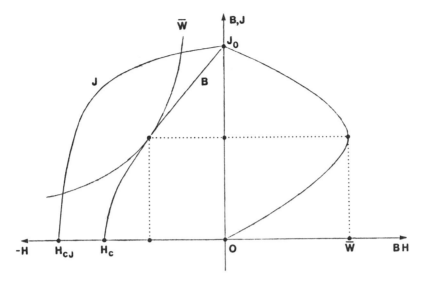

Fig. 1.7.2. Demagnetization characteristics.

as indicated in the second quadrant of Fig. 1.7.2. $\overline{W}$ is one of the basic parameters of a magnetic material. The higher the value of $\overline{W}$ is, the higher the energy stored in the magnetic material is. Thus the amount of magnetic material required to generate a given field depends primarily on the value of $\overline{W}$.

A second important characteristic of the material is the curve of the polarization density

$$J = B - \mu_0 H \qquad (1.7.2)$$

plotted in the second quadrant of Fig. 1.7.2 versus the intensity of the magnetic field. The maximum value of J is the remanence J_0. As the intensity of the demagnetizing field increases, $\vec{J}$ decreases to zero at a value $H_{c,J}$ of the intensity which corresponds to a point in the third quadrant of the cycle of Fig. 1.7.1. The intensity $H_{c,J}$ that cancels the magnetization of the material is called the intrinsic coercive force. The larger the value of $H_{c,J}$ is, the more resistant the magnetic material is to external demagnetizing forces. Thus a large value of $H_{c,J}$ is of importance in applications where the magnet is subjected to large time varying fields.

In the ideal case of a linear demagnetization curve, by virtue of Eq. 1.1.6, Eq. 1.7.2 defines the straight line

$$J = J_0 + \mu_0 \chi_m H \ . \qquad (1.7.3)$$

In the limit

$$\chi_m = 0 \qquad (1.7.4)$$

36 MAGNETOSTATIC FIELDS

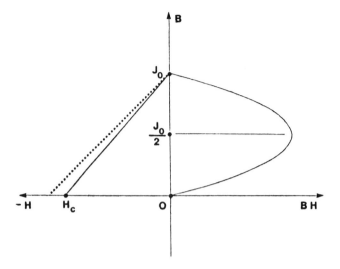

Fig. 1.7.3. Linear demagnetization characteristics.

the polarization is not affected by the demagnetizing field. Thus, limit 1.7.4 defines a rigid magnet that is equivalent to a distribution of fixed magnetic dipoles in a nonmagnetic medium. As a consequence of Eq. 1.7.4, a rigid magnetic material is perfectly transparent to the field generated by external sources.

The ideal case of a linear demagnetization curve is illustrated in Fig. 1.7.3, where the dotted line corresponds to the limit 1.7.4. The maximum value of the energy product occurs at

$$B = \frac{J_0}{2} \qquad (1.7.5)$$

independent of the value of the magnetic susceptibility, and the maximum value $\overline{W}$ is

$$\overline{W} = \frac{J_0^2}{4\mu_0} \frac{1}{1 + \chi_m} . \qquad (1.7.6)$$

Thus, $\overline{W}$ is half the value of the energy W_0 given by Eq. 1.6.11.

Examples of demagnetization characteristics of typical magnets are shown in Fig. 1.7.4 [4,5]. One observes that Alnico exhibits a larger remanence than the Nd-Fe-B rare earth alloy. However, Alnico exhibits much smaller coercive force H_c and energy product $\overline{W}$ than the Nd-Fe-B alloy. The polarization characteristic of Alnico practically coincides with its demagnetization characteristic. By contrast the quasi-linear demagnetization curve of the Nd-Fe-B alloy results in a polarization characteristic which is practically a straight line in a range of a demagnetizing field exceeding the coercive force H_c.

Typical values of the coercive force H_c and the intrinsic coercive force $H_{c,J}$ of a Nd-Fe-B alloy are

$$H_c \approx 12 KHe , \qquad H_{c,J} \approx 17 KHe . \qquad (1.7.7)$$

As a consequence a demagnetizing field much larger than the coercive force H_c is needed to cancel the magnetization of this rare earth alloy.

The third example of magnetic material shown in Fig. 1.7.4 is a ferrite which also exhibits a quasi-linear demagnetization curve. Compared to Alnico, a ferrite magnet

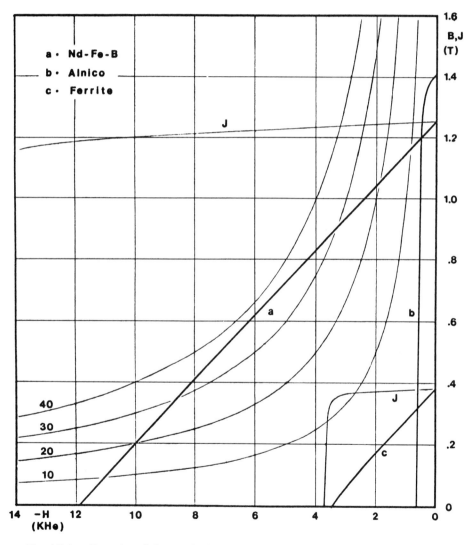

Fig. 1.7.4. Examples of demagnetization characteristics of typical magnets. The energy product values are given in MGHe (10^6 oersteds · gauss).

exhibits a much larger coercive force in spite of the fact that it has a substantially lower remanence. Compared to rare earth alloys, the energy product of a ferrite ceramic magnet is about one order of magnitude smaller.

The magnetic susceptibility χ_m within the quasi-linear range of the polarization characteristic of the Nd-Fe-B alloys shown in Fig. 1.7.4 is

$$\chi_m \approx 5 \cdot 10^{-2} . \tag{1.7.8}$$

The quasi-linear demagnetization characteristic and the transparency to a magnetic field are the most important properties that differentiate rare earth alloys and Ferrites from steels and Alnico. These properties make it possible to develop new design procedures and more efficient magnetic structures completely different from traditional magnets. These new designs and their optimization are the subject of the following chapters.

REFERENCES

[1] D. S. Jones, *The Theory of Electromagnetism*, The Macmillan Company, New York, 1964.
[2] J. A. Stratton, *Electromagnetic Theory*, McGraw-Hill Book Company, Inc, New York and London, 1941.
[3] J. D. Jackson, *Classical Electrodynamics*, John Wiley, New York, 1975.
[4] K. J. Strnat, Modern Permanent Magnets for Applications in Electro-Technology, *Proc. IEEE* 78, No. 6, (June 1990) 923-946.
[5] R. J. Parker, *Advances in Permanent Magnetism*, John Wiley, New York, 1990.

CHAPTER 2

Basic Geometries of Permanent Magnets

INTRODUCTION

Equations 1.1.16 and 1.1.18 of Chapter 1 can be solved exactly in simple geometries of uniform media separated by interfaces which coincide with coordinate surfaces of appropriately chosen orthogonal curvilinear systems such as parallel planes, coaxial cylinders or concentric spheres. This chapter reviews the solutions of the field equations in these geometries for the purpose of introducing some basic properties of the structures of permanent magnets with linear demagnetization characteristics.

2.1. ONE-DIMENSIONAL MAGNET

Consider the one-dimensional problem of a uniformly magnetized material contained between two plane parallel surfaces $y = \pm y_0$ shown in Fig. 2.1.1. The medium outside the magnetized material is assumed to be air. The magnetic susceptibility χ_m is assumed to be zero everywhere, and the remanence $\vec{J}_0$ of the magnetic material is assumed to be uniform and oriented along the y axis. Then remanence $\vec{J}_0$ satisfies the condition

$$\nabla \cdot \vec{J}_0 = 0 . \tag{2.1.1}$$

The uniformly magnetized medium is equivalent to two surface charge densities

$$\sigma = \pm J_0 , \tag{2.1.2}$$

distributed on the two planes $y = \pm y_0$, respectively. Since the value of the scalar potential Φ generated by σ at a point P depends only on the distance of P from the surfaces of the medium, one can assume that P is located on the y axis. From Eq. 1.4.13 one has

$$\Phi = \frac{J_0}{2\mu_0} \int_0^\infty \left[\frac{1}{\rho_1} - \frac{1}{\rho_2} \right] r \, dr , \tag{2.1.3}$$

39

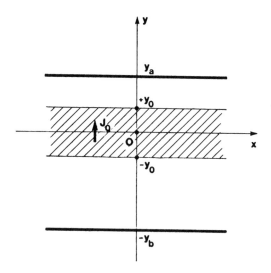

Fig. 2.1.1. One-dimensional magnet.

where

$$\rho_{1,2} = \left[r^2 + (y \mp y_0)^2 \right]^{1/2}, \qquad r = (x^2 + z^2)^{1/2}. \tag{2.1.4}$$

Equation 2.1.3 yields

$$\Phi = \begin{cases} -\dfrac{J_0}{\mu_0} y_0 & (y < -y_0) \\ \dfrac{J_0}{\mu_0} y & (|y| < y_0) \\ +\dfrac{J_0}{\mu_0} y_0 & (y > y_0) . \end{cases} \tag{2.1.5}$$

In the external regions $|y| > y_0$ the intensity of the magnetic field vanishes because the scalar potential is constant.

The linear distribution of Φ in the region $|y| < y_0$ results in a uniform intensity

$$\vec{H} = -\frac{1}{\mu_0} \vec{J}_0, \tag{2.1.6}$$

independent of the thickness $2y_0$ of the magnetized medium. By virtue of Eq. 2.1.6 the magnetic induction $\vec{B}$ inside the magnetized medium is

$$\vec{B} = \mu_0 \vec{H} + \vec{J}_0 = 0. \tag{2.1.7}$$

Thus, the magnetic induction is zero everywhere and the intensity inside the magnetic material is equal to the coercive force H_c.

A nonzero value of the intensity of the magnetic field outside the magnetized medium can be achieved only if an additional constraint is imposed on the scalar potential Φ in regions $|y| > y_0$. Assume for instance that the outside medium is limited by the two plane surfaces of a medium of infinite magnetic permeability located at

$$y = y_a, \qquad y = -y_b, \qquad (2.1.8)$$

as shown in Fig. 2.1.1. If the two surfaces close upon themselves at infinity, one can impose the boundary condition

$$\Phi(y_a) = \Phi(-y_b) = 0. \qquad (2.1.9)$$

Within the region $-y_0 < y < y_0$, the intensity of the magnetic field becomes

$$\vec{H}_1 = -\left(1 - \frac{y_0}{y_1}\right) \frac{\vec{J}_0}{\mu_0}, \qquad (2.1.10)$$

where

$$y_1 = \frac{1}{2}(y_a + y_b). \qquad (2.1.11)$$

In both regions $y_0 < y < y_a$ and $-y_b < y < -y_0$ the intensity is

$$\vec{H}_0 = \frac{y_0}{y_1} \frac{\vec{J}_0}{\mu_0}. \qquad (2.1.12)$$

One observes that $\vec{H}_0$ and $\vec{H}_1$ are uniform and independent of the position of the magnetized medium inside the region $-y_b < y < y_a$. In the limit $y_1 \to y_0$ one has

$$\lim_{y_1 \to y_0} \vec{H}_0 = \frac{\vec{J}_0}{\mu_0}, \qquad \lim_{y_1 \to y_0} \vec{H}_1 = 0. \qquad (2.1.13)$$

The parameter

$$K = \frac{\mu_0 H_0}{J_0} \qquad (2.1.14)$$

ranges from zero, in the limit of a thickness $2y_0$ of material very small compared to $2y_1$, to a maximum equal to unity in the limit of a thickness $2(y_1 - y_0)$ of the air region small compared to $2y_0$. Parameter K will be used throughout the text in the analysis of magnetic structures designed to generate a uniform field within the region of interest.

Consider a cylindrical volume with its axis parallel to y and a cross-sectional area equal to unity. By virtue of Eqs. 1.6.8 and 2.1.12, within the volume of the cylinder limited by the two planes 2.1.8 and outside the magnetized medium, the energy of the magnetic field is

$$U = \frac{J_0^2}{\mu_0} \frac{y_0^2}{y_1^2} (y_1 - y_0) . \tag{2.1.15}$$

For $y_1 = y_0$, i.e., if the volume of the nonmagnetic material vanishes, U is zero. U is also zero in the limit $y_1 \to \infty$, when the intensity of the magnetic field vanishes outside the magnetic material. In this limit the energy is confined within the magnetic material where the intensity is given by Eq. 2.1.6., and the energy per unit volume "stored" inside the magnetic material is

$$W_0 = \frac{1}{2\mu_0} J_0^2 . \tag{2.1.16}$$

It should be noted that Eq. 2.1.16 coincides with Eq. 1.6.11.

The maximum value of U occurs when

$$y_1 = 2y_0 , \tag{2.1.17}$$

i.e., when the air region between the two planes 2.1.8 has the same thickness as the magnetized material. When condition 2.1.17 is satisfied, the magnitudes of both vectors $\vec{H}_1$ and $\vec{H}_0$ are also equal to each other:

$$H_1 = H_0 = \frac{J_0}{2\mu_0} , \tag{2.1.18}$$

and the maximum of U is

$$U_{max} = \overline{W} , \tag{2.1.19}$$

where $\overline{W}$ is the maximum value of the energy product given by Eq. 1.7.6 for $\chi_m = 0$. Thus, in a one-dimensional magnet the maximum of U is attained when the magnetic material operates at the peak of the energy product curve defined in Section 1.7.

2.2 UNIFORMLY MAGNETIZED CYLINDRICAL MAGNET

One of the simplest two-dimensional magnetic structures is a geometry of cylindrical coaxial interfaces between the media. Such a geometry is shown in Fig. 2.2.1 with three interfaces of radii $r_1 < r_2 < r_3$. The medium inside the cylinder of radius r_1 is assumed to be air ($\chi_m = 0$). The medium between the cylinders of radii r_1 and r_2 is the magnetized material. The region between the cylinders of radii r_2 and r_3 is

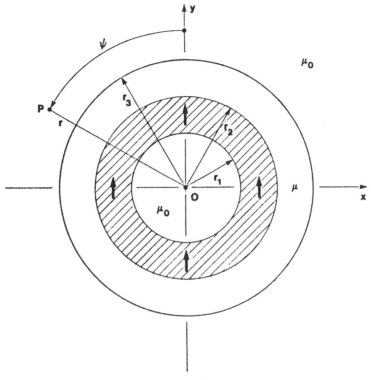

Fig. 2.2.1. Cylindrical magnet.

assumed to be a nonmagnetized medium of permeability μ, and the region outside the cylinder of radius r_3 is assumed to be free space.

Assume that the medium in the region $r_1 < r < r_2$ is magnetized with a uniform remanence $\vec{J}_0$ oriented along the y axis as shown by the heavy arrows in Fig. 2.2.1. Let us select a system of cylindrical polar coordinates r, ψ, z, where r is the distance from the z axis, ψ is the angle between r and the y axis, and z is the axial coordinate perpendicular to the plane of Fig. 2.2.1. In this system of coordinates the components of $\vec{J}_0$ are

$$J_r = J_0 \cos \psi, \qquad J_\psi = -J_0 \sin \psi, \qquad J_z = 0. \tag{2.2.1}$$

Remanence $\vec{J}_0$ is independent of z. Thus, a z-independent solution of Laplace's equation can be sought in each of the four regions of the structure of Fig. 2.2.1. Because of condition 2.2.1, the scalar potential $\Phi(r,\psi)$ satisfies Eq. 1.3.18. A particular solution consistent with remanence 2.2.1 is

$$\Phi(r,\psi) = R(r) \cos \psi. \tag{2.2.2}$$

Function $R(r)$ satisfies the equation

$$r \frac{d}{dr} \left[r \frac{dR}{dr} \right] - R = 0, \tag{2.2.3}$$

whose general solution is

$$R(r) = C_{i,1} r + C_{i,2}\frac{1}{r} . \qquad (2.2.4)$$

The subindices i identify the four regions of integration

$$\begin{aligned} i &= 0 & (r < r_1) \\ i &= 1 & (r_1 < r < r_2) \\ i &= 2 & (r_2 < r < r_3) \\ i &= 3 & (r > r_3) , \end{aligned} \qquad (2.2.5)$$

and the eight constants of integration $C_{i,1}$, $C_{i,2}$ are determined by the boundary conditions at the interfaces between the regions of integration. First, since $\Phi(r,\psi)$ must be finite everywhere,

$$C_{0,2} = C_{3,1} = 0 . \qquad (2.2.6)$$

By virtue of Eqs. 1.1.5, 2.2.2, and 2.2.4, the radial component of the magnetic induction $\vec{B}$ is

$$B_r = -\mu\frac{\partial \Phi}{\partial r} + J_r = -(\mu_0 \kappa_i C_{i,1} - \mu_0 \kappa_i C_{i,2}\frac{1}{r^2} - J_i)\cos\psi , \qquad (2.2.7)$$

where

$$J_i = \begin{cases} J_0 & (i = 1) \\ 0 & (i \neq 1) , \end{cases} \qquad (2.2.8)$$

and

$$\kappa_i = \begin{cases} \kappa_m & (i = 2) \\ 1 & (i \neq 2) . \end{cases} \qquad (2.2.9)$$

Since the magnetic susceptibility of the magnetized material is assumed to be zero and there is no surface current density on any interface, boundary conditions 1.1.35 and 1.1.41 yield the system of six linear equations

$$\begin{aligned} C_{0,1} - C_{1,1} - C_{1,2}r_1^{-2} &= 0 \\ -C_{0,1} + C_{1,1} - C_{1,2}r_1^{-2} &= \frac{J_0}{\mu_0} \\ C_{1,1} + C_{1,2}r_2^{-2} - C_{2,1} - C_{2,2}r_2^{-2} &= 0 \\ -C_{1,1} + C_{1,2}r_2^{-2} + \kappa_m C_{2,1} - \kappa_m C_{2,2}r_2^{-2} &= -\frac{J_0}{\mu_0} \\ C_{2,1} + C_{2,2}r_3^{-2} - C_{3,2}r_3^{-2} &= 0 \\ -\kappa_m C_{2,1} + \kappa_m C_{2,2}r_3^{-2} - C_{3,2}r_3^{-2} &= 0 . \end{aligned} \qquad (2.2.10)$$

The value of constant $C_{0,1}$ derived from Eq. 2.2.10 is

$$C_{0,1} = -\frac{J_0}{2\mu_0}(\kappa_m^2 - 1)\frac{\left[1 - \frac{r_1^2}{r_2^2}\right]\left[1 - \frac{r_2^2}{r_3^2}\right]}{(\kappa_m + 1)^2 - \frac{r_2^2}{r_3^2}(\kappa_m - 1)^2}. \qquad (2.2.11)$$

Hence, within region $r < r_1$, the scalar potential is

$$\Phi(r, \psi) = C_{0,1} r \cos\psi, \qquad (2.2.12)$$

and the intensity $\vec{H}_0$ of the magnetic field is

$$\vec{H}_0 = -C_{0,1}(\cos\psi\,\vec{r} - \sin\psi\,\vec{\psi}). \qquad (2.2.13)$$

Thus, the uniform remanence 2.2.1 generates a uniform magnetic field oriented in the direction of the $\psi = 0$ axis. From Eq. 2.2.11 one observes that $\vec{H}_0$ vanishes if

$$\kappa_m = 1 \quad \text{or} \quad r_3 = r_2; \qquad (2.2.14)$$

i.e., $\vec{H}_0$ vanishes in the absence of the medium of permeability $\mu \neq \mu_0$. In the limit 2.2.14 the system of Eqs. 2.2.10 yields the value of the constant of integration $C_{3,2}$,

$$C_{3,2} = \frac{J_0}{2\mu_0}(r_2^2 - r_1^2). \qquad (2.2.15)$$

Thus a uniformly magnetized hollow cylinder in free space generates a scalar potential

$$\Phi(r, \psi) = \frac{J_0}{2\mu_0}(r_2^2 - r_1^2)\frac{\cos\psi}{r} \qquad (2.2.16)$$

in the external region $r > r_2$ surrounding the magnet and no field in the region $r < r_1$.

Equation 2.2.16 can be rewritten in the form

$$\Phi(r, \psi) = \frac{1}{2\pi\mu_0}\frac{\vec{p}_l \cdot \vec{r}}{r}, \qquad (2.2.17)$$

where $\vec{r}$ is the unit vector oriented in the direction of the coordinate r and

$$\vec{p}_l = \pi(r_2^2 - r_1^2)\vec{J}_0. \qquad (2.2.18)$$

By definition $\vec{J}_0$ is the magnetic dipole moment per unit volume of the magnetized medium, and $\pi(r_2^2 - r_1^2)$ is the area of the cross-section of the medium perpendicular to the axis z. Thus Eq. 2.2.18 gives the magnetic dipole moment per unit length of the magnetized hollow cylinder and, as shown in Section 4 of Chapter 1, Eq. 2.2.17 is the scalar potential generated by the uniform distribution of $\vec{p}_l$ on the z axis.

46 BASIC GEOMETRIES OF PERMANENT MAGNETS

The vanishing of $\vec{H}_0$ in the absence of an external $\kappa_m \neq 1$ layer is the two-dimensional equivalent of the the one-dimensional problem analyzed in the preceding section in the absence of the two external planes of infinite magnetic permeability.

In the limit $\kappa_m \to \infty$, the intensity of the magnetic field $\vec{H}_0$ given by Eq. 2.2.13 becomes [1]

$$\lim_{\kappa_m \to \infty} \vec{H}_0 = \frac{1}{2\mu_0}\left[1 - \frac{r_1^2}{r_2^2}\right]\vec{J}_0, \qquad (2.2.19)$$

and the constants of integration $C_{1,1}, C_{1,2}$ reduce to

$$\lim_{\kappa_m \to \infty} C_{1,1} = \frac{J_0}{2\mu_0}\frac{r_1^2}{r_2^2}$$

$$\lim_{\kappa_m \to \infty} C_{1,2} = -\frac{J_0}{2\mu_0}r_1^2 \ . \qquad (2.2.20)$$

Thus, the scalar potential becomes

$$\Phi(r,\psi) = \begin{cases} -\dfrac{J_0}{2\mu_0}\left[1 - \dfrac{r_1^2}{r_2^2}\right] r\cos\psi & (r < r_1) \\[2ex] -\dfrac{J_0}{2\mu_0}\dfrac{r_1^2}{r_2}\left[\dfrac{r_2}{r} - \dfrac{r}{r_2}\right]\cos\psi & (r_1 < r < r_2) \\[2ex] 0 & (r > r_2) \ . \end{cases} \qquad (2.2.21)$$

2.3 HOLLOW CYLINDER OF MAGNETIC MATERIAL

The vanishing of $\vec{H}_0$ in the absence of the medium of permeability μ is the consequence of the particular type of magnetization defined by Eqs. 2.2.1. Let us consider the more general case of a nonuniform distribution of remanence

$$\vec{J} = \sum_n \left[F_n(r)\cos n\psi\, \vec{r} + G_n(r)\sin n\psi\, \vec{\psi}\right] \qquad (2.3.1)$$

of a medium within the two coaxial cylinders of radii r_1, r_2 shown in Fig. 2.3.1. The magnetic susceptibility χ_m is assumed to be zero and the media in regions $r < r_1$ and $r > r_2$ are assumed to be nonmagnetic. In Eq. 2.3.1, $F_n(r)$ and $G_n(r)$ are arbitrary functions of the radial coordinate r, independent of the axial coordinate z.

$\vec{J}$ can be written as the linear superposition of two remanences $\vec{J}_1, \vec{J}_2$

$$\vec{J} = \vec{J}_1 + \vec{J}_2, \qquad (2.3.2)$$

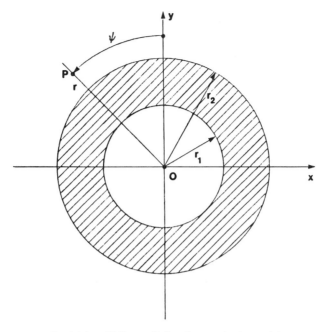

Fig. 2.3.1. Hollow cylinder of magnetized material.

where $\vec{J}_1$ is given by

$$\vec{J}_1 = \sum_n J_{1,n}(r) \left[\cos n\psi \, \vec{r} - \sin n\psi \, \vec{\psi}\right] \tag{2.3.3}$$

and $\vec{J}_2$ by

$$\vec{J}_2 = \sum_n J_{2,n}(r) \left[\cos n\psi \, \vec{r} + \sin n\psi \, \vec{\psi}\right]. \tag{2.3.4}$$

The two new functions $J_{1,n}(r)$, $J_{2,n}(r)$ are related to F_n and G_n by the equations

$$J_{1,n} = \frac{1}{2}(F_n - G_n), \qquad J_{2,n} = \frac{1}{2}(F_n + G_n). \tag{2.3.5}$$

Let us analyze the field generated by $\vec{J}_1$ and $\vec{J}_2$ independently from each other. Within the region $r_1 < r < r_2$, the scalar potential $\Phi(r, \psi)$ satisfies Poisson's equation

$$\nabla^2 \Phi = \frac{1}{r}\frac{\partial}{\partial r}\left[r \frac{\partial \Phi}{\partial r}\right] + \frac{1}{r^2}\frac{\partial^2 \Phi}{\partial \psi^2} = \frac{1}{\mu_0}\nabla \cdot \vec{J}, \tag{2.3.6}$$

where $\vec{J}$ is equal to either $\vec{J}_1$ or $\vec{J}_2$.

48 BASIC GEOMETRIES OF PERMANENT MAGNETS

Assume first $\vec{J} = \vec{J}_1$. One has

$$\nabla \cdot \vec{J}_1 = \frac{1}{r} \sum_n \left[r \frac{dJ_{1,n}}{dr} - (n-1) J_{1,n} \right] \cos n\psi. \quad (2.3.7)$$

The solution of Eq. 2.3.6 can be written in the form

$$\Phi(r, \psi) = \sum_n \Phi_{1,n}(r) \cos n\psi, \quad (2.3.8)$$

where $\Phi_{1,n}$ satisfies the equation

$$r \frac{d}{dr} \left[r \frac{d}{dr} (\Phi_{1,n} - X_n) \right] - n^2 (\Phi_{1,n} - X_n) = 0. \quad (2.3.9)$$

In Eq. 2.3.9, function X_n is related to $J_{1,n}$ by the differential equation

$$\frac{d}{dr}(r^n X_n) = \frac{1}{\mu_0} r^n J_{1,n}. \quad (2.3.10)$$

The general solution of Eq. 2.3.9 is

$$\Phi_{1,n} = X_n + C_{n,2,1} r^n + C_{n,2,2} r^{-n}, \quad (2.3.11)$$

where the values $C_{n,2,1}$, $C_{n,2,2}$ are constants of integration. The radial component of the magnetic induction $\vec{B}$ can be written in the form

$$B_r(r, \psi) = \sum_n B_{r,1,n}(r) \cos n\psi, \quad (2.3.12)$$

where, by virtue of Eqs. 2.3.10 and 2.3.11, functions $B_{r,1,n}$ are

$$B_{r,1,n} = -\mu_0 \frac{n}{r} \left[C_{n,2,1} r^n - C_{n,2,2} r^{-n} - X_n \right]. \quad (2.3.13)$$

In the two nonmagnetic regions the scalar potential is

$$\Phi(r, \psi) = \begin{cases} \sum_n C_{n,1} r^n \cos n\psi & (r < r_1) \\ \sum_n C_{n,3} r^{-n} \cos n\psi & (r > r_2), \end{cases} \quad (2.3.14)$$

where the values $C_{n,1}$, $C_{n,3}$ are additional constants of integration.

HOLLOW CYLINDER OF MAGNETIC MATERIAL

The boundary conditions at $r = r_1$ and $r = r_2$ must be identically satisfied for all values of ψ. Thus, for each value of n, the continuity of Φ and B_r results in a system of four equations in the unknown quantities $C_{n,1}$, $C_{n,2,1}$, $C_{n,2,2}$, $C_{n,3}$:

$$\begin{aligned} C_{n,1} - C_{n,2,1} - C_{n,2,2} r_1^{-2n} &= X_n(r_1) r_1^{-n} \\ -C_{n,1} + C_{n,2,1} - C_{n,2,2} r_1^{-2n} &= X_n(r_1) r_1^{-n} \\ C_{n,2,1} + C_{n,2,2} r_2^{-2n} - C_{n,3} r_2^{-2n} &= -X_n(r_2) r_2^{-n} \\ -C_{n,2,1} + C_{n,2,2} r_2^{-2n} - C_{n,3} r_2^{-2n} &= -X_n(r_2) r_2^{-n} \end{aligned} \quad (2.3.15)$$

The solution of the above system of equations is

$$\begin{aligned} C_{n,1} &= C_{n,2,1} = 0 \\ C_{n,2,2} &= -X_n(r_1) r_1^n \\ C_{n,3} &= X_n(r_2) r_2^n - X_n(r_1) r_1^n \ . \end{aligned} \quad (2.3.16)$$

Thus, by virtue of Eq. 2.3.10, function $\Phi_{1,n}(r)$ in the three regions of the cylindrical structure is

$$\Phi_{1,n}(r) = \begin{cases} 0 & (r < r_1) \\ \dfrac{1}{\mu_0 r^n} \displaystyle\int_{r_1}^{r} r^n J_{1,n} \, dr & (r_1 < r < r_2) \\ \dfrac{1}{\mu_0 r^n} \displaystyle\int_{r_1}^{r_2} r^n J_{1,n} \, dr & (r > r_2) \ . \end{cases} \quad (2.3.17)$$

The first equation of system 2.3.17 shows that the intensity of the magnetic field generated by remanence $\vec{J}_1$ inside the cylinder of radius r_1 is equal to zero, regardless of the value of r_1 and r_2. The interface $r = r_1$ in the region $r < r_1$ is a $\Phi = 0$ equipotential surface that satisfies the condition

$$B_r(r_1, \psi) = 0 \ . \quad (2.3.18)$$

The third equation of system 2.3.17 shows that remanence $\vec{J}_1$ generates a nonzero field outside the cylindrical surface $r = r_2$.

Let us consider next the field generated by the remanence $\vec{J}_2$ given by the Eq. 2.3.4. One has

$$\nabla \cdot \vec{J}_2 = \frac{1}{r} \sum_n \left[r \frac{dJ_{2,n}}{dr} + (n+1) J_{2,n} \right] . \quad (2.3.19)$$

50 BASIC GEOMETRIES OF PERMANENT MAGNETS

The solution of Eq. 2.3.6 for $\vec{J} = \vec{J}_2$ is

$$\Phi(r, \psi) = \sum \Phi_{2,n}(r) \cos n\psi , \quad (2.3.20)$$

where $\Phi_{2,n}$ satisfies the equation

$$r \frac{d}{dr}\left[r \frac{d}{dr}(\Phi_{2,n} - Y_n)\right] - n^2(\Phi_{2,n} - Y_n) = 0 . \quad (2.3.21)$$

Function Y_n is related to $J_{2,n}$ by the equation

$$\frac{d}{dr}\left[\frac{Y_n}{r^n}\right] = \frac{1}{\mu_0} \frac{J_{2,n}}{r^n} . \quad (2.3.22)$$

The solution of Eq. 2.3.21 in the three regions of the cylindrical structure is

$$\Phi_{2,n} = \begin{cases} D_{n,1} r^n & (r < r_1) \\ D_{n,2,1} r^n + D_{n,2,2} r^{-n} + Y_n & (r_1 < r < r_2) \\ D_{n,3} r^{-n} & (r > r_2) , \end{cases} \quad (2.3.23)$$

where $D_{n,1}$, $D_{n,2,1}$, $D_{n,2,2}$, $D_{n,3}$ are the new constants of integration. The value of the radial component $B_{r,2,n}$ of the magnetic induction in the region $r_1 < r < r_2$ is

$$B_{r,2,n} = -\mu_0 \frac{n}{r}\left[D_{n,2,1} r^n - D_{n,2,2} r^{-n} + Y_n\right] . \quad (2.3.24)$$

The boundary conditions at $r = r_1$, $r = r_2$ yield the system of equations

$$\begin{aligned}
D_{n,1} - D_{n,2,1} - D_{n,2,2} r_1^{-2n} &= Y_n(r_1) r_1^{-n} \\
-D_{n,1} + D_{n,2,1} - D_{n,2,2} r_1^{-2n} &= -Y_n(r_1) r_1^{-n} \\
D_{n,2,1} + D_{n,2,2} r_2^{-2n} - D_{n,3} r_2^{-2n} &= -Y_n(r_2) r_2^{-n} \\
-D_{n,2,1} + D_{n,2,2} r_2^{-2n} - D_{n,3} r_2^{-2n} &= Y_n(r_2) r_2^{-n} ,
\end{aligned} \quad (2.3.25)$$

whose solution is

$$\begin{aligned}
D_{n,1} &= Y_n(r_1) r_1^{-n} - Y_n(r_2) r_2^{-n} \\
D_{n,2,1} &= -Y_n(r_2) r_2^{-n} \\
D_{n,2,2} &= D_{n,3} = 0
\end{aligned} \quad (2.3.26)$$

and

$$\Phi_{2,n}(r) = \begin{cases} -\dfrac{r^n}{\mu_0} \displaystyle\int_{r_1}^{r_2} \dfrac{J_{2,n}}{r^n} dr & (r < r_1) \\[1em] -\dfrac{r^n}{\mu_0} \displaystyle\int_{r}^{r_2} \dfrac{J_{2n}}{r^n} dr & (r_1 < r < r_2) \\[1em] 0 & (r > r_2) \,. \end{cases} \qquad (2.3.27)$$

Thus, remanence $\vec{J}_2$ given by Eq. 2.3.4 results in the confinement of the field within the external boundary r_2 of the magnetized material. The external boundary is a $\Phi = 0$ equipotential surface that satisfies the condition

$$B_r(r_2, \psi) = 0 \,. \qquad (2.3.28)$$

Equations 2.3.17 and 2.3.27 show that the field generated by the distribution of remanence 2.3.1 can be considered as the linear superposition of two fields: one characterized by the absence of the field within the cylinder of radius r_1, and the other characterized by the absence of the field outside the cylinder of radius r_2.

If vectors $\vec{J}_1, \vec{J}_2$ are written as

$$\vec{J}_1 = \sum_n \vec{J}_{1,n}(r,\psi) \,, \qquad \vec{J}_2 = \sum_n \vec{J}_{2,n}(r,\psi) \,, \qquad (2.3.29)$$

it is of interest to compare the configurations of the lines of force of each vector $\vec{J}_{1,n}(r,\psi)$ and $\vec{J}_{2,n}(r,\psi)$. The equation of the lines of force of $\vec{J}_{1,n}$ is

$$r^n \sin n\psi = constant \,, \qquad (2.3.30)$$

and the equation of the lines of force of $\vec{J}_{2,n}$ is

$$r^{-n} \sin n\psi = constant \,. \qquad (2.3.31)$$

A plotting of the families of curves 2.3.30 and 2.3.31 is presented in Fig. 2.3.2 for the particular case $n = 1$. Fig. 2.3.2(a) is the uniform remanence that generates no field in the region $r < r_1$, while Fig. 2.3.2(b) is the remanence that generates a uniform field in the region $r < r_1$ and no field in the region $r > r_2$. In general, Eq. 2.3.31 describes the lines of force of $\vec{J}_{2,n}$ of a multipole magnet that generates the nth harmonic of the potential $\Phi(r, \psi)$ inside the cylinder of radius r_1 [1,2].

52 BASIC GEOMETRIES OF PERMANENT MAGNETS

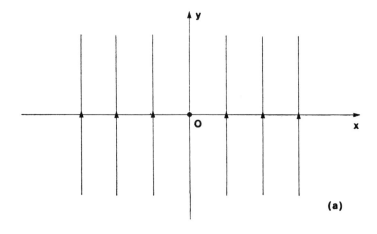

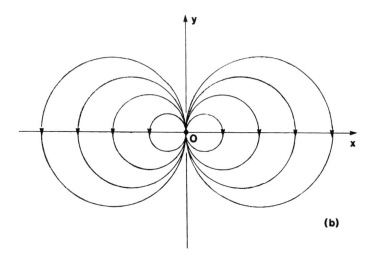

Fig. 2.3.2. (a) Lines of force of $\vec{J}_1$, (b) lines of force of $\vec{J}_2$ for $n=1$.

2.4 EXAMPLE OF FIELD CONFINEMENT

Consider the particular case of field confinement where Eq. 2.3.4 reduces to the fundamental term $n = 1$, and assume that coefficient $J_{2,1}(r)$ is a constant J_2, i.e.,

$$\vec{J}_2 = J_2 \, [\cos \psi \, \vec{r} + \sin \psi \, \vec{\psi}] \, . \tag{2.4.1}$$

The remanence $\vec{J}_2$ is oriented at an angle 2ψ with respect to the axis $\psi = 0$ as shown in Fig. 2.4.1.

By virtue of Eq. 2.3.27, the distribution of the scalar potential $\Phi(r,\psi)$ becomes [3]

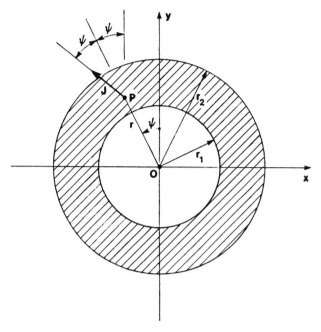

Fig. 2.4.1. Yokeless cylindrical magnet.

$$\Phi(r,\psi) = \begin{cases} -\dfrac{J_2}{\mu_0} r \ln\dfrac{r_2}{r_1} \cos\psi & (r < r_1) \\ -\dfrac{J_2}{\mu_0} r \ln\dfrac{r_2}{r} \cos\psi & (r_1 < r < r_2) \\ 0 & (r > r_2) \end{cases} \quad (2.4.2)$$

and the intensity of the magnetic field in the region $r < r_1$ is

$$\vec{H} = \frac{J_2}{\mu_0} \ln\frac{r_2}{r_1} \left[\cos\psi\, \vec{r} - \sin\psi\, \vec{\psi}\right], \tag{2.4.3}$$

i.e., the field $\vec{H}$ is uniform and oriented in the directions of the axis $\psi = 0$. The magnitude H_0 of vector $\vec{H}$ is

$$H_0 = \frac{J_2}{\mu_0} \ln\frac{r_2}{r_1}. \tag{2.4.4}$$

The remanence 2.4.1 exhibits the interesting property that parameter

$$K = \frac{\mu_0 H_0}{J_2} = \ln\frac{r_2}{r_1} \tag{2.4.5}$$

54 BASIC GEOMETRIES OF PERMANENT MAGNETS

becomes larger than unity for large values of r_2/r_1. Thus, in principle, the value of the magnetic induction within region $r < r_1$ may be larger than the remanence of the magnetized material. As r_2 increases the rate of increase of K decreases with r_2 as r_2^{-1}.

There are two basic differences between remanence 2.4.1 and the uniform magnetization 2.2.1 discussed in Section 2 of this chapter. The uniform remanence 2.2.1 requires the presence of the high permeability shield to generate the field within the region $r < r_1$ and to confine the field within the cylinder of radius r_2. The second difference concerns the value of parameter K. From Eq. 2.2.19 one obtains

$$K = \frac{\mu_0 H_0}{J_0} = \frac{1}{2}\left[1 - \frac{r_1^2}{r_2^2}\right]. \tag{2.4.6}$$

The maximum value of K is

$$\lim_{r_2 \to \infty} K = \frac{1}{2}, \tag{2.4.7}$$

i.e., in the cylindrical magnet with the high permeability shield, the upper limit of the magnetic induction generated by remanence 2.2.1 within region $r < r_1$ is half the value of remanence J_0.

Section 4 of Chapter 1 has shown that the potential generated by a distribution of remanence in free space can be computed by means of Eq. 1.4.14 by introducing a distribution of volume and surface densities of magnetic charges. The volume charge density υ is given by Eq. 1.1.17. In the case of remanence 2.4.1 one has

$$\upsilon(r, \psi) = -\nabla \cdot \vec{J}_2 = -\frac{2J_2}{r}\cos\psi. \tag{2.4.8}$$

The surface charge density σ is given by Eq. 1.4.16. On the two surfaces of radii r_1 and r_2 one has

$$\sigma(r_2,\psi) = -\sigma(r_1,\psi) = J_2\cos\psi. \tag{2.4.9}$$

The uniform remanence 2.2.1 leads to the same distribution of surface charges in the absence of the material of permeability μ. However, the uniform remanence 2.2.1 is solenoidal and, as a consequence, the equivalent volume charge density is zero.

The field confinement is not necessarily the consequence of the condition $\nabla \cdot \vec{J} \neq 0$, as can be seen from another particular case of remanence $\vec{J}_2$. Assume that Eq. 2.3.4 reduces again to the fundamental term $n = 1$, and that function $J_{2,1}(r)$ is

$$J_{2,1}(r) = \bar{J}_0 \left[\frac{r_1}{r}\right]^2, \tag{2.4.10}$$

where $\bar{J}_0$ is an arbitrary constant. The radial and angular components of $\vec{J}_2$ are

$$J_r = \bar{J}_0 \left[\frac{r_1}{r}\right]^2 \cos \psi , \qquad J_\psi = \bar{J}_0 \left[\frac{r_1}{r}\right]^2 \sin \psi , \qquad (2.4.11)$$

and

$$\nabla \cdot \vec{J}_2 = 0 . \qquad (2.4.12)$$

By virtue of Eqs. 2.3.27, the distribution of the scalar potential generated by remanence 2.4.11 is

$$\Phi(r,\psi) = \begin{cases} -\dfrac{\bar{J}_0}{2\mu_0} r \left[1 - \dfrac{r_1^2}{r_2^2}\right] \cos \psi & (r < r_1) \\ -\dfrac{\bar{J}_0}{2\mu_0} \dfrac{r_1^2}{r_2} \left[\dfrac{r_2}{r} - \dfrac{r}{r_2}\right] \cos \psi & (r_1 < r < r_2) \\ 0 & (r > r_2) . \end{cases} \qquad (2.4.13)$$

The first equation of the system 2.4.13 is the scalar potential of a uniform field of intensity

$$\vec{H} = \frac{\bar{J}_0}{2\mu_0} \left[1 - \frac{r_1^2}{r_2^2}\right] \left[\cos \psi \, \vec{r} - \sin \psi \, \vec{\psi}\right] , \qquad (2.4.14)$$

which coincides with the intensity generated by the uniform remanence 2.2.1 in the limit $\mu \to \infty$. Thus, remanence 2.4.11 results in the field confinement within the cylinder of radius r_2. As opposed to the remanence given by Eq. 2.4.1, remanence 2.4.11 yields the same upper limit of parameter K given by Eq. 2.4.7.

2.5 SPHERICAL MAGNET

Assume a sphere of radius ρ_0 magnetized with a uniform remanence $\vec{J}_0$. In a spherical system of coordinates ρ, θ, ψ with the origin at the center of the sphere and the axis $\theta = 0$ oriented in the direction of $\vec{J}_0$ shown in Fig. 2.5.1, $\vec{J}_0$ can be written in the form

$$\vec{J}_0 = J_0(\cos \theta \, \vec{\rho} - \sin \theta \, \vec{\theta}) . \qquad (2.5.1)$$

56 BASIC GEOMETRIES OF PERMANENT MAGNETS

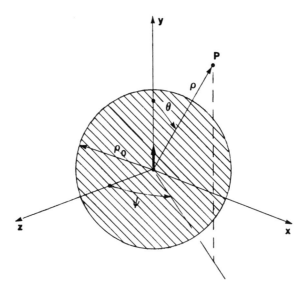

Fig. 2.5.1. Magnetized sphere.

Because of symmetry, the scalar potential Φ may be assumed to be independent of coordinate ψ, and Laplace's equation in spherical coordinates reduces to

$$\frac{\partial}{\partial \rho}\left[\rho^2 \frac{\partial \Phi}{\partial \rho}\right] + \frac{1}{\sin \theta}\frac{\partial}{\partial \theta}\left[\sin \theta \frac{\partial \Phi}{\partial \theta}\right] = 0 \ . \qquad (2.5.2)$$

Because the radial component of $\vec{J}_0$ is

$$J_r = J_0 \cos \theta \ , \qquad (2.5.3)$$

one can seek the particular solution of Eq. 2.5.2

$$\Phi(\rho, \theta) = R(\rho) \cos \theta \ , \qquad (2.5.4)$$

where R satisfies the equation

$$\frac{d}{d\rho}\left[\rho^2 \frac{dR}{d\rho}\right] - 2R = 0 \ . \qquad (2.5.5)$$

The general solution of Eq. 2.5.5 is

$$R = M_{i,1}\rho + M_{i,2}\rho^{-2} \ , \qquad (2.5.6)$$

where the values $i = 1$, $i = 2$ correspond to the regions $\rho < \rho_0$, $\rho > \rho_0$, respectively. The two constants of integration $M_{i,1}$, $M_{i,2}$ must be chosen in such a way that the scalar potential Φ is finite and continuous everywhere and the field components satisfy the boundary conditions at the surface of the sphere. These conditions are expressed by the equations

$$M_{1,2} = M_{2,1} = 0$$
$$M_{1,1} - M_{2,2}\rho_0^{-3} = 0 \qquad (2.5.7)$$
$$M_{1,1} + 2M_{2,2}\rho_0^{-3} = \frac{J_0}{\mu_0}.$$

Thus,

$$M_{1,1} = \frac{1}{3}\frac{J_0}{\mu_0}, \qquad M_{2,2} = \frac{\rho_0^3}{3}\frac{J_0}{\mu_0}, \qquad (2.5.8)$$

and the scalar potential is

$$\Phi(\rho, \theta) = \begin{cases} \dfrac{J_0}{3\mu_0}\rho\cos\theta = \dfrac{J_0}{3\mu_0}y & (\rho < \rho_0) \\ \dfrac{J_0}{3\mu_0}\dfrac{\rho_0^3}{\rho^2}\cos\theta & (\rho > \rho_0). \end{cases} \qquad (2.5.9)$$

In the region $\rho < \rho_0$, Φ is a function of the coordinate y only. Hence inside the sphere the intensity $\vec{H}$ is uniform and oriented in the opposite direction of the remanence $\vec{J}_0$,

$$\vec{H} = -\frac{J_0}{3\mu_0}\vec{y} \qquad (\rho < \rho_0), \qquad (2.5.10)$$

where $\vec{y}$ is the unit vector oriented in the direction of the y axis. One observes that the magnitude of $\vec{H}$ inside the sphere is independent of ρ_0. Outside the sphere, Φ is the scalar potential of a magnetic dipole of moment

$$\vec{p} = \frac{4}{3}\pi\rho_0^3\vec{J}_0. \qquad (2.5.11)$$

Assume now that the material magnetized with the uniform remanence $\vec{J}_0$ is confined between two concentric spheres of radii ρ_1, ρ_2 and that the medium inside the

58 BASIC GEOMETRIES OF PERMANENT MAGNETS

sphere of radius ρ_1 is nonmagnetic. The scalar potential Φ generated by the magnetized hollow sphere can be computed as the linear superposition of the potential generated by a sphere of radius ρ_2 magnetized with the polarization $\vec{J}_0$ and the potential generated by a sphere of radius ρ_1 magnetized with a polarization $-\vec{J}_0$. As a consequence, the distribution of Φ is

$$\Phi = \begin{cases} 0 & (\rho < \rho_1) \\ \dfrac{J_0}{3\mu_0}\left[\rho - \dfrac{\rho_1^3}{\rho^2}\right]\cos\theta & (\rho_1 < \rho < \rho_2) \\ \dfrac{J_0}{3\mu_0}\left[\rho_2^3 - \rho_1^3\right]\dfrac{\cos\theta}{\rho^2} & (\rho > \rho_2). \end{cases} \quad (2.5.12)$$

Equation 2.5.12 shows that a uniformly magnetized hollow sphere generates no field within its cavity. Section 2.2 has shown that this property is also exhibited by the uniformly magnetized hollow cylinder. As in the case of the hollow cylinder, a finite field can be generated inside the uniformly magnetized hollow sphere if one alters the boundary condition at $\rho = \rho_2$. Assume that the sphere is bound by a nonmagnetized medium of high magnetic permeability. Let M_3 denote the constant of integration within the high permeability medium outside of the sphere of radius ρ_2. The boundary conditions at the two surfaces $\rho = \rho_1$, $\rho = \rho_2$ yield the equations

$$\begin{aligned} M_{1,1} - M_{2,1} - M_{2,2}\rho_1^{-3} &= 0 \\ M_{1,1} - M_{2,1} + 2M_{2,2}\rho_1^{-3} &= -\dfrac{J_0}{\mu_0} \\ M_{2,1} + M_{2,2}\rho_2^{-3} - M_3\rho_2^{-3} &= 0 \\ M_{2,1} - 2M_{2,2}\rho_2^{-3} + 2M_3\kappa_m\rho_2^{-3} &= \dfrac{J_0}{\mu_0}. \end{aligned} \quad (2.5.13)$$

Thus the constants of integration are

$$\begin{aligned} M_{1,1} &= -\dfrac{2J_0}{3\mu_0}\dfrac{\kappa_m - 1}{2\kappa_m + 1}\left[1 - \dfrac{\rho_1^3}{\rho_2^3}\right] \\ M_{2,1} &= \dfrac{J_0}{3\mu_0}\dfrac{1}{2\kappa_m + 1}\left[3 + 2(\kappa_m - 1)\dfrac{\rho_1^3}{\rho_2^3}\right] \\ M_{2,2} &= -\dfrac{J_0}{3\mu_0}\rho_1^3 \\ M_3 &= \dfrac{J_0}{\mu_0}\dfrac{\rho_2^3}{2\kappa_m + 1}\left[1 - \dfrac{\rho_1^3}{\rho_2^3}\right] \end{aligned} \quad (2.5.14)$$

and the scalar potential in the three regions of the spherical magnet is

$$\Phi(\rho,\theta) = \begin{cases} -\dfrac{2J_0}{3\mu_0}\dfrac{\kappa_m - 1}{2\kappa_m + 1}\left[1 - \dfrac{\rho_1^3}{\rho_2^3}\right]\rho\cos\theta & (\rho < \rho_1) \\[2ex] \dfrac{J_0}{3\mu_0}\dfrac{\rho\cos\theta}{2\kappa_m+1}\left[3 + 2(\kappa_m - 1)\dfrac{\rho_1^3}{\rho_2^3} - (2\kappa_m + 1)\dfrac{\rho_1^3}{\rho^3}\right] & (\rho_1 < \rho < \rho_2) \quad (2.5.15) \\[2ex] \dfrac{J_0}{\mu_0}\dfrac{1}{2\kappa_m + 1}\left[1 - \dfrac{\rho_1^3}{\rho_2^3}\right]\dfrac{\rho_2^3}{\rho^2}\cos\theta & (\rho > \rho_2) . \end{cases}$$

In the limit $\kappa_m \to \infty$, the scalar potential vanishes outside the sphere of radius ρ_2, which becomes an equipotential surface $\Phi = 0$. The field inside the sphere of radius ρ_1 is always uniform, and in the limit $\kappa_m \to \infty$ its intensity becomes

$$\lim_{\kappa_m \to \infty}\vec{H} = +\frac{1}{3}\left[1 - \frac{\rho_1^3}{\rho_2^3}\right]\frac{\vec{J_0}}{\mu_0} \qquad (\rho < \rho_1) . \qquad (2.5.16)$$

By comparing this result with Eqs. 2.1.12 (one-dimensional geometry) and Eq. 2.1.15 (two-dimensional geometry), one observes that the upper limit of the intensity generated by a uniform remanence $\vec{J_0}$ in an l-dimensional geometry is

$$(\vec{H}_0)_{max} = \frac{\vec{J_0}}{l\,\mu_0} , \qquad l = 1, 2, 3. \qquad (2.5.17)$$

For a finite value of κ_m, the scalar potential is zero over a spherical surface of radius ρ given by

$$\frac{\rho_1^3}{\rho_2^3} \leq \frac{\rho^3}{\rho_2^3} = \frac{2\kappa_m + 1}{2(\kappa_m - 1) + 3\dfrac{\rho_2^3}{\rho_1^3}} \leq 1 . \qquad (2.5.18)$$

As expected, in the limit $\kappa_m = 1$, the scalar potential vanishes over the sphere of radius ρ_1. A plotting of Φ versus the radial coordinate ρ along the axis $\theta = 0$ is shown in Fig. 2.5.2.

A uniform field within the cavity of the hollow sphere can be generated without the need of an external medium of high magnetic permeability by a distribution of magnetization of the type defined by Eq. 2.3.4 for the cylindrical magnet. Assume again that the medium outside the external surface $\rho = \rho_2$ is air and consider the distribution of remanence

$$\vec{J} = J_0[\cos\theta\,\vec{\rho} + \sin\theta\,\vec{\theta}] . \qquad (2.5.19)$$

60 BASIC GEOMETRIES OF PERMANENT MAGNETS

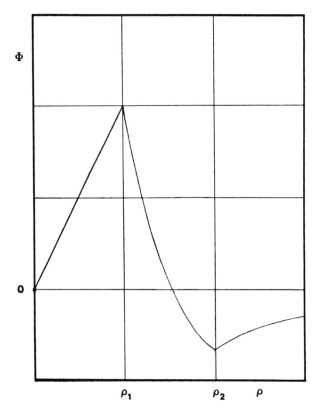

Fig. 2.5.2. Distribution of Φ along the axis $\theta = 0$ in the spherical magnet.

The magnitude of $\vec{J}$ is uniform within the region $\rho_1 < \rho < \rho_2$ and $\vec{J}$ is oriented in the planes $\psi = constant$ at an angle 2θ with respect to the $\theta = 0$ axis. Thus $\vec{J}$ depends on θ in the same way as remanence 2.4.1 depends on the angular coordinate ψ of the cylindrical magnet.

The divergence of remanence 2.5.19 is

$$\nabla \cdot \vec{J} = \frac{4J_0}{\rho} \cos\theta . \tag{2.5.20}$$

Since Eq. 2.5.19 is independent of coordinate ψ, Poisson's equation in the region $\rho_1 < \rho < \rho_2$ reduces to

$$\frac{\partial}{\partial \rho}\left[\rho^2 \frac{\partial \Phi}{\partial \rho}\right] + \frac{1}{\sin\theta} \frac{\partial}{\partial \theta}\left[\sin\theta \frac{\partial \Phi}{\partial \theta}\right] = \frac{4}{\mu_0} \rho J_0 \cos\theta . \tag{2.5.21}$$

The general solution of Eq. 2.5.21 is

$$\Phi(\rho, \theta) = \left[N_{2,1}\rho + N_{2,2}\frac{1}{\rho^2} + \frac{4J_0}{3\mu_0}\rho \ln\rho\right] \cos\theta , \tag{2.5.22}$$

where $N_{2,1}, N_{2,2}$ are the constants of integration. The boundary conditions at $\rho = \rho_1$ and $\rho = \rho_2$ yield the following distribution of the scalar potential in the three spherical regions:

$$\Phi(\rho, \theta) = \begin{cases} -\dfrac{4J_0}{3\mu_0} \rho \ln\dfrac{\rho_2}{\rho_1} \cos\theta & (\rho < \rho_1) \\ -\dfrac{4J_0}{3\mu_0} \left[\rho \ln\dfrac{\rho_2}{\rho} + \dfrac{\rho}{12}\left[1 - \dfrac{\rho_1^3}{\rho^3}\right]\right] \cos\theta & (\rho_1 < \rho < \rho_2) \\ -\dfrac{J_0}{9\mu_0} \dfrac{\rho_2^3}{\rho^2}\left[1 - \dfrac{\rho_1^3}{\rho_2^3}\right] \cos\theta & (\rho > \rho_2). \end{cases} \quad (2.5.23)$$

Thus the intensity of the magnetic field within the cavity of the spherical magnet is

$$\vec{H} = \frac{4}{3} \frac{J_0}{\mu_0} \ln\frac{\rho_2}{\rho_1} \vec{y}_0 \qquad (2.5.24)$$

and

$$K = \frac{4}{3} \ln\frac{\rho_2}{\rho_1}. \qquad (2.5.25)$$

As in the case of the hollow cylinder, K increases logarithmically with the ratio ρ_2/ρ_1. However, the remanence 2.5.19 of the hollow sphere does not confine the field within the external boundary of the magnetized material. One observes that the value of K given by Eq. 2.5.25 is larger than the value 2.4.5 obtained for the cylindrical magnet.

Equation 2.5.23 show that in the region $\rho > \rho_2$ Φ is the potential of a magnetic dipole located at the origin. The dipole moment $\vec{p}_0$ is equal to the integral of polarization 2.5.19 over the volume of the magnetized material:

$$\vec{p}_0 = J_0 \vec{y}_0 \int_0^{2\pi} d\psi \int_0^{\pi} d\theta \int_{\rho_1}^{\rho_2} \cos 2\theta \sin\theta \, \rho^2 \, d\rho = -\frac{4}{9}\pi(\rho_2^3 - \rho_1^3) J_0 \vec{y}_0. \quad (2.5.26)$$

To achieve the field confinement, the remanence of the magnetized material must be modified in such a way as to eliminate the dipole moment $\vec{p}_0$ given by Eq. 2.5.26. This can be accomplished by superimposing to remanence 2.5.19 a uniform remanence

$$\vec{J}' = J'_0(\cos\theta \, \vec{\rho} - \sin\theta \, \vec{\theta}). \qquad (2.5.27)$$

As shown by Eq. 2.5.12, the potential generated by $\vec{J}'$ is zero inside the sphere of

62 BASIC GEOMETRIES OF PERMANENT MAGNETS

radius ρ_1. Outside the sphere of radius ρ_2 it is equal to the potential of a dipole of moment $\vec{p}'_0$,

$$\vec{p}'_0 = \frac{4}{3}\pi(\rho_2^3 - \rho_1^3)J'_0\vec{y}_0 . \qquad (2.5.28)$$

The two moments $\vec{p}'_0$ and $\vec{p}_0$ cancel each other if

$$J'_0 = \frac{1}{3}J_0 . \qquad (2.5.29)$$

Thus the uniform field within the cavity of the spherical magnet and the confinement of the field are achieved with a distribution of remanence

$$\vec{J} = \frac{2J_0}{3}(2\cos\theta\,\vec{\rho} + \sin\theta\,\vec{\theta}) . \qquad (2.5.30)$$

An alternate way of achieving the field confinement is to enclose the sphere of radius ρ_2 inside a spherical layer of thickness $\rho_3 - \rho_2$ magnetized with remanence 2.5.27. In this case the intensity outside the sphere of radius r_3 vanishes if J'_0 and r_3 satisfy the condition

$$J'_0(\rho_3^3 - \rho_2^3) = \frac{J_0}{3}(\rho_2^3 - \rho_1^3) . \qquad (2.5.31)$$

2.6 MAGNETIC MATERIAL WITH NONZERO SUSCEPTIBILITY

The properties of the magnetic structures analyzed in the preceding sections are based on the assumption of zero magnetic susceptibility of the magnetic material. Assume now a linear demagnetization characteristic with $\chi_m \neq 0$ and consider again the case of the hollow cylinder in air, with the distribution of magnetization given by Eq. 2.3.4. The scalar potential in the region of the magnetized material ($r_1 < r < r_2$) is the solution of Eq. 2.3.21, where Y_n is replaced by a function Y'_n which satisfies the equation

$$\frac{d}{dr}\left[\frac{Y'_n}{r^n}\right] = \frac{1}{\mu_0(1+\chi_m)}\frac{J_{2,n}}{r^n} . \qquad (2.6.1)$$

Functions $\Phi_{2,n}$ in Eq. 2.3.23 become

$$\Phi_{2,n} = \begin{cases} D'_{n,1}\,r^n & (r < r_1) \\ D'_{n,2,1}\,r^n + D'_{n,2,2}\,r^{-n} + Y'_n & (r_1 < r < r_2) \\ D'_{n,3}\,r^{-n} & (r > r_2) . \end{cases} \qquad (2.6.2)$$

MAGNETIC MATERIAL WITH NONZERO SUSCEPTIBILITY

The new constants of integration are again determined by the boundary conditions at $r = r_1$ and $r = r_2$. In particular $D'_{n,1}$ and $D'_{n,3}$ satisfy the two equations

$$\left[1 - \frac{\chi_m f_n}{2(1 + \chi_m)}\right] D'_{n,1} - r_2^{-2n} D'_{n,3} = -\frac{1}{\mu_0(1 + \chi_m)} \int_{r_1}^{r_2} \frac{J_{2,n}}{r^n} dr \qquad (2.6.3)$$

$$\left[1 + \frac{\chi_m f_n}{2}\right] D'_{n,1} + r_2^{-2n} D'_{n,3} = -\frac{1}{\mu_0} \int_{r_1}^{r_2} \frac{J_{2,n}}{r^n} dr,$$

where

$$f_n = 1 - \left[\frac{r_1}{r_2}\right]^{2n}. \qquad (2.6.4)$$

Equation 2.6.3 yields

$$D'_{n,1} = -\frac{2}{\mu_0} \frac{2 + \chi_m}{4(1 + \chi_m) + \chi_m^2 f_n} \int_{r_1}^{r_2} \frac{J_{2,n}}{r^n} dr \qquad (2.6.5)$$

and

$$D'_{n,3} = -\frac{2}{\mu_0} \frac{\chi_m r_1^{2n}}{4(1 + \chi_m) + \chi_m^2 f_n} \int_{r_1}^{r_2} \frac{J_{2,n}}{r^n} dr. \qquad (2.6.6)$$

In general $D'_{n,3} \neq 0$; i.e. the field confinement within the $r < r_2$ region can be achieved with remanence 2.3.4 only if $\chi_m = 0$.

In the limit $\chi_m \ll 1$, functions $\Phi_{2,n}$ reduce to

$$\Phi_{2,n}(r) = \begin{cases} -\left[1 - \frac{\chi_m}{2}\right] \frac{r^n}{\mu_0} \int_{r_1}^{r_2} \frac{J_{2,n}}{r^n} dr & (r < r_1) \\ \\ -\frac{\chi_m}{2\mu_0} \frac{r_1^{2n}}{r^n} \int_{r_1}^{r_2} \frac{J_{2,n}}{r^n} dr & (r > r_2). \end{cases} \qquad (2.6.7)$$

The field confinement can be restored by modifying the distribution of the remanence, in a way similar to the procedure described in Section 2.5. Thus one can

64 BASIC GEOMETRIES OF PERMANENT MAGNETS

assume that a remanence 2.3.3, designed to generate a field in the region $r > r_2$ is superimposed to remanence 2.3.4 in the region $r_1 < r < r_2$. In the limit $\chi_m \ll 1$, the magnitude of $\vec{J}_2$ required to cancel the field generated by $\vec{J}_1$ outside the magnet may be assumed to be small compared to the magnitude of $\vec{J}_1$. The field generated by $\vec{J}_1$ can be computed in the limit $\chi_m = 0$. Thus the potential $\Phi_{1,n}(r)$ in the $r > r_2$ region is given by Eq. 2.3.17 and the field outside the magnetic structure cancels if $J_{1,n}$ satisfies the equation

$$\int_{r_1}^{r_2} r^n J_{1,n} dr = \frac{\chi_m}{2} r_1^{2n} \int_{r_1}^{r_2} \frac{J_{2,n}}{r^n} dr . \tag{2.6.8}$$

In particular, for $n=1$, if $J_{1,n}$ and $J_{2,n}$ are constants, Eq. 2.6.8 reduces to

$$J_{1,1} = \chi_m \frac{r_1^2}{r_2^2 - r_1^2} \ln\frac{r_2}{r_1} J_{2,1} . \tag{2.6.9}$$

In this case the field confinement is achieved with a nonuniform distribution of remanence

$$\vec{J} = \vec{J}_{2,1} + \vec{J}_{1,1} \tag{2.6.10}$$

whose magnitude is

$$|\vec{J}| = J_{2,1}\left[1 + \chi_m \frac{r_1^2}{r_2^2 - r_1^2} \ln\frac{r_2}{r_1} \cos 2\psi\right] , \tag{2.6.11}$$

and $\vec{J}$ is oriented with respect to the axis $\psi = 0$ at an angle

$$\phi = 2\psi - \chi_m \frac{r_1^2}{r_2^2 - r_1^2} \ln\frac{r_2}{r_1} \sin 2\psi . \tag{2.6.12}$$

In the absence of compensation with the additional distribution of remanence $\vec{J}_{2,1}$, the flux Ψ_e of $\vec{B}$ per unit length in the z direction, crossing the external boundary $r = r_2$ within the angular interval $-\pi/2 < \psi < \pi/2$ is

$$\Psi_e = \chi_m \frac{r_1^2}{r_2} J_0 \ln\frac{r_2}{r_1} . \tag{2.6.13}$$

If one compares Eq. 2.6.13 with the flux Ψ_i of $\vec{B}$ per unit length in the z direction, within the cylinder of radius r_1 where the uniform intensity is

$$\vec{H}_y \approx \left[1 - \frac{\chi_m}{2}\right] \frac{J_0}{\mu_0} \ln\frac{r_2}{r_1} , \tag{2.6.14}$$

one has

$$\frac{\Psi_e}{\Psi_i} \approx \frac{1}{2} \chi_m \frac{r_1}{r_2} \ll 1 ; \qquad (2.6.15)$$

i.e., the flux leakage due to the nonzero magnetic susceptibility of the magnetic material is small as long as $\chi_m \ll 1$.

Consider now the effect of $\chi_m \neq 0$ on the yoked cylindrical magnet in the limit of infinite magnetic permeability of the medium in region $r > r_2$.

With the uniform magnetization 2.2.1 of the region $r_1 < r < r_2$, the general solution of the scalar potential is again given by Eqs. 2.2.2 and 2.2.4. The radial component B_r of the magnetic induction in region $r_1 < r < r_2$ is

$$B_r = -\mu_0(1 + \chi_m) \frac{\partial \Phi}{\partial r} + J_r$$

$$= -\mu_0(1 + \chi_m) \left[C_{1,1}'' - C_{1,2}'' \frac{1}{r^2} \right] \cos \psi + J_0 \cos \psi , \qquad (2.6.16)$$

where $C_{1,1}''$, $C_{1,2}''$ are the new constants of integration. The boundary conditions at $r = r_1$ and $r = r_2$ yield the system of equations

$$\begin{aligned} C_{0,1}'' \quad &- C_{1,1}'' \quad - C_{1,2}'' r_1^{-2} = 0 \\ -C_{0,1}'' + (1+\chi_m) C_{1,1}'' &- (1+\chi_m) C_{1,2}'' r_1^{-2} = \frac{J_0}{\mu_0} \\ C_{1,1}'' \quad &+ C_{1,2}'' r_2^{-2} = 0 . \end{aligned} \qquad (2.6.17)$$

In the region $r < r_1$ the value of the new constant of integration is

$$C_{0,1}'' = -\frac{J_0}{2\mu_0} \left[1 - \frac{r_1^2}{r_2^2} \right] \frac{1}{1 + \frac{\chi_m}{2}\left[1 + \frac{r_1^2}{r_2^2}\right]} , \qquad (2.6.18)$$

and the intensity is

$$\vec{H} = \frac{1}{2\mu_0} \left[1 - \frac{r_1^2}{r_2^2} \right] \frac{\vec{J}_0}{1 + \frac{\chi_m}{2}\left[1 + \frac{r_1^2}{r_2^2}\right]} . \qquad (2.6.19)$$

If Eq. 2.6.19 is compared with Eq. 2.2.19, one observes that the only effect of the nonzero susceptibility of the magnetic material on the field within the region $r < r_1$ of the uniformly magnetized cylindrical magnet is a reduction of the magnitude of the uniform field.

REFERENCES

[1] K. Halbach, Proceedings of the Eighth International Workshop on Rare Earth-Cobalt Permanent Magnets and Their Applications, Dayton, Ohio, May 5-9, 1985 p. 103.
[2] K. Halbach, Design of Permanent Multipole Magnets with Oriented Rare Earth-Cobalt Material, *Nuclear Instruments and Methods* 169 (1980) 1-10.
[3] M.G. Abele, Yokeless Permanent Magnets, Technical Report No. 14, New York University, Nov. 1, 1986.

CHAPTER 3

Generation of a Uniform Field in Prismatic Cavities

INTRODUCTION

The field configurations analyzed in Chapter 2 have been computed by a direct solution of Poisson and Laplace's equations in structures defined by their geometry and by the remanences and permeabilities of the different media.

This chapter is devoted to the inverse problem, where the magnetic field is assumed to be known within a closed surface of assigned geometry, and one is seeking a magnetic structure outside this surface capable of generating the assigned field. This is the classical problem of the magnet designer, who must satisfy the design requirements by choosing the appropriate magnetic material and the geometry of the structure. In theory, an infinity of different geometries of magnetized materials can generate the same magnetic field within the assigned closed surface. In practice, the uniqueness of the solution can be approached by formulating a number of conditions and design criteria to be satisfied by the magnetic structure.

A simple example of inverse problem is the design of a magnetic structure capable of generating a uniform magnetic field within a cylindrical surface of radius r_1 and infinite length. If the uniform field is perpendicular to the axis of the cylinder, Chapter 2 has shown that a variety of distributions of the remanence of the magnetized material contained between two coaxial cylindrical surfaces of radii r_1 and r_2 can generate the same uniform field. By specifying, for instance, that a particular solution of the inverse problem is sought where the material outside of the cylinder of radius r_1 is uniformly magnetized and the cylinder of radius r_2 is the boundary of a medium of infinite permeability, the design problem is solved by considering Eq. 2.2.19 as the equation to be satisfied by r_2 and remanence $\vec{J}_0$.

To be of practical interest, the solution of the inverse problem must be based on available materials that can be magnetized with standard magnetization procedures. One cannot expect to implement designs based on complex distributions of magnetization, like, for instance, the configuration depicted in Fig. 2.3.2(b). As a consequence, the designer is forced to accept some degree of approximation in the implementation of the ideal design requirements.

A special situation is found in the design of magnets capable of generating a uniform field in the region of interest. This chapter will show the existence of an exact solution of the design of these magnets based on structures of uniformly magnetized blocks. The formulation of the design of these structures and their basic properties is the subject of the following sections.

68 GENERATION OF A UNIFORM FIELD IN PRISMATIC CAVITIES

3.1 EXISTENCE OF A UNIFORM FIELD SOLUTION

Consider a plane surface limited by two lines $\tau = \pm r_0$ in the $\eta = 0$ plane of the frame of reference shown in Fig. 3.1.1. Assume a uniform surface charge density σ induced on this surface. By virtue of Eq. 1.4.16, σ may be the result of a uniform magnetization of the region $|\tau|<r_0$, $\eta>0$ with a remanence oriented along the η axis. Assume also that the magnetic susceptibility is zero everywhere.

By virtue of Eq. 1.4.15, an element of charge $\sigma d\xi$ on the surface of Fig. 3.1.1 generates at point $P(\tau, \eta, \zeta)$ a scalar potential

$$d\Phi(\tau, \eta) = -\frac{\sigma d\xi}{2\pi\mu_0} \ln\frac{r}{\bar{r}}, \qquad (3.1.1)$$

where

$$r = \left[\eta^2 + (\xi - \tau)^2\right]^{\frac{1}{2}} \qquad (-r_0 < \xi < +r_0) \qquad (3.1.2)$$

and $\bar{r}$ is an arbitrary constant. By integrating Eq. 3.1.1 and applying the gradient to the resulting scalar potential, one obtains the intensity $\vec{H}$ of components

$$H_\tau = -\frac{\sigma}{4\pi\mu_0}\ln\frac{(\tau-r_0)^2 + \eta^2}{(\tau+r_0)^2 + \eta^2}$$

$$H_\eta = \frac{\sigma}{2\pi\mu_0}\left[\arctan\frac{r_0-\tau}{\eta} + \arctan\frac{r_0+\tau}{\eta}\right] \qquad (3.1.3)$$

$$H_\zeta = 0.$$

Let r_1 and r_2 denote the distances of a point P in Fig. 3.1.1 from the lines $\tau = \pm r_0$, $\eta = 0$, respectively. One has

$$r_{1,2} = \left[(\tau \mp r_0)^2 + \eta^2\right]^{\frac{1}{2}}. \qquad (3.1.4)$$

In the proximity of lines $\tau = \pm r_0$, i.e., for either r_1 or r_2 small compared to r_0, the H_τ component in Eq. 3.1.3 reduces to

$$H_\tau = \begin{cases} -\dfrac{\sigma}{2\pi\mu_0}\ln\dfrac{r_1}{2r_0} & (r_1 \ll r_0) \\[2mm] +\dfrac{\sigma}{2\pi\mu_0}\ln\dfrac{r_2}{2r_0} & (r_2 \ll r_0) \end{cases} \qquad (3.1.5)$$

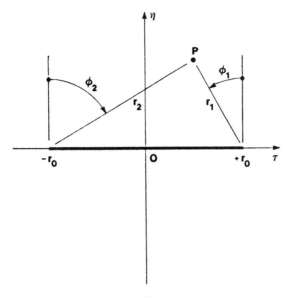

Fig. 3.1.1. Strip of a uniform surface charge density.

and the H_η component reduces to

$$H_\eta = \begin{cases} \pm \dfrac{\sigma}{2\pi\mu_0}\left[\dfrac{\pi}{2}+\phi_1\right] & (r_1 \ll r_0) \\[2ex] \pm \dfrac{\sigma}{2\pi\mu_0}\left[\dfrac{\pi}{2}+\phi_2\right] & (r_2 \ll r_0) \end{cases} \quad (3.1.6)$$

for $\eta \gtreqless 0$, respectively. Angles ϕ_1 and ϕ_2 are the angles between the axis η and r_1, r_2 respectively, as indicated in Fig. 3.1.1. Equations 3.1.5 and 3.1.6 show that in a proximity of an edge of the strip, the field component parallel to the strip is a function of the distance from the edge only, and the field component perpendicular to the strip is a function of the angular orientation of the distance from the edge only. In the limit $\eta \to 0$, angles ϕ_1, ϕ_2 are equal to $\pm\pi/2$ and Eq. 3.1.6 reduces to

$$H_\eta = \begin{cases} \pm \dfrac{\sigma}{2\mu_0} & (\eta \gtreqless 0,\ -r_0 < \tau < r_0) \\[2ex] 0 & (|\tau| > r_0) . \end{cases} \quad (3.1.7)$$

The value 3.1.7 of H_η close to the surface of the strip reflects the fact that a uniform charge distribution on an infinite planar surface generates a uniform field, as in the case of the one-dimensional magnetic structure discussed in Section 2.1.

70   GENERATION OF A UNIFORM FIELD IN PRISMATIC CAVITIES

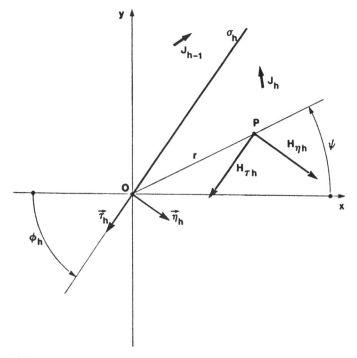

Fig. 3.1.2. Interface between two prisms magnetized with remanences $\vec{J}_{h-1}$ and $\vec{J}_h$.

Equation 3.1.5 shows that the component of the field parallel to the strip has a singularity at both edges $r_1 = 0$ and $r_2 = 0$. The cancellation of this singularity is the basic condition to be satisfied by a magnetic structure designed to generate a uniform field.

Assume a structure of uniformly magnetized prisms whose edges are parallel to the axis z of the cartesian frame of Figure 3.1.2. Assume also that the axis z is the common edge of n magnetized prisms. Thus, the z axis is common to n planar interfaces between prisms. The remanences $\vec{J}_h$ of the prisms are assumed to be perpendicular to the z axis. Figure 3.1.2 shows the interface between two prisms with remanences $\vec{J}_{h-1}, \vec{J}_h$. The orientation of the interface relative to the x axis is given by the angle ϕ_h shown in the figure.

By virtue of Eq. 1.4.16, the surface charge density σ_h induced on the interface between prisms with remanences $\vec{J}_{h-1}, \vec{J}_h$ is

$$\sigma_h = (\vec{J}_{h-1} - \vec{J}_h) \cdot \vec{\eta}_h , \qquad (3.1.8)$$

where $\vec{\eta}_h$ is a unit vector perpendicular to the hth interface and oriented from the prism with remanence $\vec{J}_{h-1}$ to the prism with remanence $\vec{J}_h$.

Select in Fig. 3.1.2 a point P of polar coordinates r, ψ and assume that P is located within the prism of remanence $\vec{J}_h$. The intensity generated at P by the prisms is the sum of the intensities generated by each surface charge density σ_h. If the remanences $\vec{J}_h$ are chosen arbitrarily, the intensity has a singularity on the z axis and, as a consequence, the field generated by the prisms is not uniform.

The field at point P can be analyzed by separating in Eq. 3.1.3 the contribution of the edges of the interfaces that coincide with the z axis from the contribution of the other edges. By virtue of Eqs. 3.1.3 the field generated at P by the full structure of magnetized prisms can be written in the form

$$\vec{H} = -\frac{1}{2\pi\mu_0} \sum_{h=1}^{n} \sigma_h \left[\ln r \, \vec{\tau}_h - \left[\frac{\pi}{2} - \phi_h + \psi \right] \vec{\eta}_h \right] + \vec{H}_e(P), \quad (3.1.9)$$

where $\vec{H}_e$ is the sum of the contribution of the other edges of the n interfaces and the field intensity generated by the surface charges on all the other interfaces between the prisms of the magnetic structure. As shown in Fig. 3.1.2, unit vector $\vec{\tau}_h$, parallel to the interface, is pointing away from the interface.

Assume that geometry and remanence of each prism of the full magnetic structure are such that vector $\vec{H}_e$ in Eq. 3.1.9 is independent of position within the prism of remanence $\vec{J}_h$. In general, even if $\vec{H}_e$ is uniform, vector $\vec{H}$ is a function of the position of point P and it suffers a singularity at $x = y = 0$.

Assume that charges σ_h and orientations $\vec{\tau}_h$ of the n interfaces satisfy the condition

$$\sum_{h=1}^{n} \sigma_h \vec{\tau}_h = 0. \quad (3.1.10)$$

Thus the x and y components of vector $\vec{H}$ are

$$H_x = +\frac{1}{2\pi\mu_0} \sum_{h=1}^{n} \sigma_h \left[\frac{\pi}{2} - \phi_h + \psi \right] \sin\phi_h + H_{e,x}$$

$$H_y = -\frac{1}{2\pi\mu_0} \sum_{h=1}^{n} \sigma_h \left[\frac{\pi}{2} - \phi_h + \psi \right] \cos\phi_h + H_{e,y} . \quad (3.1.11)$$

where $H_{e,x}$, $H_{e,y}$ are the x and y components of the uniform field intensity $\vec{H}_e$.

By virtue of Eq. 3.1.10, one has

$$\sum_{h=1}^{n} \sigma_h \sin\phi_h = \sum_{h=1}^{n} \sigma_h \cos\phi_h = 0, \quad (3.1.12)$$

and Eqs. 3.1.11 reduce to

$$H_x = -\frac{1}{2\pi\mu_0} \sum_{h=1}^{n} \sigma_h \phi_h \sin\phi_h + H_{e,x}$$

$$H_y = +\frac{1}{2\pi\mu_0} \sum_{h=1}^{n} \sigma_h \phi_h \cos\phi_h + H_{e,y} . \quad (3.1.13)$$

72 GENERATION OF A UNIFORM FIELD IN PRISMATIC CAVITIES

The two components 3.1.13 are independent of position. Consequently the field generated within the region of remanence $\vec{J}_h$ is uniform. If the derivation of Eqs. 3.1.13 is extended to each intersection of interfaces of the structure, the basic condition for the generation of a uniform field is expressed by the following theorem [1]:

Theorem 1. In a structure of uniformly magnetized prisms designed to generate a uniform field, the remanences of the prisms and their geometries must satisfy condition 3.1.10 at each point of the interfaces between prisms.

Across the interface between the prisms with remanences $\vec{J}_{h-1}, \vec{J}_h$, by virtue of Eq. 3.1.7, the component of the intensity perpendicular to the interface suffers a discontinuity given by

$$\frac{\sigma_h}{\mu_0}. \qquad (3.1.14)$$

Thus the difference between the x and y components of intensities $\vec{H}_{h-1}$ and $\vec{H}_h$ within the prisms of remanences $\vec{J}_{h-1}$ and $\vec{J}_h$ is given by

$$(\vec{H}_{h-1} - \vec{H}_h)_x = -\frac{\sigma_h}{\mu_0} \sin \phi_h$$

$$(\vec{H}_{h-1} - \vec{H}_h)_y = +\frac{\sigma_h}{\mu_0} \cos \phi_h . \qquad (3.1.15)$$

It is worthwhile mentioning that the interfaces of the magnetic structure include the interfaces between the magnet prismatic cavity and the magnetic material, as well as the interfaces between the magnetic structure and an external nonmagnetic medium.

In the limit of only two interfaces, Eq. 3.1.10 reduces to

$$\sigma_1 \vec{\tau}_1 = -\sigma_2 \vec{\tau}_2 . \qquad (3.1.16)$$

Thus $\vec{\tau}_1$ and $\vec{\tau}_2$ must be oriented in opposite directions to each other and

$$\sigma_1 = \sigma_2 ; \qquad (3.1.17)$$

i.e., the two interfaces are coplanar and the surface charge density on the interface between the magnetized media is uniform.

As an example consider the situation of three surfaces oriented at an angle $2\pi/3$ with respect to each other, as shown on Fig. 3.1.3, and let $\sigma_1, \sigma_2, \sigma_3$ be the three surface charge densities. The axis x of the frame of reference is oriented in the positive direction of unit vector $\vec{\tau}_1$. With the definition of angular orientations given by Fig. 3.1.2 one has

$$\phi_1 = \pi, \qquad \phi_2 = \frac{\pi}{3}, \qquad \phi_3 = 2\pi - \frac{\pi}{3}, \qquad (3.1.18)$$

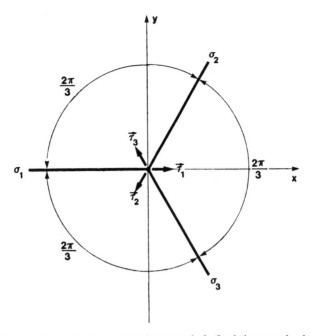

Fig. 3.1.3. Three interfaces oriented at an angle $2\pi/3$ relative to each other.

and Eqs. 3.1.12 yield

$$\sigma_1 - \frac{1}{2}\sigma_2 - \frac{1}{2}\sigma_3 = 0$$
$$\sigma_2 - \sigma_3 = 0 \; ; \quad (3.1.19)$$

i.e.,

$$\sigma_1 = \sigma_2 = \sigma_3 \, . \quad (3.1.20)$$

Because the half planes in Fig. 3.1.3 are the only interfaces of the magnetic structure, the contributions of the interface edges at infinity cancel each other, and one can assume

$$\vec{H}_e = 0 \quad (3.1.21)$$

everywhere. Thus, within the wedge limited by the interfaces $\vec{\tau}_2$, $\vec{\tau}_3$, Eq. 3.1.13 yields the components of the intensity

$$H_{1,x} = \frac{\sigma}{\sqrt{3}\,\mu_0}$$
$$H_{1,y} = 0 \, . \quad (3.1.22)$$

By virtue of Eqs. 3.1.15 and 3.1.22 the components of the intensity within the two wedges limited by $\vec{\tau}_1, \vec{\tau}_2$ and $\vec{\tau}_3, \vec{\tau}_1$ are

$$H_{3,x} = -\frac{\sigma}{2\sqrt{3}\,\mu_0}, \qquad H_{2,x} = -\frac{\sigma}{2\sqrt{3}\,\mu_0},$$
$$H_{3,y} = \frac{\sigma}{2\,\mu_0}, \qquad H_{2,y} = -\frac{\sigma}{2\,\mu_0}, \qquad (3.1.23)$$

respectively. Thus, in the three wedges the intensities have equal magnitude and are oriented at an angle $2\pi/3$ relative to each other. The lines of force of the intensity in the three wedges are shown in Fig. 3.1.4.

Equation 3.1.8 provides the relationship between the remanences $\vec{J}_1, \vec{J}_2, \vec{J}_3$ of the three wedges that generate the uniform surface charge densities σ. By solving the system of equations 3.1.8 at the three interfaces, and by virtue of Eq. 3.1.20, the components of $\vec{J}_1, \vec{J}_2, \vec{J}_3$ are

$$J_{1,x} = \frac{\sigma}{2\sqrt{3}} + J_{0,x}, \quad J_{2,x} = -\frac{\sigma}{\sqrt{3}} + J_{0,x}, \quad J_{3,x} = \frac{\sigma}{2\sqrt{3}} + J_{0,x},$$
$$J_{1,y} = -\frac{\sigma}{2} + J_{0,y}, \quad J_{2,y} = J_{0,y}, \quad J_{3,y} = \frac{\sigma}{2} + J_{0,y}, \qquad (3.1.24)$$

where $J_{0,x}$ and $J_{0,y}$ are the components of an arbitrary vector $\vec{J}_0$. Equations 3.1.24 reflect the indeterminacy of the solution of the problem of finding the distribution of magnetization capable of generating the intensity given by Eqs. 3.1.22 and 3.1.23. Two particular solutions are of interest. If the magnitude of vector $\vec{J}_0$ is assumed to be

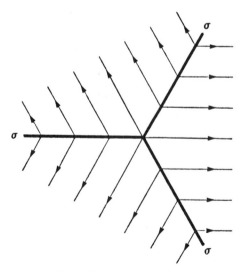

Fig. 3.1.4. Lines of force of the intensity in the structure of three prisms.

zero, the three vectors $\vec{J}_1, \vec{J}_2, \vec{J}_3$ have equal magnitude

$$|\vec{J}_1| = |\vec{J}_2| = |\vec{J}_3| = \frac{\sigma}{\sqrt{3}} \tag{3.1.25}$$

and are oriented at an angle $2\pi/3$ relative to each other, as shown in Fig. 3.1.5(a). By virtue of Eqs. 3.1.22, 3.1.23, and 3.1.24 the magnetic induction $\vec{B}$ in the three wedges of magnetized material is

$$\vec{B} = \vec{J} + \mu_0 \vec{H} = 0, \tag{3.1.26}$$

i.e., no flux of $\vec{B}$ is generated in the magnetized wedges.

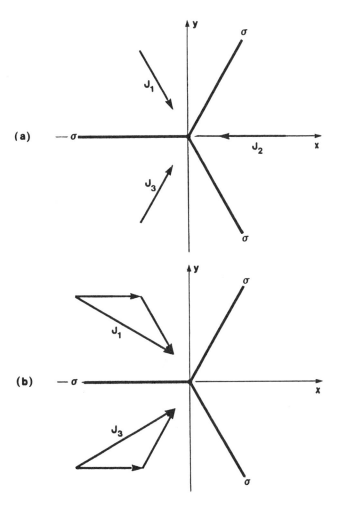

Fig. 3.1.5. Two special cases of the three interfaces: (a) identical remanences in the three prisms, (b) one of the prisms has no magnetic material.

Figure 3.1.5(b) corresponds to a vector $\vec{J}_0$ of components

$$J_{0,x} = \frac{\sigma}{\sqrt{3}}$$
$$J_{0,y} = 0. \qquad (3.1.27)$$

Equation 3.1.24 yields

$$J_{1,x} = J_{3,x} = -\frac{\sqrt{3}\sigma}{2}$$
$$J_{1,y} = -J_{3,y} = \frac{\sigma}{2} \qquad (3.1.28)$$

and

$$J_{2,x} = J_{2,y} = 0. \qquad (3.1.29)$$

The distribution of remanences given by Eqs. 3.1.28 and 3.1.29 corresponds to the particular case of a nonmagnetic material in the wedge limited by the interfaces $\vec{\tau}_2, \vec{\tau}_3$. In this case the magnetic induction $\vec{B}$ is equal to

$$\vec{B} = \frac{\sigma}{\sqrt{3}}\vec{\tau}_1 \qquad (3.1.30)$$

everywhere; i.e., the two remanences $\vec{J}_1, \vec{J}_3$ generate a uniform flux of $\vec{B}$ in the direction of the x axis.

3.2 STRUCTURES OF UNIFORMLY MAGNETIZED PRISMS

Let us analyze the development of two-dimensional structures of uniformly magnetized prisms that satisfy the condition expressed by Eq. 3.1.10. Under ideal conditions of linear demagnetization characteristics and zero magnetic susceptibility, one is seeking the geometries and the distributions of magnetization capable of generating a uniform magnetic field of intensity $\vec{H}_0$ within a prism of infinite length whose cross-section is a polygon s_a of n sides, as indicated in Fig. 3.2.1. The medium inside s_a is assumed to be nonmagnetic (air in particular) and the intensity $\vec{H}_0$ is assumed to be uniform and oriented in a direction perpendicular to the z axis, which is parallel to the axis of the prism.

The magnetic material outside s_a is assumed to be distributed in prisms, uniformly magnetized in a direction perpendicular to the z axis. These prisms interface with the region of nonmagnetic material through the n sides of the polygon s_a. The methodology of the calculation of geometry and magnetization of each prism can be developed by considering first the wedge of magnetized material shown in Fig. 3.2.2. The frame

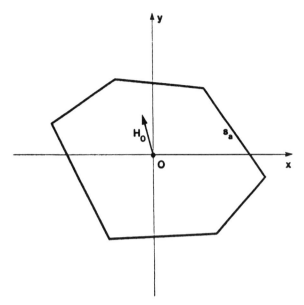

Fig. 3.2.1. Cross-section of a prismatic cavity.

of reference x, y, z of Fig. 3.2.2 is such that the edge P of the wedge is located in the plane $y = 0$ and is parallel to the z axis. A solution is sought in the region $y > 0$, which satisfies the condition that the scalar potential Φ has a reference value $\Phi = 0$ on the plane $y = 0$.

The direct boundary value problem would be formulated by assigning the geometry of the wedge, i.e., the angles α_i, α_e, between the x axis and the surfaces s_i, s_e of the wedge. Furthermore one would specify the magnetic properties of the media of Fig. 3.2.2, the boundary conditions on the surface s_e, and the value of the remanence of the medium of the wedge.

The inverse problem is the "design" of the wedge, i.e., the calculation of its geometry and its magnetization capable of generating the uniform intensity $\vec{H}_0$ in the region limited by s_i and the plane $y = 0$. The inverse problem is formulated by assigning the value of $\vec{H}_0$ and the angle α_i.

Assume that the boundary conditions at interface s_i are the only constraints of the design. The component H_s of the intensity in the wedge parallel to s_i and the component B_n of the induction in the wedge perpendicular to the s_i must be equal to

$$H_s = H_0 \sin \alpha_i , \qquad B_n = \mu_0 H_0 \cos \alpha_i . \qquad (3.2.1)$$

Since any remanence $\vec{J}$ that satisfies conditions 3.2.1 is a solution to the design problem, the calculation of $\vec{J}$ is indeterminate.

Let us analyze in vector form the solution of the design problem for a selected value J_0 of the magnitude of $\vec{J}$. The vector diagram of Fig. 3.2.3 shows the line parallel to s_i and the line n_i perpendicular to s_i drawn from the tip N_0 of vector $\vec{H}_0$. For simplicity, the constant μ_0 is assumed to be equal to unity in Fig. 3.2.3 as well as in

78 GENERATION OF A UNIFORM FIELD IN PRISMATIC CAVITIES

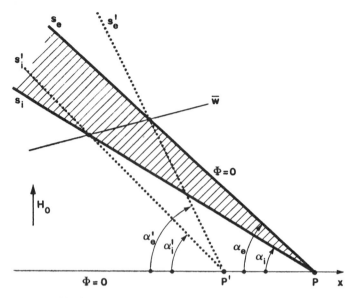

Fig. 3.2.2. Wedge of uniformly magnetized material.

all the vector diagrams used throughout the text of the book. Any remanence $\vec{J}$ of magnitude J_0 drawn from a point N_n of line n_i with the tip N_s on line s_i satisfies conditions 3.2.1. As indicated in Fig. 3.2.3, point A is the common origin of intensity $\vec{H}_0$ within the cavity and of intensity $\vec{H}$ and induction $\vec{B}$ within the wedge.

The equipotential surfaces within the wedge are perpendicular to $\vec{H}$. In particular

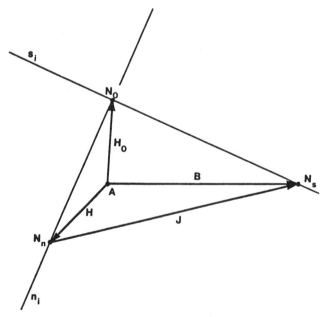

Fig. 3.2.3. Vector diagram of the field generated by the uniformly magnetized wedge.

STRUCTURES OF UNIFORMLY MAGNETIZED PRISMS 79

the $\Phi = 0$ equipotential surface must intersect surface s_i on the line of plane $y = 0$ which contains point P as shown in Fig. 3.2.2. Should the $\Phi = 0$ equipotential surface intercept the $y = 0$ plane on a different line, at $y = 0$ vector $\vec{H}$ would be perpendicular to the plane $y = 0$, and the field would not be uniform. The design of the wedge may be completed at this point by assuming that the external surface s_e of the wedge is the surface of a medium of infinite magnetic permeability. Thus s_e must coincide with the zero equipotential surface whose orientation is derived from the vector diagram of Fig. 3.2.3.

Assume now that the vector diagram of Fig. 3.2.3 is repeated to design a second wedge of magnetized material with surface s_i' oriented at an angle α_i' with respect to the x axis as shown in Fig. 3.2.2. Assume that the magnitude of the remanence $\vec{J}'$ of the second wedge is also equal to J_0 and that both wedges are designed to generate the same intensity $\vec{H}_0$. Any remanence $\vec{J}'$ with end points on the lines parallel and perpendicular to s_i' drawn from point N_0 of Fig. 3.2.2 satisfies the boundary conditions across s_i'. The selection of $\vec{J}'$ determines the position of the s_e' surface of the second wedge where $\Phi = 0$. Again s_e', s_i' must intersect each other on the line of plane $y = 0$ as shown by point P' of Fig. 3.2.2.

Assume the orientation of $\vec{J}$ and $\vec{J}'$ indicated in Fig. 3.2.4. Remanences $\vec{J}, \vec{J}'$ generate in the two wedges intensities $\vec{H}, \vec{H}'$ and inductions $\vec{B}, \vec{B}'$, respectively. Let us determine how the two wedges can be combined in such a way that the two surfaces s_i, s_i' form the interface with the region of nonmagnetic material where the uniform field has intensity $\vec{H}_0$. As long as the transition between the two regions of remanences $\vec{J}, \vec{J}'$ satisfies Eq. 3.1.10, the field generated by the combined structure is uniform.

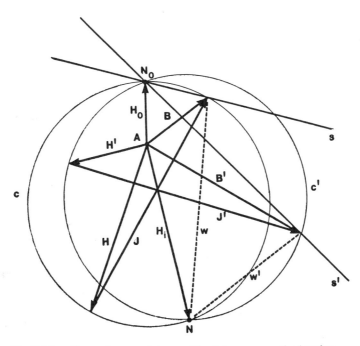

Fig. 3.2.4. Vector diagram of the transition between magnetized wedges.

80 GENERATION OF A UNIFORM FIELD IN PRISMATIC CAVITIES

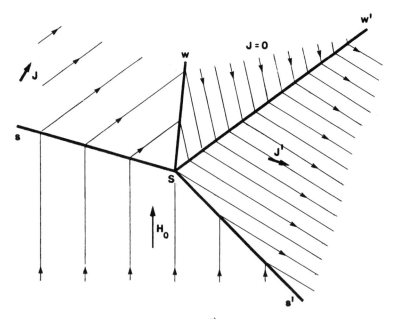

Fig. 3.2.5. Lines of flux of $\vec{B}$ in the transition region.

In general, condition 3.1.10 is not satisfied by a combined structure where the interface between the two regions of remanences $\vec{J}, \vec{J}'$ is the line $\overline{w}$ shown in Fig. 3.2.2, formed by the intersection of lines s_i, s_i' and the intersection of lines s_e, s_e'.

Consider the two circles c, c' in Fig. 3.2.4, with diameters $\vec{J}, \vec{J}'$. The two circles intersect each other at points N_0 and N. Let w, w' be the lines joining point N and the tips of the two vectors $\vec{J}, \vec{J}'$. The boundary conditions are satisfied everywhere across the transition between the two media of remanences $\vec{J}, \vec{J}'$ if w, w' are the boundaries of a wedge of nonmagnetic material inserted between the two media, with the vertex at the point of intersection of lines s, s', as indicated in Fig. 3.2.5. The intensity $\vec{H}_i$ within the nonmagnetic wedge is the vector with origin at point A and tip at point N, as shown in Fig. 3.2.4. Figure 3.2.5 shows the lines of flux of the magnetic induction in the two media and within the nonmagnetic wedge. Obviously, the induction $\vec{B}_i$ within the nonmagnetic wedge is

$$\vec{B}_i = \mu_0 \vec{H}_i . \tag{3.2.2}$$

The equipotential lines in the structure of Fig. 3.2.5 are shown in Fig. 3.2.6. Condition 3.1.10 is satisfied, as shown by the vector diagram of Fig. 3.2.7 that verifies the cancellation of the sum of the four vectors $\sigma_h \vec{\tau}_h$ at the point S of intersection of interfaces s, s', w, w'. The notation used in Fig. 3.2.7 corresponds to the selection of four unit vectors $\vec{n}_1, \vec{n}_2, \vec{n}_3, \vec{n}_4$ perpendicular to the interfaces of Fig. 3.2.6 oriented in a counterclockwise direction. The surface charge densities on the interfaces are

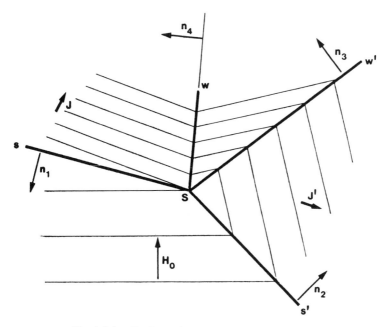

Fig. 3.2.6. Equipotential lines in the transition region.

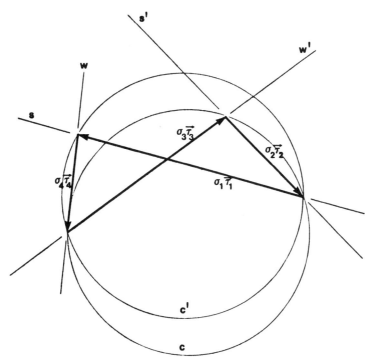

Fig. 3.2.7. Cancellation of the field singularity at the intersection of interfaces s, s', w, w'.

$$\sigma_1 = +\vec{J} \cdot \vec{n}_1$$
$$\sigma_2 = -\vec{J}' \cdot \vec{n}_2$$
$$\sigma_3 = +\vec{J}' \cdot \vec{n}_3 \quad (3.2.3)$$
$$\sigma_4 = -\vec{J} \cdot \vec{n}_4 .$$

The boundary conditions on each interface do not depend upon which side of the interfaces the field components are computed. In the schematic of Fig. 3.2.5, interfaces s, s' form a concave region of uniform intensity $\vec{H}_0$. Let us assume now the dual distribution of magnetic field, i.e., let the uniform intensity $\vec{H}_0$ be assigned in the convex region formed by s, s' indicated in Fig. 3.2.8. In both cases s and s' are the interfaces between the region of nonmagnetic material where the intensity is $\vec{H}_0$ and the two regions of remanences $\vec{J}$ and $\vec{J}'$ computed from the same vector diagram of Fig. 3.2.4.

The difference between the dual configurations of Figs. 3.2.5 and 3.2.8 lies in the transition between the two regions of remanences $\vec{J}$ and $\vec{J}'$. The role of the two lines w, w' drawn from the point of intersection of s and s' is reversed in the two configurations. In Fig. 3.2.8 line w' becomes the boundary of the medium of remanence $\vec{J}$ and line w becomes the boundary of the medium of remanence $\vec{J}'$. Thus, the boundary conditions on w and w' are not satisfied by the intensity $\vec{H}_i$ computed in the vector diagram of Fig. 3.2.4, and the region confined by lines w and w' cannot be a wedge of nonmagnetic material.

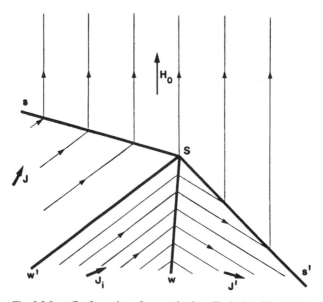

Fig. 3.2.8. Configuration of magnetized media dual to Fig. 3.2.5.

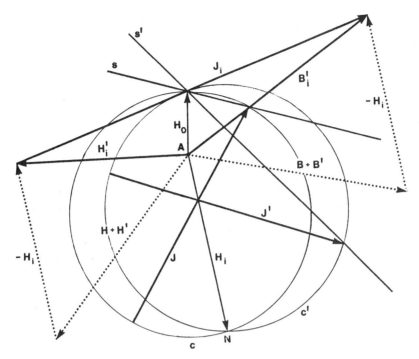

Fig. 3.2.9. Computation of remanences of the transition wedge between the magnetized media of Fig. 3.2.8.

The continuity of the tangential component of the intensity across boundaries w and w' in Fig. 3.2.8 is satisfied if the intensity $\vec{H}_i'$ within the wedge of Fig. 3.2.8 is

$$\vec{H}_i' = \vec{H} + \vec{H}' - \vec{H}_i , \qquad (3.2.4)$$

where $\vec{H}$ is the intensity in the medium limited by s and w'. Consequently the continuity of the tangential components of $\vec{H}$ and $\vec{H}_i'$ across line w' reduces to the continuity of the tangential components of $\vec{H}'$ and $\vec{H}_i$ as established by the vector diagram of Fig. 3.2.4. Similarly, the continuity of the tangential components of $\vec{H}'$ and $\vec{H}_i'$ across line w reduces to the continuity of the tangential components of $\vec{H}$ and $\vec{H}_i$ established in Fig. 3.2.4. Intensity $\vec{H}_i'$ is shown in the vector diagram of Fig. 3.2.9.

By virtue of the vector diagram of Fig. 3.2.4 the continuity of the normal components of the induction B_i' across boundaries w and w' in Fig. 3.2.8 is satisfied if the magnetic induction within the wedge is

$$\vec{B}_i' = \vec{B} + \vec{B}' - \mu_0 \vec{H}_i , \qquad (3.2.5)$$

also shown in the vector diagram of Fig. 3.2.9.

84 GENERATION OF A UNIFORM FIELD IN PRISMATIC CAVITIES

To generate the intensity $\vec{H}_i'$ and the magnetic induction $\vec{B}_i'$, the region between w and w' must be a wedge of a magnetic material with remanence

$$\vec{J}_i = \vec{B}_i' - \mu_0 \vec{H}_i' , \qquad (3.2.6)$$

as shown in the diagram of Fig. 3.2.9. By virtue of Eqs. 3.2.4 and 3.2.5 one has

$$\vec{J}_i = \vec{J} + \vec{J}' . \qquad (3.2.7)$$

The need of a wedge of magnetized material between the two media is not necessarily a consequence of a convex boundary of the region of uniform intensity $\vec{H}_0$, but rather depends on the selection of remanences $\vec{J}$ and $\vec{J}'$. Assume, for instance, that a structure of magnetic material is sought such that the region of intensity $\vec{H}_0$ is bound by media with remanences of equal magnitude and orientation. The vector diagram of Fig. 3.2.10 provides the value of intensities $\vec{H}, \vec{H}'$ and magnetic inductions $\vec{B}, \vec{B}'$ in the two media that interface with the region of intensity $\vec{H}_0$ across s, s', respectively. The two circles c, c' of diameter $\vec{J}$ and $\vec{J}'$ intersect each other at point N, yielding the

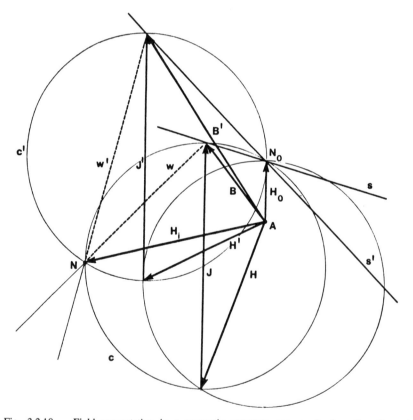

Fig. 3.2.10.   Field computation in a magnetic structure composed of media of equal remanences in magnitude and orientation.

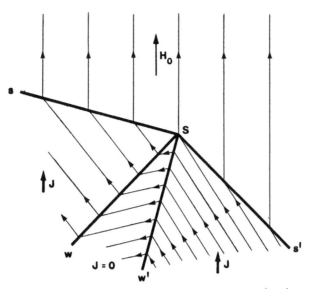

Fig. 3.2.11. Lines of flux of the magnetic induction for $\vec{J}' = \vec{J}$.

two boundaries w, w' of the transition region between the two media indicated in Fig. 3.2.10. One observes that the relative position of w' and w in Fig. 3.2.10 is inverted with respect to their relative position in Fig. 3.2.4. As a consequence, because of the assumption

$$\vec{J}' = \vec{J}, \qquad (3.2.8)$$

the region confined by w, w' is now a wedge of nonmagnetic material on the concave side of the boundary formed by s and s'. The region confined by w, w' is a wedge of magnetized material on the convex side of the boundary formed by s and s', opposite to the situation illustrated by Figs. 3.2.5 and 3.2.8.

Figure 3.2.11 shows the lines of flux of the magnetic induction generated by condition 3.2.8 when the two media located on the concave side of the boundary are separated by the wedge of nonmagnetic material. When the two media are located on the convex side of the boundary, the remanence 3.2.7 of the wedge of magnetic material between w and w' becomes

$$\vec{J}_i = 2\vec{J}, \qquad (3.2.9)$$

as shown in the vector diagram of Fig. 3.2.12. The lines of flux of the magnetic induction generated by condition 3.2.9 are shown in Fig. 3.2.13.

The arbitrary selection of vectors $\vec{J}$ and $\vec{J}'$ which satisfy the boundary conditions across interfaces s and s' is equivalent to the selection of circles c and c' of diameters equal to the magnitude J_0 of $\vec{J}$ and $\vec{J}'$. Assume the selection of circles c, c' shown in Fig. 3.2.14 and let N_s be the intersection of s with circle c. Points N_s, N define the

86 GENERATION OF A UNIFORM FIELD IN PRISMATIC CAVITIES

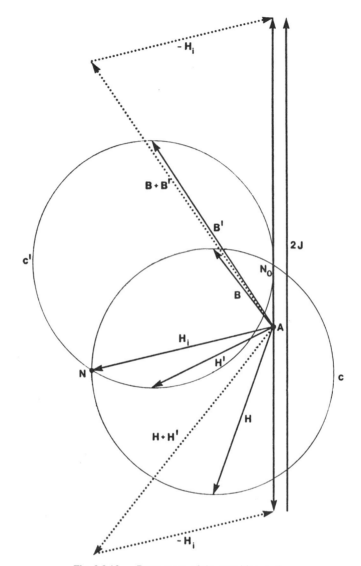

Fig. 3.2.12. Remanence of the transition wedge.

orientation of line w. A particular orientation of s' exists whose intersection N'_s with circle c' lies on the line w. In this particular situation line w' coincides with line w and it becomes the interface between the two media of remanences $\vec{J}$ and $\vec{J'}$. No wedge of either non-magnetic or magnetic material is necessary to establish the transition between the two media, independent of what side of the boundary the two media are located. As shown in the vector diagram of Fig. 3.2.14, the geometry that eliminates the wedge between the two media is defined by the angular orientation θ_0 of s' with respect to s. If s' is oriented at an angle $\theta < \theta_0$ with respect to s, the two media are separated by a wedge of magnetized material, if they are located on the concave side of the boundary. If $\theta > \theta_0$, then the wedge of nonmagnetic material

STRUCTURES OF UNIFORMLY MAGNETIZED PRISMS 87

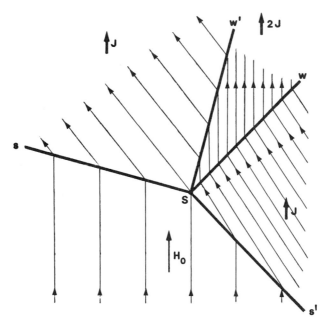

Fig. 3.2.13. Lines of flux of the magnetic induction in the transition region with remanence $\vec{J}_i = 2\vec{J}$.

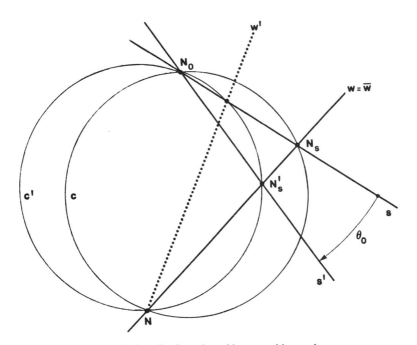

Fig. 3.2.14. Configuration with no transition wedges.

88 GENERATION OF A UNIFORM FIELD IN PRISMATIC CAVITIES

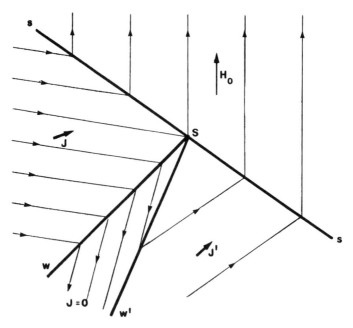

Fig. 3.2.15. Limit of s and s' aligned with each other.

separates the two media located on the convex side, and a wedge of magnetic material separates the two media located on the concave side.

A design approach based on the selection of the two independent circles c, c' makes it necessary to insert a transition wedge between the two media even if s and s' are collinear. This situation is shown in Fig. 3.2.15, where the interfaces w and w' are derived from the vector diagram of Fig. 3.2.14. In this configuration w, w' define a wedge of nonmagnetic material, whereas w, w' would define a wedge of magnetized material if the media were located on the opposite side of the boundary formed by lines s, s'.

The need for the transition wedge in the geometry of Fig. 3.2.15 is avoided if the same circle c is used to compute the field distribution in all the magnetized elements that interface with the nonmagnetic region of uniform field with intensity $\vec{H}_0$. Obviously, one can choose any circle of diameter J_0 that passes through the tip N_0 of $\vec{H}_0$. Assume that a single circle c_0 is chosen with center on the line that contains vector $\vec{H}_0$. In this case, the field distribution across interfaces s and s' is determined by the vector diagram of Fig. 3.2.16. In particular, the remanence of a medium whose boundary is perpendicular to $\vec{H}_0$ is parallel to $\vec{H}_0$.

In the vector diagram of Fig. 3.2.16 the angle between $\vec{H}_0$ and the remanence is always twice the angle between $\vec{H}_0$ and the line perpendicular to s, regardless of the orientation of s. Thus, in a magnetic structure designed on the basis of the vector diagram of Fig. 3.2.16, intensity $\vec{H}_0$ within the polygonal cavity s_a and remanence $\vec{J}$ of a prism that interface with s_a satisfy the following theorem [2]:

Theorem 2. Intensity $\vec{H}_0$ and remanence $\vec{J}$ are oriented at equal and opposite angles with respect to the vector perpendicular to the interface between the magnet cavity and the magnetic material.

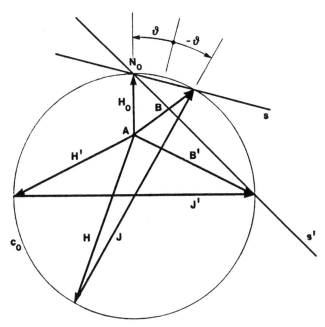

Fig. 3.2.16. Design approach with the same circular diagram for all components of the structure.

3.3 YOKED TWO-DIMENSIONAL PRISMATIC STRUCTURES

Let us apply the vector diagram defined by the single circle of Fig. 3.2.16 to the computation of a structure composed of n prisms of magnetic material that enclose a prismatic cavity. As shown in the preceding section, the point N of intersection of each pair of independent circles selected for adjacent sides of the cavity provides the geometry and the magnetization of each transition region between prisms of magnetic material.

The problem of computing the transition region becomes indeterminate if all circles coincide. In principle, n points D_h can be arbitrarily chosen on the circle of Fig. 3.2.16 to compute the n transition regions.

Assume that circle c_0 in Fig. 3.3.1 of diameter J_0 is used to compute the remanences $\vec{J}_h, \vec{J}_{h+1}$ of the prisms that interface with the cavity through the sides s_h, s_{h+1}. Points N_h and N_{h+1} are the intersections of lines s_h and s_{h+1} with the circle c_0. Select a point D_h on the circle c_0 and let $w_{h,1}, w_{h,2}$ be the lines that have point D_h in common and contain points N_h and N_{h+1}, respectively. Let $H_{h,i}$ be the vector with the tip at D_h and the origin at A, and assume that $H_{h,i}$ is the intensity in a wedge limited by $w_{h,1}$ and $w_{h,2}$ with the edge at the vertex S_h common to sides s_h, s_{h+1} of the prismatic cavity.

Regardless of the position of D_h on circle c_0, the angle formed by interfaces s_{h+1} and s_h between cavity and media of remanences $\vec{J}_h, \vec{J}_{h+1}$ has the same sign as the angle between $w_{h,1}$ and $w_{h,2}$. Thus, by virtue of the analysis of the preceding section, if the n-sided polygon is convex, the wedges between the prisms of remanences

90 GENERATION OF A UNIFORM FIELD IN PRISMATIC CAVITIES

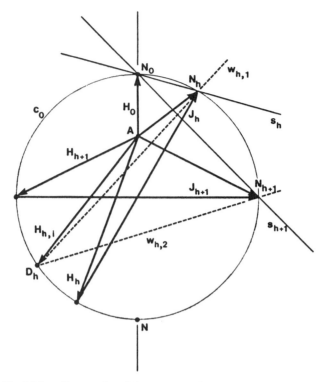

Fig. 3.3.1. Computation of the geometry of a nonmagnetic wedge.

$\vec{J}_h, \vec{J}_{h+1}$ are regions of nonmagnetic material. As a consequence, the uniform intensity within a cavity with an arbitrary convex polygonal contour can always be generated with a structure of prisms of magnetic material, all magnetized with remanences having the same magnitude J_0.

Figure 3.3.2 shows the lines of flux of the magnetic induction $\vec{B}$ in the two media of remanences $\vec{J}_h, \vec{J}_{h+1}$ and in the wedge of nonmagnetic material formed by lines $w_{h,1}$ and $w_{h,2}$. The equipotential lines in the three regions of the magnetic structure of Fig. 3.3.2 are perpendicular to the intensities $\vec{H}_h, \vec{H}_{h,i}, \vec{H}_{h+1}$. As shown in Section 3.2, the scalar potential reaches the reference value

$$\Phi = 0 \qquad (3.3.1)$$

on the equipotential lines u_h, u_{h+1} indicated in Fig. 3.3.2. The two lines intersect $w_{h,1}, w_{h,2}$ at points $U_{h,1}, U_{h,2}$. Because $H_{h,i}$ satisfies the boundary conditions on the two lines $w_{h,1}, w_{h,2}$, the line joining points $U_{h,1}, U_{h,2}$ must also be a $\Phi = 0$ equipotential line. Consequently, there exists a closed $\Phi = 0$ equipotential line u_0 that completely encloses the n-sided polygonal cavity.

By definition, Φ is the scalar potential relative to the potential of the external nonmagnetic medium surrounding the magnetic structure, provided that no magnetic field exists in the external medium. As shown in Fig. 3.3.2, in general the lines of flux of

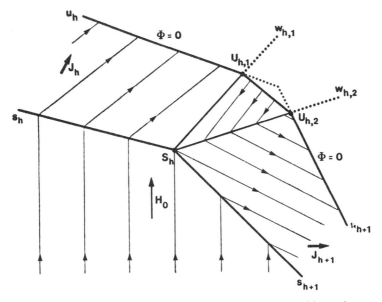

Fig. 3.3.2. Lines of flux of the magnetic induction in the transition region.

$\vec{B}$ cross the $\Phi = 0$ equipotential line. Therefore line u_0 cannot be the interface between the magnetic structure and an external nonmagnetic medium of uniform scalar potential. The magnetic structure can be truncated by the equipotential line u_0 only if u_0 is the interface between the structure and an external medium of infinite magnetic permeability. The external medium itself channels the flux of the magnetic induction generated by the magnetic structure, without perturbing the field configuration within the structure.

The external medium of infinite magnetic permeability is an ideal model of a perfect yoke, and the schematic of Fig. 3.3.2 defines the general approach of the design of yoked prismatic magnets based on the vector diagram of Fig. 3.3.1.

While a change of position of point D_h on the circle of Fig. 3.3.1 does not alter the position of the lines u_h, u_{h+1}, it affects the geometry of the wedge of nonmagnetic material with vertex S_h. In particular, point D_h can be selected to coincide with the origin of either vector $\vec{J}_h$ or vector $\vec{J}_{h+1}$. The resulting geometries of the transition region are shown in Fig. 3.3.3. The geometry of Fig. 3.3.3(a) corresponds to the selection of D_h at the origin of $\vec{J}_h$ and Fig. 3.3.3(b) corresponds to the selection of D_h at the origin of $\vec{J}_{h+1}$. Both structures have the same external boundary with the medium of infinite magnetic permeability, which is an n-sided polygon formed by the external boundaries u_h of the n elements of magnetized material. In both cases one of the interfaces of the wedge is the line joining the vertices S_h, V_h of the internal and external polygons. The other interface is parallel to $\vec{J}_h$ in Fig. 3.3.3(a) and it is parallel to $\vec{J}_{h+1}$ in Fig. 3.3.3(b).

Figure 3.3.4 shows the geometry of the transition region if point D_h coincides with point N in Fig. 3.3.1, which is the second point of intersection of circle c_0 and the line that contains $\vec{H}_0$. In this case the intensity $\vec{H}_{h,i}$ within the wedge is parallel and opposite to $\vec{H}_0$. As a consequence, the interface between the wedge and the yoke of infinite

92 GENERATION OF A UNIFORM FIELD IN PRISMATIC CAVITIES

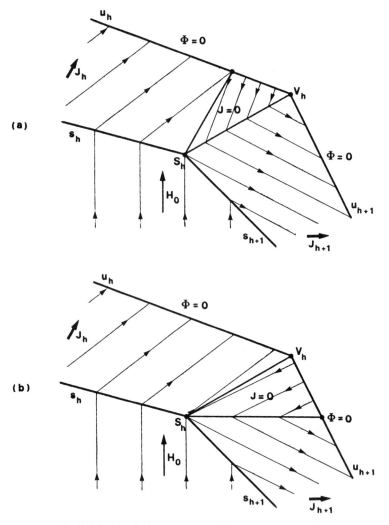

Fig. 3.3.3. Particular geometries of the nonmagnetic wedges.

permeability is perpendicular to $\vec{H}_0$. If point D_h is made to coincide with the point N_0 in the vector diagram of Fig. 3.3.1, lines $w_{h,1}$ and $w_{h,2}$ coincide with lines s_h and s_{h+1}, respectively, and $\vec{H}_{h,i}$ reduces to $\vec{H}_0$. Thus the selection of point D_h at N_0 causes the elimination of either the prism of magnetized material with remanence $\vec{J}_{h+1}$ or the prism of magnetized material with remanence $\vec{J}_h$.

If the point D_h is selected in such a way that line $w_{h,2}$ is perpendicular to $\vec{H}_{h+1}$, as indicated in the diagram of Fig. 3.3.5, the line $w_{h,2}$ becomes parallel to the outside boundary u_{h+1}. In this extreme case the volume of magnetized material in the transition region diverges.

The above examples show that a variety of wedges can establish the transition between magnetized prisms. Since one of the major goals of a designer is to minimize the amount of magnetic material, the indeterminacy of the selection of the transition

YOKED TWO-DIMENSIONAL PRISMATIC STRUCTURES 93

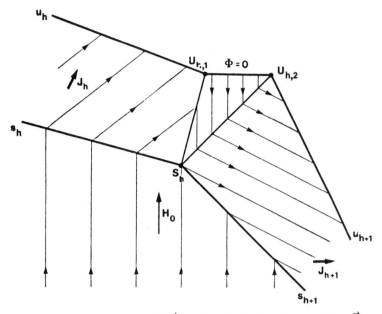

Fig. 3.3.4. Transition region with $\vec{H}_{h,i}$ oriented in the direction opposite to $\vec{H}_0$.

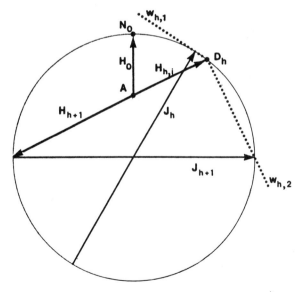

Fig. 3.3.5. Transition region with $w_{h,2}$ perpendicular to $\vec{H}_{h+1}$.

94 GENERATION OF A UNIFORM FIELD IN PRISMATIC CAVITIES

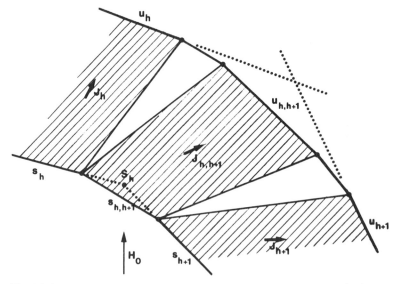

Fig. 3.3.6. Magnetized prism inserted between the prisms of remanences $\vec{J}_h, \vec{J}_{h+1}$.

wedge can be eliminated by imposing additional constraints, such as the criterion of minimum cross-sectional area of the magnetic structure.

It is of importance to point out that the transitions discussed so far are the result of an implicit assumption that S_h is the vertex of a transition region involving a single wedge of nonmagnetic material. If this restriction is eliminated, the transition can be accomplished with an infinity of solutions which consist of a multiplicity of wedges of magnetized material separated from each other by wedges of nonmagnetic material with the same common vertex S_h.

Let us view point S_h as the limit of a side $s_{h,h+1}$ whose length converges to zero, as shown in Fig. 3.3.6. Side $s_{h,h+1}$ becomes the interface between the cavity and a magnetized medium whose remanence $\vec{J}_{h,h+1}$ is provided by the vector diagram of Fig. 3.3.7. The geometry of the transitions between the medium of remanence $\vec{J}_{h,h+1}$ and the two media of remanences $\vec{J}_h$ and $\vec{J}_{h+1}$ corresponds to the selection of two points $D_{h,1}$ and $D_{h,2}$ on the circle of the vector diagram, as indicated in Fig. 3.3.7.

If the length of side $s_{h,h+1}$ reduces to zero, point S_h becomes the vertex of a wedge of remanence $\vec{J}_{h,h+1}$, as shown in the schematic of Fig. 3.3.8. The wedge is separated from the two prisms of remanences $\vec{J}_h, \vec{J}_{h+1}$ by the two wedges of nonmagnetic material determined by the selection of points $D_{h,1}$ and $D_{h,2}$ in the vector diagram of Fig. 3.3.7. The external boundary of the wedge of remanence $\vec{J}_{h,h+1}$ is the line $u_{h,h+1}$ shown in Fig. 3.3.8. Line $u_{h,h+1}$ becomes a new side of the external polygonal contour where $\Phi = 0$.

The procedure can be extended to any number of points D_h. In the limit of an infinite number of points, the external boundary of the transition region becomes a continuous curve, whose envelope is formed by the lines $u_{h,h+1}$ corresponding to each position of points $D_{h,i}$.

YOKED TWO-DIMENSIONAL PRISMATIC STRUCTURES 95

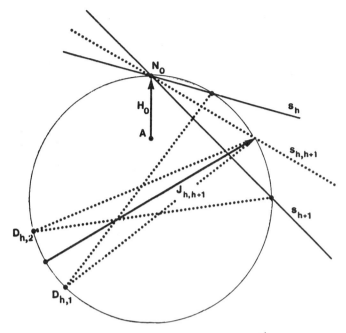

Fig. 3.3.7. Computation of the remanence $\vec{J}_{h,h+1}$.

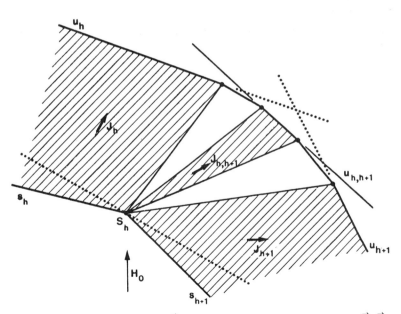

Fig. 3.3.8. Wedge of remanence $\vec{J}_{h,h+1}$ between the media of magnetizations $\vec{J}_h$, $\vec{J}_{h+1}$.

3.4 YOKELESS TWO-DIMENSIONAL PRISMATIC STRUCTURES

The ideal yoke of infinite magnetic permeability is essential to generate a uniform field intensity with magnetized prisms confined within the equipotential surface u_0. The yoke can be eliminated only if it can be replaced by a structure of magnetized material whose geometry and magnetization satisfy two requirements: (a) confining the flux of the magnetic induction and (b) satisfying the condition of field uniformity within each region of the magnetic structure. To be consistent with the methodology developed in the preceding section, such a yokeless structure is sought with the condition that its magnetic components have all equal remanence.

The relationship between the values of $\vec{B}$ and $\vec{H}$ in the region of magnetized material that interfaces with the prismatic cavity was provided by the vector diagram 3.3.1. That relationship depends solely on the boundary conditions at the interface between the cavity and the magnetic material; i.e., the vector diagram in Fig. 3.3.1 is not affected by the boundary conditions at the external surface of the magnetic structure. One can extend that relationship to the external medium where the field is to vanish. If the vector diagram of Fig. 3.3.1 is applied to the particular value

$$\vec{H}_0 = 0, \tag{3.4.1}$$

point A coincides with point N_0 and $\vec{B}$ and $\vec{H}$ are perpendicular to each other, i.e.,

$$\vec{B} \cdot \vec{H} = 0, \tag{3.4.2}$$

regardless of the orientation of the interface between external medium and magnetic material and regardless of the geometry of the material.

The basic schematic of a yokeless structure is shown in Fig. 3.4.1. The interface s_1 separates the magnet cavity where the intensity is $\vec{H}_0$ from the region V_1 of magnetized material whose remanence is $\vec{J}_1$. The vectors $\vec{B}_1$ and $\vec{H}_1$ are provided by the vec-

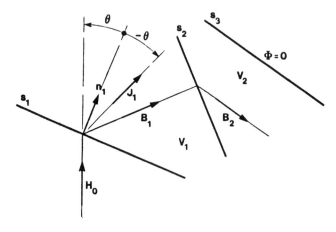

Fig. 3.4.1. Basic structure of a yokeless magnet.

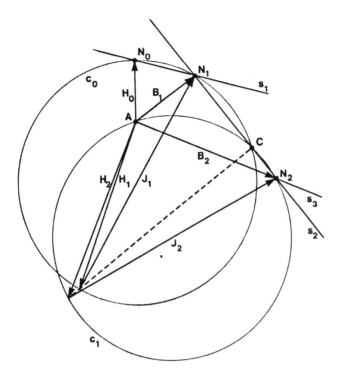

Fig. 3.4.2. Computation of the field components in a yokeless structure.

tor diagram inscribed in circle c_0 of Fig. 3.4.2. Line s_2 is the interface between regions V_1 and V_2 of magnetized material designed to confine the field. Line s_3 is the interface between V_2 and the surrounding medium where the scalar potential is assumed to be uniform and equal to zero. Thus s_3 must be an equipotential line, and by virtue of Eq. 3.4.2, the lines of flux of $\vec{B}_2$ in region V_2 must be parallel to s_3, as indicated in Fig. 3.4.1 [3,4].

If the orientation of the segment s_2 is given, the values of $\vec{J}_2, \vec{B}_2, \vec{H}_2$ in the region V_2 are provided by a second vector diagram inscribed in circle c_1, shown in Fig. 3.4.2.

To determine the relative position of the vector diagrams inscribed in the two circles c_0 and c_1, consider the line parallel to interface s_2 drawn from the tip N_1 of remanence $\vec{J}_1$. Line s_2 in Fig. 3.4.2 intersects circle c_0 at point C.

Consider circle c_1 with a diameter equal to the magnitude J_0 of remanence $\vec{J}_1$ that passes through point C and the origin A of vector $\vec{H}_0$. If the vector diagram inscribed in circle c_1 is to provide the values of $\vec{B}_2$ and $\vec{H}_2$ that satisfy Eq. 3.4.2, the origin of both vectors must be on circle c_1.

Assume that the tip of remanence $\vec{J}_2$ is selected to coincide with the second point N_2 of intersection of line s_2 and circle c_1. The boundary conditions at interface s_2 are satisfied if the origins of $\vec{B}_2$ and $\vec{H}_2$ coincide with the origin A of intensity $\vec{H}_0$, as shown in Fig. 3.4.2. Furthermore, by virtue of Eq. 3.4.2, the boundary condition at the interface s_3 between the region V_2 and the external nonmagnetic medium is satisfied if s_3 passes through points A and N_2. Thus, Fig. 3.4.2 provides the basic vector diagram for the computation of geometry and magnetization of a yokeless structure.

98 GENERATION OF A UNIFORM FIELD IN PRISMATIC CAVITIES

Since in general two circles of diameter J_0 can be drawn through points A, C, two solutions are found for the field vectors in region V_2. These solutions lead to two different geometries of region V_2. The criterion that controls the choice between the two solutions in an actual magnet design will be defined in the following chapter.

As in the case of a yoked structure, by virtue of Theorem 2 of Section 3.2, vectors $\vec{H}_0$ and $\vec{J}_1$ are oriented at equal and opposite angles with respect to the unit vector $\vec{n}_1$ perpendicular to interface s_1, as shown in Fig. 3.4.1. In general, no such simple relationship holds for the orientation of $\vec{J}_2$ in the yokeless structure.

A particular case worth mentioning is a yokeless magnet designed for

$$K = \frac{\mu_0 H_0}{J_0} = \frac{1}{2}. \qquad (3.4.3)$$

This choice results in point A coinciding with the center of circle c_0, as shown in the vector diagram of Fig. 3.4.3. In this case the center O_1 of circle c_1 is on the circle c_0

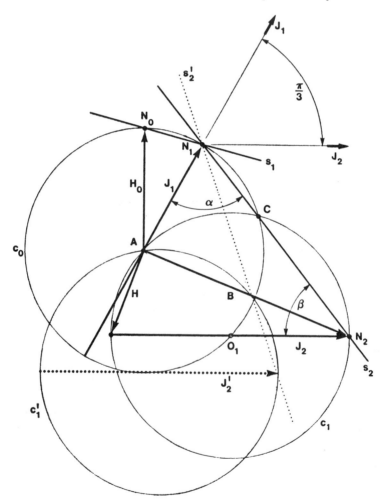

Fig. 3.4.3. Yokeless structure designed for $H_0 = J_0/2\mu_0$.

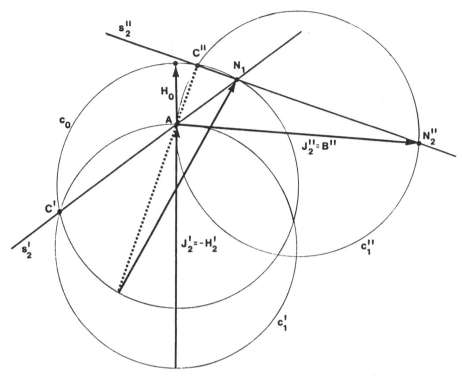

Fig. 3.4.4. Limit of s_2 parallel to $\vec{B}_1$ and s_2 perpendicular to $\vec{H}_1$.

and points A, C, O_1 form an equilateral triangle. In the vector diagram of Fig. 3.4.3, angle α between line s_2 and vector $\vec{J}_1$ and angle β between line s_2 and vector $\vec{J}_2$ satisfy the condition

$$\alpha + \beta = \frac{2\pi}{3}. \tag{3.4.4}$$

Thus, in this particular case, vectors $\vec{J}_1$ and $\vec{J}_2$ are oriented at an angle $\pi/3$ relative to each other, independent of the orientation of interface s_2 as indicated in Fig 3.4.3. The dotted line in Fig. 3.4.3 shows the position of vector $\vec{J}'_2$, parallel to $\vec{J}_2$, which corresponds to a different interface s'_2.

Another particular field configuration in region V_2 corresponds to an orientation of interface s_2 either parallel to magnetic induction $\vec{B}_1$ or perpendicular to intensity $\vec{H}_1$. Assume in vector diagram 3.4.2 an interface s'_2 parallel to $\vec{B}_1$. Point C moves to point C' as indicated in Fig. 3.4.4 and point N_2 coincides with the origin A of vector $\vec{H}_0$. As a consequence, A is the tip of the remanence $\vec{J}''_2$ in region V_2, and by virtue of Eq. 3.4.2 vectors $\vec{B}'_2$, $\vec{H}'_2$ are given by

$$\vec{B}'_2 = 0, \qquad \vec{H}'_2 = -\vec{J}''_2; \tag{3.4.5}$$

i.e., there is no flux of the magnetic induction in the component of magnetized material that interfaces with the surrounding medium.

A second particular case occurs when s_2 becomes s_2'', perpendicular to $\vec{H}_1$, which is oriented along the dotted line of Fig. 3.4.4. Point C moves to the new position C'' and point N_2 moves to the new position N_2'' as indicated in the figure. Point N_2'' belongs to the diameter of circle c_1'' whose other end coincides with point A. As a consequence, point A is the origin of the remanence $\vec{J}_2''$ in region V_2. Again by virtue of Eq. 3.4.2, vectors $\vec{B}_2''$, $\vec{H}_2''$ are given by

$$\vec{B}_2'' = \vec{J}_2'', \qquad \vec{H}_2'' = 0 ; \qquad (3.4.6)$$

i.e., region V_2 is an equipotential region whose potential coincides with the potential of the surrounding nonmagnetic medium.

The distribution of $\vec{J}, \vec{B}, \vec{H}$ in each element of magnetized material that interfaces with the cavity changes from element to element. Thus the interface between region V_1 and region V_2 in Fig. 3.4.1 must emerge from a vertex of the polygonal contour of the cavity. Furthermore, because the field configuration in region V_1 does not satisfy condition 3.4.2, region V_1 must interface with the surrounding medium at a single point where the scalar potential in region V_1 becomes equal to zero. Thus, in general, the cross-section of region V_1 is a triangle as shown in Fig. 3.4.5. In that figure S_0

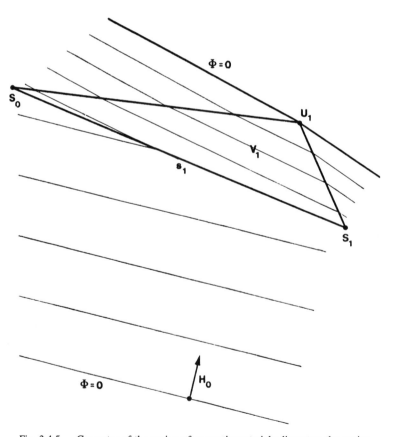

Fig. 3.4.5. Geometry of the region of magnetic material adjacent to the cavity.

and S_1 are the two vertices of the side s_1 of the polygonal cavity and point U_1 is the vertex of the triangle of the medium of remanence $\vec{J}_1$ where $\Phi = 0$.

3.5 INVARIANCE PROPERTIES OF YOKELESS STRUCTURES

The value of the scalar potential Φ at each point of the yokeless structure of Fig. 3.4.5 is measured relative to a reference equipotential line $\Phi = 0$ and is a function of the geometry and the distribution of remanence J_0. For a given orientation of intensity $\vec{H}_0$ and a given value of J_0, the orientation of the interfaces provided by the vector diagram of Fig. 3.4.2 depends only on the position of point A on the diagonal of circle c_0. Thus, the orientation of the interfaces is a function of parameter

$$K = \frac{\mu_0 H_0}{J_0}. \tag{3.5.1}$$

Assume the geometry that corresponds to the position of the reference line $\Phi = 0$ indicated in Fig. 3.4.5. In general, a different geometry results from a shift or a rotation of the reference line due to a change of orientation of $\vec{H}_0$. Assume, however, that $\vec{H}_0$ and the $\Phi = 0$ line both rotate by an angle α about a fixed point F that lies on the line perpendicular to s_1 that passes through the vertex U_1 of region V_1 as indicated in Fig. 3.5.1.

The effect of the rotation of $\vec{H}_0$ by an angle $+\alpha$ can be computed in the vector diagram of Fig. 3.4.2 by assuming that interfaces s_1, s_2 are rotated by an angle $-\alpha$ with respect to $\vec{H}_0$. As shown in the vector diagram of Fig. 3.5.2, points N_1 and N_2 move to points N_1' and N_2', respectively. Points N_1', N_2' are the tips of the new vectors $\vec{J}_1'$, $\vec{J}_2'$ in regions V_1, V_2, respectively. Thus, vectors $\vec{J}_1'$, $\vec{J}_2'$ are rotated by an angle -2α with respect to $\vec{J}_1$, $\vec{J}_2$.

Points A, N_2' in Fig. 3.5.2 are the end points of the new vector $\vec{B}_2'$ in region V_2 which is rotated by an angle $-\alpha$ with respect to $\vec{B}_2$. Thus, $\vec{B}_2'$ is still parallel to the interface between volume V_2 and the surrounding air. On the other hand, because point A is located on circle c_1, and $\vec{J}_2'$ is a diameter of c_1, the new intensity $\vec{H}_2'$ is perpendicular to vector $\vec{B}_2'$.

Let ξ_0, ξ_1 be the distances of points F and U_1 from the interface s_1 as indicated in Fig. 3.5.1. The difference between the scalar potentials at points F and U_1 is zero. Thus intensity $\vec{H}_1$ in the region V_1 satisfies the condition

$$(\xi_0 \vec{H}_0 + \xi_1 \vec{H}_1) \cdot \vec{n}_1 = 0, \tag{3.5.2}$$

where $\vec{n}_1$ is a unit vector perpendicular to s_1, as shown in Fig. 3.5.1. By virtue of Eq. 3.5.1 and the vector diagram of Fig 3.4.2, Eq. 3.5.2 reduces to

$$[K\xi_0 - (1-K)\xi_1] J_0 \cos\theta = 0, \tag{3.5.3}$$

where θ is the angle between $\vec{H}_0$ and $\vec{n}_1$. Thus the relationship between ξ_0 and ξ_1

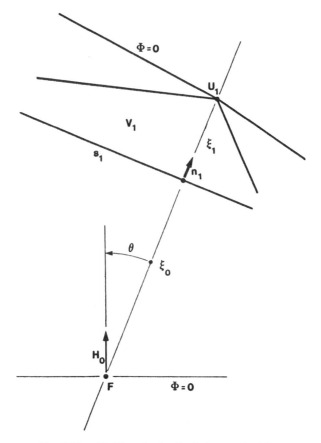

Fig. 3.5.1. Position of point F relative to point U_1.

given by Eq. 3.5.3 is independent of θ. Moreover, the scalar potential at point U_1 is zero, independent of the orientation of $\vec{H}_0$.

In conclusion, the external boundary of the yokeless magnet is always a $\Phi = 0$ surface and the field confinement is maintained for each orientation of $\vec{H}_0$. The geometry of the magnetic structure satisfies the invariance theorem:

Theorem 3. The geometry of the yokeless magnetic structure is a function of parameter K and point F only, and it is independent of the orientation of $\vec{H}_0$.

The angular orientation of the remanences in each element of the magnetic structure satisfies the theorem [5]:

Theorem 4. For a constant value of K and a fixed position of point F, a rotation of intensity $\vec{H}_0$ by an angle α results in a rotation by an angle $-\alpha$ of remanences $\vec{J}$ in all components of the magnetic structure.

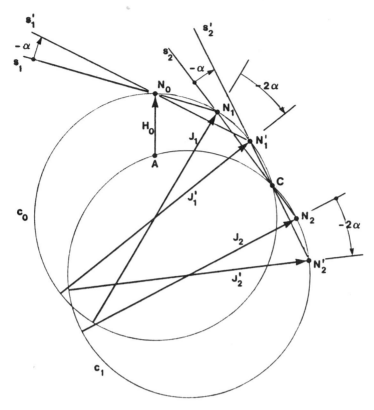

Fig. 3.5.2. Rotation of the equipotential line $\Phi = 0$.

The vector diagram of Fig. 3.5.2 has been derived with the assumption of a given orientation of interface s_2 between the two regions V_1 and V_2 of the magnetized material. The following chapter will show how s_2 is computed in the design of a yokeless magnet. Once side s_1 of the cavity is assigned, the orientation of s_2 is determined by the vertex U_1 of region V_1 in the schematic of Fig. 3.4.5. Thus, for a given orientation of $\vec{H}_0$, interface s_2 may depend upon both parameters K and F. In the particular case $K = 0.5$, Section 3.4 has shown that the angle between $\vec{J}_1$ and $\vec{J}_2$ is always equal to $\pi/3$, independent of the position of point F.

Consider the particular situation depicted in Fig. 3.5.3 where the line $\Phi = 0$ within the cavity intersects side s_1 of the cavity polygonal contour. In the regions of magnetic material that interface with the surrounding medium, the scalar potential is zero only on their external boundaries. Then in region V_1 the $\Phi = 0$ line must contain the vertex U_1, as shown in Fig. 3.5.3. Also shown in the figure are two equipotential lines $\Phi \neq 0$. The values of $\vec{J}_1, \vec{B}_1, \vec{H}_1$ in region V_1 are given by the vector diagram of Fig. 3.5.4.

Assume now that vector $\vec{H}_0$ is rotated by an angle $\pi/2$ around point F. By virtue of theorem 4, the remanence $\vec{J}_1$ will rotate by the angle π with respect to $\vec{H}_0$. Thus the new remanence of region V_1 is

$$\vec{J}_1' = -\vec{J}_1 \tag{3.5.4}$$

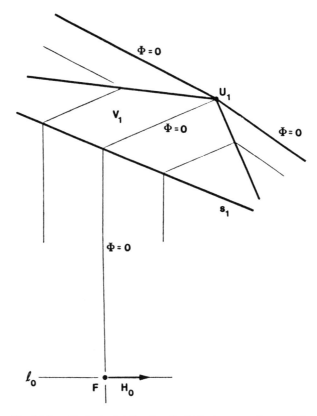

Fig. 3.5.3. Equipotential line $\Phi = 0$ within the magnetic material.

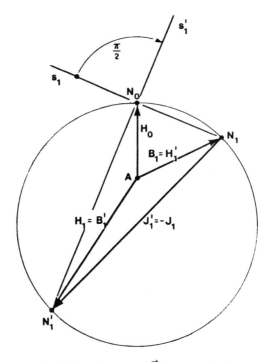

Fig. 3.5.4. Rotation of $\vec{H}_0$ by an angle $\pi/2$.

and the new vectors $\vec{B}_1^{\,\prime}$ and $\vec{H}_1^{\,\prime}$ are

$$\vec{B}_1^{\,\prime} = \vec{H}_1, \qquad \vec{H}_1^{\,\prime} = \vec{B}_1, \qquad (3.5.5)$$

as shown in the diagram of Fig. 3.5.4. Vector $\vec{B}_1^{\,\prime}$ is oriented in the directions of the equipotential lines of field $\vec{H}_1$ and vector $\vec{B}_1$ is oriented in the direction of the equipotential lines of field $\vec{H}_1^{\,\prime}$. Thus, the rotation of $\pi/2$ of vector $\vec{H}_0$ results in the exchange of the equipotential lines with the lines of flux of the magnetic induction throughout the magnetic structure.

The distribution of the lines of flux of $\vec{B}$ resulting from the rotation of $\vec{H}_0$ by an angle $\pi/2$ is shown in Fig. 3.5.5. The $\Phi = 0$ equipotential line of Fig. 3.5.3 transforms into the line of flux l_0 of $\vec{B}$ which passes through point F and terminates at point U_1 of the external boundary. As a consequence, line l_0 divides the flux of $\vec{B}$ in two paths flowing in opposite directions within the magnetic structure as indicated in Fig. 3.5.5.

Thus, point F can be defined as the point of intersection of the $\Phi = 0$ equipotential line and the line l_0 dividing the flux within the magnetic structure. The position of point F is one of the basic parameters which determine the geometry of the magnetic structure, and its selection will be discussed in detail in the analysis of the yokeless magnets presented in the following chapter.

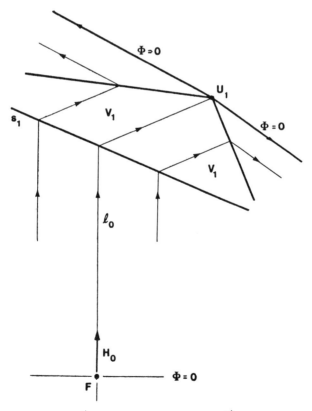

Fig. 3.5.5. Lines of flux of $\vec{B}$ corresponding to a rotation of $\vec{H}_0$ by an angle $\pi/2$ relative to the orientation in Fig. 3.5.3.

REFERENCES

[1] M. G. Abele, Design of Yokeless Rare Earth Magnets for NMR Medical Imaging, Proceedings of the Tenth International Workshop on Rare Earth Magnets and Their Applications, Kyoto, Japan, May 16-19, 1989, pp. 121-130.

[2] M. G. Abele, Some Considerations about Permanent Magnet Design for NMR, Technical Report No. 13, New York University, February 1, 1986.

[3] M. G. Abele, Linear Theory of Yokeless Permanent Magnets, *Journal of Magnetization and Magnetic Materials*, 83 (1990), pp. 276-278.

[4] M. G. Abele, Design of Two-Dimensional Magnets without Magnetic Yoke, Technical Report No. 15, New York University, March 1987.

[5] K. Halbach, Design of Permanent Multipole Magnets with Oriented Rare Earth-Cobalt Material, *Nuclear Instruments and Methods*, 169 (1980), pp. 1-10.

CHAPTER 4

Yokeless Magnets

INTRODUCTION

The initial steps in calculating the geometry and the magnetization of a magnetic structure are the same for both yoked and yokeless magnets. The basic characteristics of both types of structures are determined by the vector diagrams introduced in Sections 3.4 and 3.5. The methodology defined in Chapter 3 will now be applied to the full calculation of a yokeless magnet designed to generate a uniform field inside a closed cavity. The methodology will be applied first to two-dimensional structures and then extended to a three-dimensional yokeless magnet with the assumption of ideal linear demagnetization characteristics with zero magnetic susceptibility.

4.1 TWO-DIMENSIONAL, SINGLE LAYER YOKELESS MAGNETS

Assume a two-dimensional prismatic cavity, whose cross-section is an arbitrary polygon s_a as indicated in the schematic of Fig. 4.1.1. Assume that the magnetic

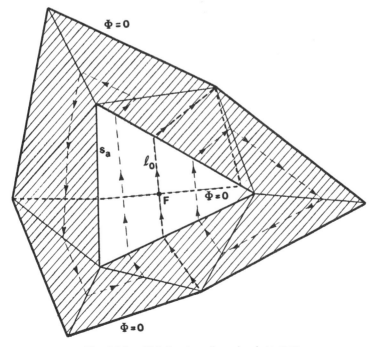

Fig. 4.1.1. Yokeless two-dimensional structure.

structure surrounding the cavity is designed to generate a uniform intensity $\vec{H}_0$ within the cavity, oriented in the direction of the arrows in Fig. 4.1.1.

The first parameter to be selected is the value of the magnitude J_0 of the remanence that is assumed to be uniform throughout the magnetic structure. Equivalently, the first design parameter is

$$K = \frac{\mu_0 H_0}{J_0} \qquad (4.1.1)$$

defined in Section 3.5.

The selection of $\vec{H}_0$ and K is not sufficient to determine the structure of the magnetic material. Chapter 3 has shown that the field configuration within the cavity is defined by the position of a characteristic point F, common to the equipotential line $\Phi = 0$ and the line of flux l_0 which divides the flux of the magnetic induction in the two independent closed paths indicated in Fig. 4.1.1.

Once the position of point F has been selected, the magnetic structure can be determined by taking full advantage of the geometric invariance theorem of Section 3.5, which states that the geometry of a two-dimensional yokeless structure is independent of the orientation of $\vec{H}_0$. The process of designing the structure is defined by the schematic of Fig. 4.1.2 and the vector diagram of Fig. 4.1.3.

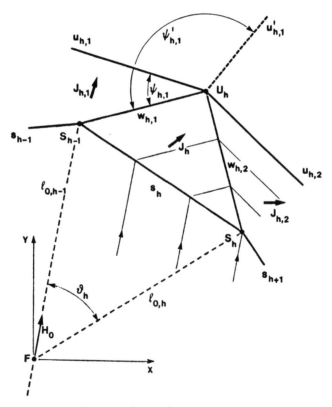

Fig. 4.1.2. Geometry of magnetic structure across interface s_h.

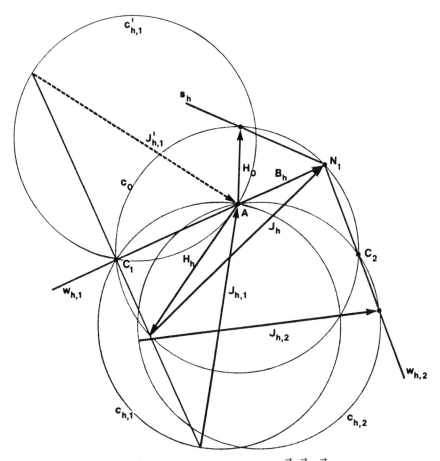

Fig. 4.1.3. Computation of remanences $\vec{J}_h, \vec{J}_{h,1}, \vec{J}_{h,2}$.

Let s_h in Fig. 4.1.2 be one side of the polygonal contour of the cavity. Assume that $\vec{H}_0$ is oriented along line $l_{0,h-1}$ that connects point F and the end point S_{h-1} of interface s_h between the cavity and a uniformly magnetized prism of remanence $\vec{J}_h$. The magnitude of $\vec{J}_h$ is J_0 and its orientation is given by the vector diagram inscribed in circle c_0 of Fig. 4.1.3. By virtue of Theorem 2 of Section 3.2, vectors $\vec{J}_h$ and $\vec{H}_0$ are oriented at equal and opposite angles with respect to a line perpendicular to s_h. Shown in Fig. 4.1.3 are the intensity $\vec{H}_h$ and the magnetic induction $\vec{B}_h$ generated by remanence $\vec{J}_h$.

Consider the line of flux $w_{h,1}$ of vector $\vec{B}_h$ which emerges from point S_{h-1}. By virtue of Eq. 3.5.2, the scalar potential on line $w_{h,1}$ vanishes at a point U_h of $w_{h,1}$ whose distances from S_{h-1} is given by

$$-\vec{H}_h \cdot \vec{w}_{h,1} = H_0 l_{0,h-1}, \qquad (4.1.2)$$

where $l_{0,h-1}$ is the distance of S_{h-1} from point F. Vector $\vec{w}_{h,1}$ is oriented in the same direction as $\vec{B}_h$ and its magnitude is equal to the distance of U_h from S_{h-1}.

The segment of line $w_{h,1}$ is the interface between the medium of remanence $\vec{J}_h$ and a medium whose remanence $\vec{J}_{h,1}$ is provided by the vector diagram inscribed in circle $c_{h,1}$ of Fig. 4.1.3. Following the methodology of Section 3.4, circle $c_{h,1}$ passes through point A and point C_1 where the line $w_{h,1}$ drawn from point N_1 intersects circle c_0. Point A is the second point of intersection of line $w_{h,1}$ and circle $c_{h,1}$. Thus, point A is the tip of the remanence $\vec{J}_{h,1}$ of the medium across interface $w_{h,1}$, and as a consequence, the intensity $\vec{H}_{h,1}$ and the magnetic induction $\vec{B}_{h,1}$ generated by $\vec{J}_{h,1}$ are

$$\vec{J}_{h,1} = -\mu_0 \vec{H}_{h,1}, \qquad \vec{B}_{h,1} = 0. \qquad (4.1.3)$$

Because of Eq. 4.1.3, line $u_{h,1}$ in Fig. 4.1.2 drawn from point U_h in a direction perpendicular to $\vec{J}_{h,1}$ is a $\Phi = 0$ equipotential line. Furthermore, because $\vec{B}_{h,1}$ is zero, line $u_{h,1}$ can be the interface between the medium of remanence $\vec{J}_{h,1}$ and the surrounding air.

As shown in Fig. 4.1.3, two solutions are possible because two circles $c_{h,1}$, $c'_{h,1}$ of diameter J_0 can be drawn through points A, C_1, yielding two possible interfaces $u_{h,1}$, $u'_{h,1}$ with the surrounding air. The uncertainty is eliminated by choosing the solution with the smallest angle $\psi_{h,1}$ between interface $u_{h,1}$ and line w_{h-1}. Because of Eq. 4.1.3, the distance of line $u_{h,1}$ from point S_{h-1} is

$$r_{h-1} = K \, l_{0,h-1}. \qquad (4.1.4)$$

Once the position of point U_h has been determined, the segment of line $w_{h,2}$ in Fig. 4.1.2 completes the boundary of the medium with remanence $\vec{J}_h$. Segment $w_{h,2}$ is the interface between the medium with remanence $\vec{J}_h$ and a medium whose remanence $\vec{J}_{h,2}$ is computed by means of the vector diagram inscribed in circle $c_{h,2}$ of Fig. 4.1.3. Vectors $\vec{B}_{h,2}$, $\vec{H}_{h,2}$ in the medium of remanence $\vec{J}_{h,2}$ are perpendicular to each other, thereby satisfying condition 3.4.2 for the field confinement within the magnetic structure.

Line $u_{h,2}$ in Fig. 4.1.2 drawn from point U_h parallel to vector $\vec{B}_{h,2}$ is a $\Phi = 0$ equipotential line and becomes a side of the external boundary of the magnetic structure. By virtue of the invariance theorem, position and orientation of line $u_{h,2}$ could have been computed by assuming vector $\vec{H}_0$ oriented along the segment of line $l_{0,h}$ between F and S_h. As a consequence, in a way similar to the computation of line $u_{h,1}$ which led to Eq. 4.1.4, the distance of line $u_{h,2}$ from point S_h is

$$r_h = K \, l_{0,h}. \qquad (4.1.5)$$

Assume now that $\vec{H}_0$ is oriented in the direction perpendicular to side s_h and let $l_{1,h}$ be the distance of point F from s_h. In this case, in the medium across interface s_h, vector $\vec{H}_h$ is also perpendicular to s_h and its magnitude is given by

$$\mu_0 H_h = (1 - K) J_0. \qquad (4.1.6)$$

The external boundary of the magnetic structure is a $\Phi = 0$ equipotential line regardless of the orientation of $\vec{H}_0$. By virtue of Eq. 3.5.3, point U_h belongs to the line perpendicular to s_h drawn from point F, and the distance $\xi_{1,h}$ of U_h from s_h is

$$\xi_{1,h} = \frac{K}{1-K} \xi_{0,h}, \qquad (4.1.7)$$

where $\xi_{0,h}$ is the distance of F from s_h.

The geometry of the magnetic structure is fully determined by applying Eqs. 4.1.4, 4.1.5, and 4.1.7 to all sides s_h of the cavity polygonal contour. If the cavity boundary is a polygon of n sides, the external boundary of the magnetic structure is a polygon of $2n$ sides with vertices U_h defined by Eq. 4.1.7 and T_h determined by the intersection of lines $u_{h,2}$ and $u_{h+1,1}$ drawn from U_h and U_{h+1}, respectively, tangent to circle of radius 4.1.5 and center at S_h, as shown in Fig. 4.1.4. The segment of line t_h between points S_h and T_h is the interface between the component of the magnetic structure with remanence $\vec{J}_{h,2}$ and the component with remanence $\vec{J}_{h+1,1}$. Thus the magnetic structure of a yokeless two-dimensional magnet with a cavity consisting of a polygonal contour of n sides is composed of $3n$ prisms of triangular cross-section: n prisms interface with the cavity and $2n$ prisms interface with the external nonmagnetic medium or air.

The geometry of Fig. 4.1.4 has been determined by selecting the orientation of $\vec{H}_0$ independently for each side of the cavity. Obviously the actual distribution of the

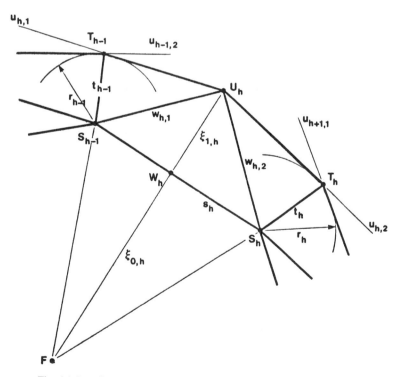

Fig. 4.1.4. Computation of the external boundary of a yokeless magnet.

112 YOKELESS MAGNETS

remanence in each prism must be consistent with an assigned orientation of $\vec{H}_0$. Assume in Fig. 4.1.3 that the actual orientation of $\vec{H}_0$ is at an angle α_{h-1} with respect to line $l_{0,h-1}$. Thus, again by virtue of the invariance theorem, the true orientation of the remanence in the three prisms with sides $(s_h, w_{h,1}, w_{h,2})$, $(t_{h-1}, u_{h,1}, w_{h,1})$, and $(t_h, u_{h,2}, w_{h,2})$ is obtained by rotating $\vec{J}_h, \vec{J}_{h,1}, \vec{J}_{h,2}$ by an angle $-\alpha_{h-1}$.

Equation 4.1.7 shows that a solution of the magnet design problem exists as long as one selects a value of parameter K smaller than unity. Thus $\mu_0 H_0 = J_0$ is the upper limit of the intensity that can be achieved in the cavity of a yokeless magnet consisting of a single layer of the magnetic structure of Fig. 4.1.2.

As an example, consider the regular hexagonal contour of the cavity cross-section shown in Fig. 4.1.5. Assume a value of parameter K

$$K = 0.25 \tag{4.1.8}$$

and point F coincident with the center of the hexagon.

Let us apply the vector diagram of Fig. 4.1.3 to determine the magnetic structure that interfaces with the hexagonal cavity through side s_1, as indicated in Fig. 4.1.5. By

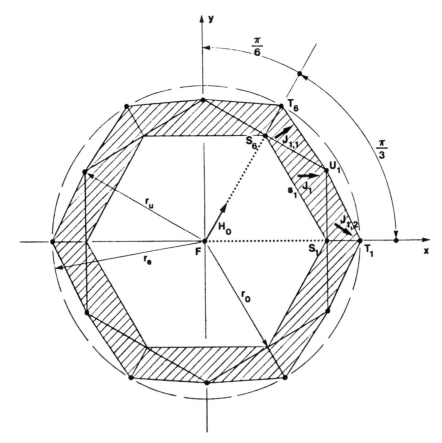

Fig. 4.1.5. Yokeless magnet with regular hexagonal cavity.

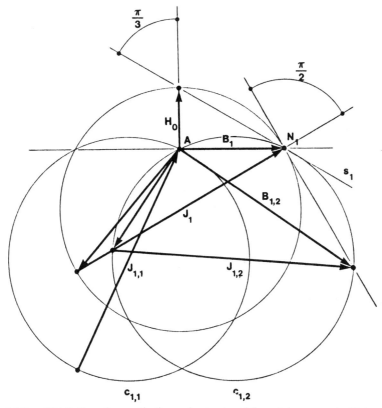

Fig. 4.1.6. Distribution of magnetization and geometry of the yokeless magnet with hexagonal cavity.

selecting the orientation of $\vec{H}_0$ from point F to vertex S_6, the geometry and the distribution of magnetization of the magnet components within the angle $\pi/3$ indicated in Fig. 4.1.5 are provided by the vector diagram of Fig. 4.1.6. Let r_0 be the radius of the circle which inscribes the hexagonal contour of the cavity. Point U_1 is located on the radial line perpendicular to s_1, and by virtue of Eqs. 4.1.7 and 4.1.8, the distance of (U_1, F) is

$$r_u = \frac{r_0}{1-K} \cos\frac{\pi}{6} = \frac{r_0}{\cos\frac{\pi}{6}}. \qquad (4.1.9)$$

Because of symmetry, points T_6, T_1 are located on the radial lines that pass through S_6, S_1, respectively. By virtue of Eqs. 4.1.7 and 4.1.8, points T_6, T_1 are located on a circle of center F and radius

$$r_e = \left[1 + \frac{1}{\sqrt{13}}\right] r_0. \qquad (4.1.10)$$

114 YOKELESS MAGNETS

The orientation of the three remanences $\vec{J}_1$, $\vec{J}_{1,1}$, and $\vec{J}_{1,2}$ shown in Fig. 4.1.5 is provided by the vector diagram of Fig. 4.1.6. Vector $\vec{J}_1$ is oriented at an angle $\pi/3$ with respect to $\vec{H}_0$. Vector $\vec{J}_{1,1}$ is perpendicular to side $(T_6 U_1)$ of the external boundary, and by virtue of Eq. 4.1.3, intensity $\vec{H}_{1,1}$ is given by

$$\mu_0 \vec{H}_{1,1} = -\vec{J}_{1,1} . \quad (4.1.11)$$

Circle $c_{1,2}$ provides the orientation of vector $\vec{J}_{1,2}$ and the values of intensity $\vec{H}_{1,2}$ and magnetic induction $\vec{B}_{1,2}$ in region $(S_1 T_1 U_1)$ of the magnetic structure.

By virtue of symmetry and because of the geometric invariance theorem of yokeless magnets, the geometries of the magnetic structures in the six angular intervals of the hexagonal cavity are all identical to each other. Assume that the desired orientation of the intensity within the cavity is parallel to the y axis. Thus, as vector $\vec{H}_0$ is rotated by an angle $+\pi/6$, vectors $\vec{J}_1, \vec{J}_{1,1}$, and $\vec{J}_{1,2}$ must be rotated by an angle $-\pi/6$. To obtain the correct angular relationship between all the remanences of the components of the magnetic structure and the intensity $\vec{H}_0$ oriented along the y axis, the remanences computed independently in the angular intervals s_h ($h = 2, 3, 4, 5, 6$) must be rotated by the angles

$$\alpha_2 = -\frac{\pi}{2} \quad (s_2)$$

$$\alpha_3 = -\frac{5}{6}\pi \quad (s_3)$$

$$\alpha_4 = -\pi - \frac{\pi}{6} \quad (s_4) \quad (4.1.12)$$

$$\alpha_5 = -\frac{3}{2}\pi \quad (s_5)$$

$$\alpha_6 = +\frac{\pi}{6} \quad (s_6) .$$

The structure of Fig. 4.1.5 has a dodecagonal external boundary. As previously stated, if n is the number of sides of the polygonal cavity, the external boundary of the magnetic structure is a polygon of $2n$ sides. In the particular case of a regular n-sided polygonal boundary of the cavity, a value of K is found that results in a regular polygon of $2n$ sides for the external boundary. In this case points T_{h-1}, U_h, T_h in Fig. 4.1.4 are found at the same distance r_e from the center of the polygonal cavity. Figure 4.1.7 shows such a magnetic structure within a $2\pi/n$ arc. In Fig. 4.1.7, r_0 is the radius of the circle that inscribes the polygonal cavity and point F is at the center of the polygon.

By virtue of Eq. 4.1.4, radii r_0 and r_e are related to each other by the equation

$$r_e = \left[1 + K \cos^{-1} \frac{\pi}{2n}\right] r_0 , \quad (4.1.13)$$

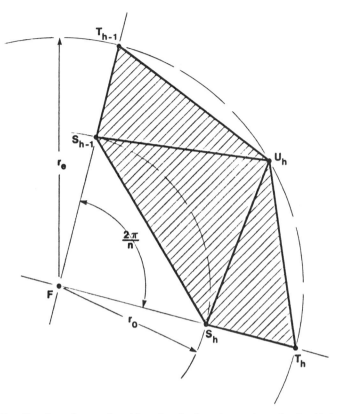

Fig. 4.1.7. Regular polygon of n sides of cavity boundary and regular $2n$-sided external boundary.

and by virtue of Eq. 4.1.7, r_0 and r_e must also satisfy the condition

$$r_e = \frac{r_0}{1-K}\cos\frac{\pi}{n}. \tag{4.1.14}$$

In order to satisfy both Eq. 4.1.13 and Eq. 4.1.14, parameter K must be a solution of the equation

$$K^2 - \left[1 - \cos\frac{\pi}{2n}\right]K - \cos\frac{\pi}{2n}\left[1 - \cos\frac{\pi}{n}\right] = 0, \tag{4.1.15}$$

i.e.,

$$K = \frac{1}{2}\left[1 - \cos\frac{\pi}{2n}\right] + \left[\frac{1}{4}(1 + \cos\frac{\pi}{2n})^2 - \cos\frac{\pi}{2n}\cos\frac{\pi}{n}\right]^{1/2}. \tag{4.1.16}$$

A listing of the values of K and r_e/r_0 is presented in Table 4.1.1 for several lowest values of n. As n increases, K decreases, and as Eq. 4.1.16 shows, $K = 0$ in the limit $n \to \infty$.

Table 4.1.1. Values of K and r_e/r_0 of yokeless structures with regular internal and external boundaries.

n	3	4	5	6	7	8	9	10
K	0.728	0.560	0.451	0.377	0.323	0.283	0.251	0.226
$\dfrac{r_e}{r_0}$	1.84	1.61	1.47	1.39	1.33	1.29	1.25	1.23

4.2 MAGNETIC STRUCTURES WITH SIMILAR POLYGONAL CONTOURS OF EXTERNAL AND INTERNAL BOUNDARIES

Consider a cavity whose boundary is a regular polygon of n sides and assume that the characteristic point F coincides with the center of the polygon. Let us determine the values of K that result in similar geometries of both external and internal boundaries of the magnetic structure.

One side of the n-sided polygonal cavity is shown in Fig. 4.2.1 where point F is the center of the polygon. Within the angular interval $2\pi/n$, external and internal boundaries of the structure shown in Fig. 4.2.1 are parallel to each other when the values of r_h, r_{h-1}, and $\xi_{1,h}$ in Eqs. 4.1.4, 4.1.5, and 4.1.7 are equal to each other. In the particular case of a regular polygon one has

$$\xi_{1,h} = (\cos^{-1}\frac{\pi}{n} - 1)\,\xi_{0,h}\;, \tag{4.2.1}$$

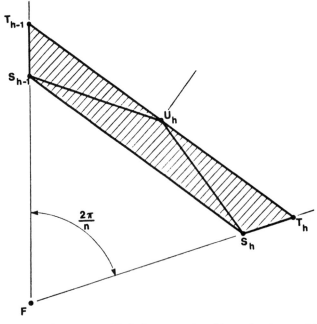

Fig. 4.2.1. Magnetic structure with similar geometries of internal and external boundaries.

where $\xi_{0,h}$ is the distance of the side of the cavity from the center of the polygon. Thus by virtue of Eq. 4.1.7 the condition of similar internal and external polygons is satisfied for the value of parameter K,

$$K = 1 - \cos\frac{\pi}{n} \, . \tag{4.2.2}$$

If the internal polygon is inscribed in a circle of radius r_0, the external polygon is inscribed in a circle of radius

$$r_e = \frac{r_0}{\cos\dfrac{\pi}{n}} \, . \tag{4.2.3}$$

As shown by Eq. 4.2.2, the value of K decreases as n increases. Figure 4.2.2 shows the values of $\vec{H}_0$ within the cavity and the values of remanence $\vec{J}_{h,2}$, intensity $\vec{H}_{h,2}$, and magnetic induction $\vec{B}_{h,2}$ within the region $(S_h \, T_h \, U_h)$ of Fig. 4.2.1 when $\vec{H}_0$ is oriented in the direction perpendicular to the side of the polygon.

The lowest value $n = 3$ results in the structure of equilateral triangles shown in Fig. 4.2.3 with a value

$$K = 0.5 \, , \tag{4.2.4}$$

which results in a 2:1 ratio between dimensions of external and internal boundaries. The computation of vectors $\vec{J}, \vec{H}, \vec{B}$ in the regions of the magnetic structure of Fig.

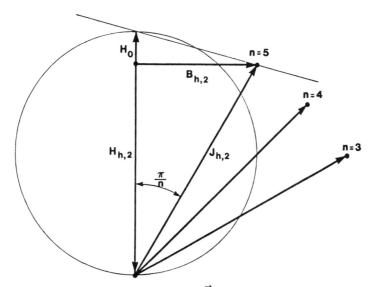

Fig. 4.2.2. Calculation of $\vec{H}_0$ as a function of n.

118 YOKELESS MAGNETS

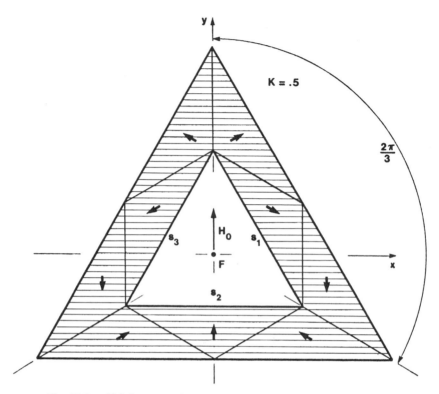

Fig. 4.2.3. Yokeless magnetic structure with similar triangular boundaries.

4.2.3 within the angular interval $2\pi/3$ is provided by the vector diagram of Fig. 4.2.4 which corresponds to $\vec{H}_0$ oriented along the axis y. In Fig. 4.2.3, the perpendicular to side s_1 is oriented at an angle $\pi/3$ with respect to the axis y. Thus, by virtue of Theorem 2 of Section 3.2, remanence J_1 and intensity $\vec{H}_0$ are oriented at an angle $2\pi/3$ with respect to each other. Because of the particular value 4.2.4 of parameter K, the angle between remanences $\vec{J}_1$ and $\vec{J}_{1,2}$ and the angle between remanences $\vec{J}_{2,1}$ and $\vec{J}_1$ are both equal to $\pi/3$, in agreement with the property of structures designed for $K = 0.5$, as discussed in Section 3.4. The configuration of equipotential lines within the $2\pi/3$ angular interval of the magnetic structure is shown in Fig. 4.2.5(a).

Assume that $\vec{H}_0$ in Fig. 4.2.3 is rotated by $-\pi/2$ to become aligned with the positive direction of the axis x of Fig. 4.2.3. By virtue of Theorem 4 of Section 3.5 the remanences in Fig. 4.2.5(a) must be rotated by the angle $+\pi/2$. The rotation results in the configuration of equipotential lines shown in Fig. 4.2.5(b). Triangle $(S_3 U_1 T_3)$, which is a region of zero magnetic induction in Fig. 4.2.5(a), becomes an equipotential region as shown by Fig. 4.2.5(b).

The $n = 4$ value results in the square cross-sectional magnet shown in Fig. 4.2.6 with a value

$$K = 1 - \frac{1}{\sqrt{2}} \qquad (4.2.5)$$

and a $\sqrt{2}:1$ ratio between dimensions of external and internal squares. The computation

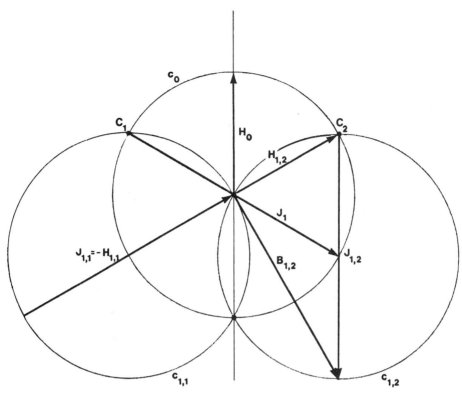

Fig. 4.2.4. Computation of the field in the triangular cross-sectional magnet.

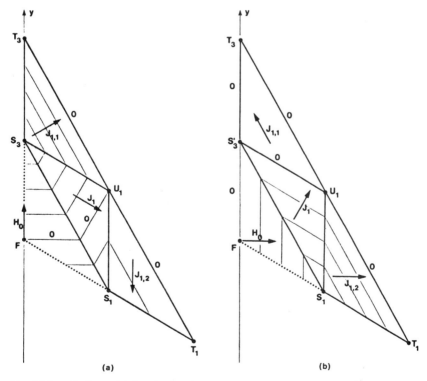

Fig. 4.2.5. Equipotential lines in the triangular cross-sectional magnet with $\vec{H}_0$ oriented (a) along the y axis, and (b) along the x axis.

119

120 YOKELESS MAGNETS

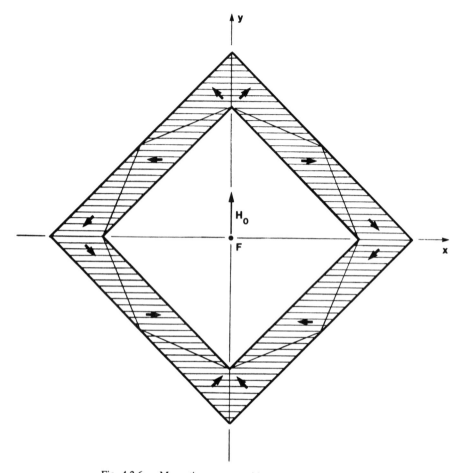

Fig. 4.2.6. Magnetic structure with a square cross-section.

of vectors $\vec{J}, \vec{H}, \vec{B}$ in the first quadrant of Fig. 4.2.6 is provided by the vector diagram of Fig. 4.2.7 which corresponds to $\vec{H}_0$ oriented along a diagonal of the square. The configuration of equipotential lines is shown in Fig. 4.2.8. In the two regions (S_4, U_1, T_4) and (S_1, T_1, U_1) the remanence is perpendicular and parallel to the side of the square, respectively, i.e., the magnetic induction is zero within triangle (S_4, U_1, T_4) and the intensity is zero within triangle (S_1, T_1, U_1).

The computation of the geometry of yokeless magnetic structures by means of the vector diagram of Fig. 4.1.3 is based on the assumption of zero magnetic susceptibility of the magnetic material. This ideal limit makes the magnetic structure perfectly transparent to the field generated by other sources of magnetic field. It is then possible to increase the field inside the prismatic cavity by adding the field generated by a number of coaxial yokeless magnets.

A multiplicity of m coaxial yokeless magnets yields within the cavity a value

$$\vec{H}_0 = \sum_{i=1}^{m} \vec{H}_{0,i} , \qquad (4.2.6)$$

MAGNETIC STRUCTURES WITH SIMILAR POLYGONAL CONTOURS **121**

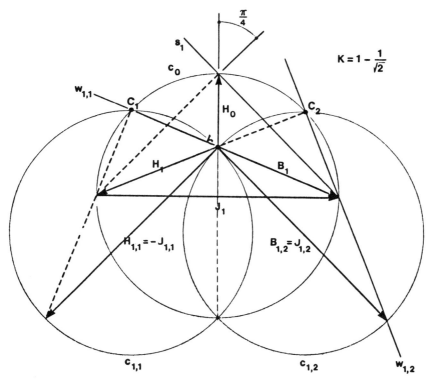

Fig. 4.2.7. Computation of the field in the square cross-sectional magnet.

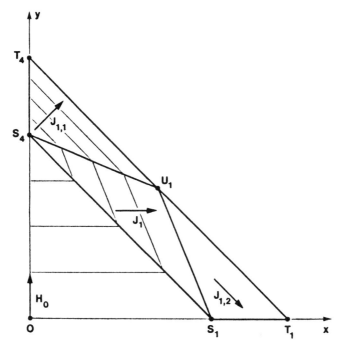

Fig. 4.2.8. Equipotential lines in the square cross-sectional magnet.

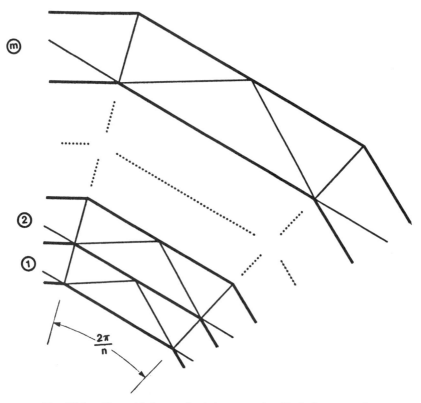

Fig. 4.2.9.   Concentric layers of yokeless magnets with similar geometries.

where $\vec{H}_{0,i}$ is the intensity generated by each magnet. In particular, if the interfaces between the m coaxial magnets are similar n-sided regular polygons as schematically shown in Fig. 4.2.9, and if all individual vectors $\vec{H}_{0,i}$ have the same orientation, the total intensity $\vec{H}_0$ within the cavity of the multilayer structure is given by

$$K = \frac{\mu_0 H_0}{J_0} = m\left(1 - \cos\frac{\pi}{n}\right) \tag{4.2.7}$$

if the remanence has a uniform value J_0 throughout the entire structure.

By virtue of Eq. 4.2.3, if no air gap is left between individual layers, the external boundary of the mth layer is a polygon inscribed in a circle of radius

$$r_e = \frac{r_0}{\cos^m \frac{\pi}{n}} . \tag{4.2.8}$$

In the limit $n \to \infty$, the polygonal interfaces between the individual magnets reduce to concentric circles and Eq. 4.2.7 yields

$$\lim_{n \to \infty} \frac{K}{m} = 0 . \tag{4.2.9}$$

If the ratio r_e/r_0 is constant, in the limit $n \to \infty$ the number of layers m also becomes infinitely large, and Eqs. 4.2.7 and 4.2.8 yield

$$\ln \frac{r_e}{r_0} = -m \ln(1 - \frac{K}{m}), \qquad (4.2.10)$$

and, in the limit $m \to \infty$, $n \to \infty$, Eq. 4.2.10 yields

$$K = \ln \frac{r_e}{r_0}, \qquad (4.2.11)$$

i.e., K coincides with the value 2.4.5 obtained in Section 2.4 for two coaxial cylinders with the distribution of remanence given by Eq. 2.4.1.

Equations 4.2.7 and 4.2.11 show that a multilayer structure of yokeless magnets removes the upper limit $K = 1$ which characterizes a single layer two-dimensional magnet. In a practical application, as long as the linear approximation of the demagnetization characteristic is valid, a multilayer structure makes it possible to achieve a field larger than the remanence of the magnetic material. This is the case, for instance, of a magnet designed with the Nd-Fe-B rare earth alloy illustrated in Section 1.7, whose intrinsic coercive force $H_{c,1}$ is larger than the coercive force H_c.

An example is the two-layer structure of two square cross-sectional magnets, shown in Fig. 4.2.10. The side of the square cross-sectional external boundary is twice the side of the cavity and the value of K is twice the value given by Eq. 4.2.5 [1,2].

Equations 4.2.2 and 4.2.3 define the geometries of magnetic structures where the sides of the similar regular polygons of internal and external boundaries are parallel to each other. A second solution where both internal and external boundaries are still similar regular polygons is shown in Fig. 4.2.11. The difference between Figs. 4.2.11 and 4.2.1 is that the two external polygons of the structure of Fig. 4.2.11 are rotated by an angle π/n with respect to each other. The distance r_u of vertex U_h from F is

$$r_u = \frac{r_0}{1 - K} \cos \frac{\pi}{n}, \qquad (4.2.12)$$

and if side $(U_h T_{h-1})$ is perpendicular to line $(F S_{h-1})$ the distance r_t of point T_{h-1} from S_{h-1} is

$$r_t = K r_0. \qquad (4.2.13)$$

On the other hand, the distance of T_{h-1} from F is

$$r_0 + r_t = r_u \cos \frac{\pi}{n}. \qquad (4.2.14)$$

Thus, by virtue of Eqs. 4.2.12, 4.2.13 and 4.2.14, the external boundary of the magnetic structure is also a regular polygon of n sides if

$$K = \sin \frac{\pi}{n}, \qquad (4.2.15)$$

124 YOKELESS MAGNETS

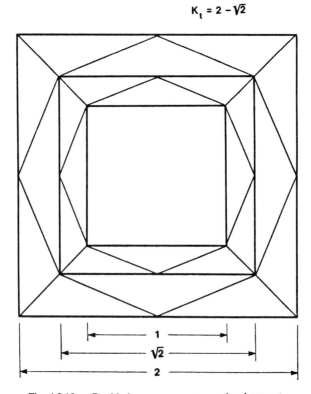

Fig. 4.2.10. Double layer square cross-sectional magnet.

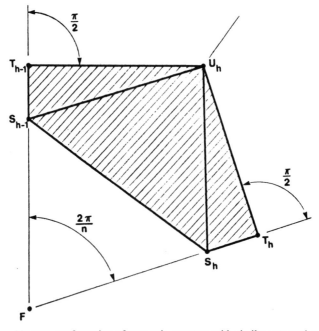

Fig. 4.2.11. Alternate configuration of magnetic structure with similar geometries of internal and external boundaries.

and the magnetic structure is inscribed in a circle of radius

$$r_e = r_u = \frac{\cos\frac{\pi}{n}}{1 - \sin\frac{\pi}{n}} r_0. \qquad (4.2.16)$$

Again the geometries governed by Eqs. 4.2.15 and 4.2.16 lend themselves to the generation of multilayered structures of magnets with similar regular polygonal boundaries with no air gap in between. The total value of K generated by m layers is

$$K = m \sin\frac{\pi}{n}, \qquad (4.2.17)$$

and the mth layer is inscribed in a circle of radius

$$r_e = \frac{\cos^m \frac{\pi}{n}}{\left[1 - \sin\frac{\pi}{n}\right]^m} r_0. \qquad (4.2.18)$$

In the limit $n \to \infty$, the interfaces between individual magnets reduce again to concentric circles and the value of K given by Eq. 4.2.17 satisfies the same limit 4.2.9. For n and m infinitely large, Eqs. 4.2.18 and 4.2.17 yield

$$\ln\frac{r_e}{r_0} = m \ln\left[1 + \sin\frac{K}{m}\right], \qquad (4.2.19)$$

and again the ratio r_e/r_0 and the value of K of the multilayered structure satisfy the limit given by Eq. 4.2.11.

4.3 FIGURE OF MERIT OF YOKELESS MAGNETS

The magnetic structures developed in the preceding sections are the result of an arbitrary selection of parameter K and point F. Once the geometry of the prismatic cavity is assigned, different values of K and different positions of F result in different geometries of the magnetic structure and, in particular, in different ratios of the cross-sectional area of the magnetic material and the cross-sectional area of the cavity.

Thus, K and F control the amount of magnetic material required to generate the assigned field, and the designer of a yokeless magnet must determine the optimum combination of these two parameters that makes the most efficient use of the magnetic material.

In the equivalent magnetic circuit model used by the designer of a traditional magnet, the optimum design condition is achieved by selecting the operating point of the demagnetization curve that maximizes the energy product of the magnetic material. This is still the case of the ideal one-dimensional magnet discussed in Section 2.1 where, by virtue of Eq. 2.1.14, the magnetic material operating at the maximum of the energy product curve maximizes the intensity of the field. The same criterion, however, does not apply to the type of two-dimensional structures developed in the preceding sections, where the intensity of the field varies in magnitude and orientation throughout the magnetic material. This is still the case of the ideal one-dimensional magnet discussed in Section 2.1 where, by virtue of Eq. 2.1.14, the magnetic material operating at the maximum of the energy product curve maximizes the intensity of the field. The same criterion, however, does not apply to the type of two-dimensional structures developed in the preceding sections, where the intensity of the field varies in magnitude and orientation throughout the magnetic material.

The quality of the design of a permanent magnet can be expressed in a quantitative way by its figure of merit M, defined as the fraction of the energy stored in the magnetic material, which is used to generate the field inside the cavity. For a two-dimensional magnet, with a uniform field within the cavity and a uniform value of the remanence throughout the material, by virtue of the definition of K and Eqs. 1.6.7 and 1.6.10, the figure of merit is

$$M = K^2 \frac{A_c}{A_m}, \qquad (4.3.1)$$

where A_c and A_m are the cross-sectional areas of the cavity and the magnetic material, respectively, and K is given by Eq. 4.1.1.

Keeping in mind that the contribution of each element of magnetic material decreases rapidly with its distance from the cavity, a regular polygonal geometry should result in an efficient magnet design. Consider again the n-sided regular polygonal contours of the cavity analyzed in the preceding section and, in particular, the single layer structures with similar internal and external polygons that satisfy Eq. 4.2.2.

In the multilayer structures of similar geometries defined by Eqs. 4.2.7 and 4.2.8, the ratio of the area of magnetic material to cavity area is

$$\frac{A_m}{A_c} = \cos^{-2m}\frac{\pi}{n} - 1, \qquad (4.3.2)$$

and Eq. 4.3.1 yields

$$M = \frac{K^2}{\cos^{-2m}\frac{\pi}{n} - 1}. \qquad (4.3.3)$$

In the limit of $m \to \infty$, $n \to \infty$, by virtue of Eq. 4.2.11, Eq. 4.3.3 reduces to

$$M = \frac{K^2}{e^{2K} - 1}. \qquad (4.3.4)$$

Thus, M vanishes at $K = 0$ and $K = \infty$. M attains a maximum at the value of K which satisfies the transcendental equation

$$(1 - K)e^{2K} - 1 = 0 \tag{4.3.5}$$

whose solution is

$$K_0 \approx 0.797, \quad M_0 \approx 0.162. \tag{4.3.6}$$

M_0 is the upper limit of the figure of merit of either a single or multilayer two-dimensional yokeless magnet of uniform remanence. For values of $K > 1$, Eq. 4.3.4 shows that M decreases rapidly as K increases. Thus, in multilayered structures large values of K are achieved at the cost of less efficient use of the magnetic material.

In the multilayer structures defined by Eqs. 4.2.17 and 4.2.18, the ratio of area A_m to the area A_c is

$$\frac{A_m}{A_c} = \left[1 - \sin\frac{\pi}{n}\right]^{-2m} \cos^{2m}\frac{\pi}{n} - 1, \tag{4.3.7}$$

and the figure of merit is

$$M = \frac{m^2 \sin^2\frac{\pi}{n}}{\left[1 - \sin\frac{\pi}{n}\right]^{-2m} \cos^{2m}\frac{\pi}{n} - 1}. \tag{4.3.8}$$

By virtue of Eq. 4.2.19, in the limit $m \to \infty$, $n \to \infty$, Eq. 4.3.8 reduces again to Eq. 4.3.4.

Each n-sided regular polygonal cavity is characterized by an optimum value of $M < M_0$ which is a function of the number m of n-sided concentric layers of similar geometries. Table 4.3.1 lists the values of K and M given by Eqs. 4.2.7 and 4.3.3 for the lowest integers m, n. One observes that the optimum value of M of a triangular geometry is achieved with a single layer. Increasing values of n require an increasing number of layers to achieve the optimum value of M. A plotting of M versus the number of layers for large values of n is shown in Fig. 4.3.1. Regardless of the values of m, n, the optimum value of M is always found at $K < 1$.

In the multilayer structures defined by Eqs. 4.2.17 and 4.2.18, Table 4.3.2 lists the values of K and M given by Eqs. 4.2.17 and 4.3.8. The obvious difference between Table 4.3.1 and Table 4.3.2 is that in the structures defined by Eqs. 4.2.12 and 4.2.15 the same value of K is attained with a smaller number of layers, and the difference becomes more and more pronounced as n increases. On the other hand the figures of merit listed in Table 4.3.2 are smaller than the figures listed in Table 4.3.1, although the difference becomes negligible as n increases.

128 YOKELESS MAGNETS

Table 4.3.1. Values of M and K of multilayer structures with regular and similar polygonal contours. m denotes the number of layers and n the number of sides.

n	3		4		5		6		7	
m	K	M	K	M	K	M	K	M	K	M
1	0.50	0.083	0.29	0.086	0.19	0.069	0.13	0.054	0.10	0.042
2	1.00	0.067	0.59	0.114	0.38	0.109	0.27	0.092	0.20	0.076
3	1.50	0.036	0.88	0.110	0.57	0.128	0.40	0.118	0.30	0.102
4	2.00	0.016	1.17	0.092	0.76	0.131	0.54	0.133	0.40	0.120
5			1.46	0.069	0.95	0.124	0.67	0.140	0.49	0.133
6			1.76	0.049	1.15	0.112	0.80	0.140	0.59	0.141
7			2.05	0.033	1.34	0.097	0.94	0.135	0.69	0.145
8					1.53	0.081	1.07	0.128	0.79	0.146
9					1.72	0.067	1.21	0.118	0.89	0.144
10					1.91	0.053	1.34	0.107	0.99	0.139
11					2.10	0.042	1.47	0.096	1.09	0.133
12							1.61	0.085	1.19	0.126
13							1.74	0.074	1.29	0.118
14							1.88	0.064	1.39	0.110
15							2.01	0.055	1.48	0.101

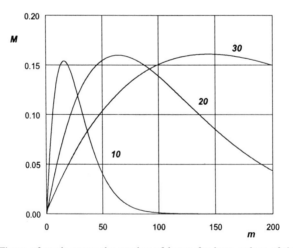

Fig. 4.3.1. Figure of merit versus the number of layers for large values of the number of sides of the polygonal boundaries.

Table 4.3.2. Values of M and K of alternate geometries for multilayer magnets with similar regular polygonal cavities.

n	3		4		5		6		7	
m	K	M	K	M	K	M	K	M	K	M
1	0.87	0.058	0.08	0.104	0.59	0.121	0.50	0.125	0.43	0.123
2	1.73	0.015	1.41	0.061	1.18	0.100	1.00	0.125	0.87	0.139
3					1.76	0.055	1.50	0.087	1.30	0.111

In general, a magnet design starts from a given geometry of the cavity which is not necessarily a regular polygon. As an example, consider the quadrilateral contour of the magnet cavity shown in Fig. 4.3.2. The geometry of the single layer magnet of Fig. 4.3.2 corresponds to the value

$$K = 0.3 , \qquad (4.3.9)$$

with point F selected at the origin of the system of coordinates x, y which pass through the vertices of the quadrangle. The values of M and area ratio A_m/A_c of the structure of Fig. 4.3.2 are

$$M \approx 0.077 , \qquad A_m/A_c \approx 1.16 . \qquad (4.3.10)$$

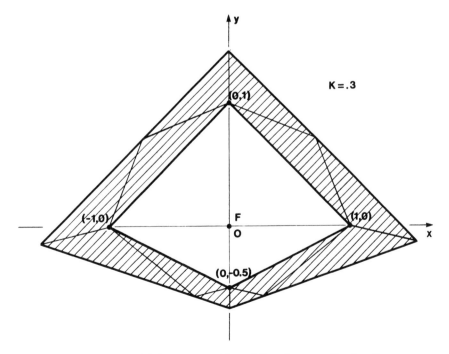

Fig. 4.3.2. Yokeless magnet with quadrilateral cross-sectional cavity.

The determination of the optimum single layer magnetic structure around the cavity of Fig. 4.3.2 results from the calculation of the figures of merit of magnet geometries computed for values of K in the range $0 < K < 1$ and values of the coordinates x_F, y_F of point F within the area of the cavity. The optimum value of M is found to be

$$M \approx 0.110 \qquad (4.3.11)$$

for the values

$$K \approx 0.590, \quad x_F = 0, \quad y_F = 0.149. \qquad (4.3.12)$$

The optimum magnet geometry is shown in Fig. 4.3.3. The effect of a change of the position of point F is shown in Fig. 4.3.4, where the lines of constant values of the figure of merit are plotted in the plane of the cavity for the value $K = 0.590$. One observes that M suffers a substantial decrease as F is moved away from its optimum position to points close to the boundary of the cavity.

For symmetric cavities the optimum position of point F is the center of symmetry and the optimization of the magnet structure reduces to the calculation of the optimum value of parameter K [3]. For regular polygonal contours of the cavity and F at the center of the cavity, the values of M are plotted versus K in Fig. 4.3.5 for $m = 1$ and

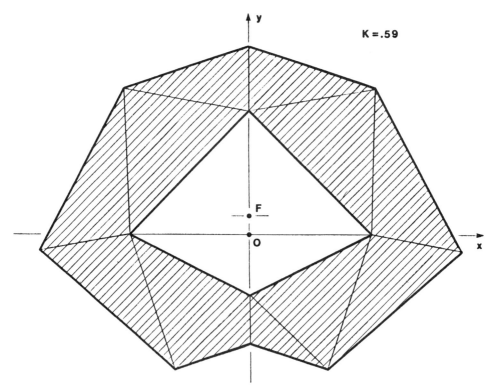

Fig. 4.3.3. Optimum magnetic structure with the quadrilateral cavity of Fig. 4.3.2.

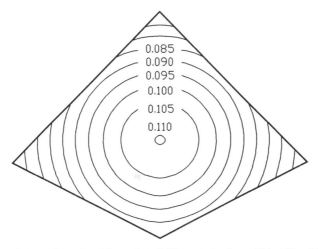

Fig. 4.3.4. Lines of positions of F yielding equal values of M at $K = 0.590$.

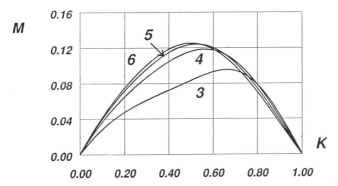

Fig. 4.3.5. Figure of merit versus K of single layer magnets with regular polygonal cavities.

the lowest values of n. The triangular cavity exhibits the lowest maximum of the figure of merit. However, for K of the order or larger than 0.8, Fig. 4.3.5 shows that the triangular cavity provides a better figure of merit than the other polygonal cavities with larger values of n. The optimum values of M and K are shown in Table 4.3.3 for the lowest values of n.

In general a change of position of point F affects the geometry of the magnetic structure and the distribution of remanences as well. An example is the magnetic structure designed around the equilateral triangular cavity of Fig. 4.2.3 for the particular case $K = 0.5$. If point F is located at the center of the triangle, the geometry of the structure is shown in Fig. 4.2.3, and the area ratio is

$$\frac{A_m}{A_c} = 3, \qquad (4.3.13)$$

132 YOKELESS MAGNETS

Table 4.3.3. Optimum values of M and K of single layer magnets with regular polygonal cavities.

n	3	4	5	6	7	8
K	0.66	0.56	0.52	0.50	0.48	0.47
M	0.095	0.118	0.124	0.125	0.124	0.123

which yields the figure of merit

$$M_{y_F=0} \approx 0.083, \qquad (4.3.14)$$

as shown by Table 4.3.1 for $n = 3$ and $m = 1$.

The effect of a change of position of F is illustrated in Fig. 4.3.6, where the figure of merit is plotted versus the the coordinate y_F. As expected, the maximum value of M is found when F coincides with the center of the triangle.

The change of the magnet geometry caused by the change of position of point F on the axis y is shown in Fig. 4.3.7. The geometry shown in Fig. 4.3.7(a) corresponds to the coordinates of F

$$x_F = 0, \qquad y_F = 0.25. \qquad (4.3.15)$$

In the limit of F coinciding with vertex S_3 of the triangle, the magnetic structure transforms into the geometry of Fig. 4.3.7(b). By comparing Fig. 4.3.7(b) with Fig. 4.2.5, one observes that the shift of point F to point S_3 results in the elimination from Fig. 4.2.5 of the two regions $(S_3U_1T_3)$ and $(S_3S_1U_1)$ of magnetized material. The area of the cross-section of magnetized material in Fig. 4.3.7(b) is five times the area of the triangular cavity. Thus, the figure of merit of the magnetic structure of Fig.

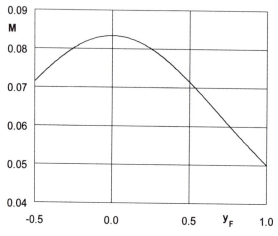

Fig. 4.3.6. Plotting of the figure of merit versus the position of point F, at $K = 0.5$.

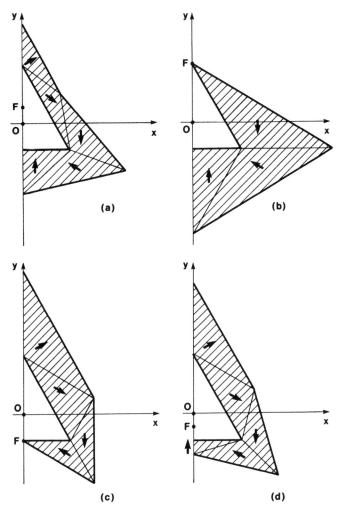

Fig. 4.3.7. Effect of position of point F on the magnet geometry: (a) $x_F = 0$, $y_F = 0.25$; (b) F at the center of the triangle; (c) $x_F = 0$, $y_F = -0.5$; and (d) $x_F = 0$, $-0.5 < y_F < 0$.

4.3.7(b) reduces to

$$M_{y_F=1} = 0.050 \ . \tag{4.3.16}$$

A shift of the position of point F to the center point of the side of the triangle at $y = -0.5$ yields the geometry shown in Fig. 4.3.7(c), whose figure of merit is

$$M_{y_F=-0.5} \approx 0.071 \ . \tag{4.3.17}$$

The magnet geometry for a value $0 > y_F > -0.5$ is shown in Fig. 4.3.7(d).

One observes that the orientation of the remanences of all four structures shown in Fig. 4.3.7 is the same as in Fig. 4.2.5(a), in agreement with the invariance of $\vec{J}$ with the position of F.

4.4 FIGURE OF MERIT OF MULTILAYERED MAGNETS

The properties of the multilayered structures catalogued in Table 4.3.1 are based on the assumption that the cross-sections of cavity boundary and interfaces between individual magnets are similar regular polygons. For large values of n, as Table 4.3.1 and Fig. 4.3.1 show, a large number of layers is required to achieve the optimum value of the figure of merit. For the lowest values of n, the maxima of M listed in Table 4.3.1 do not necessarily represent the optimum magnetic structure. In the case of the triangular cavity, for instance, Fig. 4.3.5 and Table 4.3.2 show that the optimum value of M is achieved at $K \approx 0.66$ with a single layer magnetic structure with a hexagonal external boundary. In the case of a square cross-sectional cavity ($n = 4$), Table 4.3.3 shows that the maximum value of M achieved also with a single layer structure at $K = 0.56$ is higher than the figure of merit of the double layer structure listed in Table 4.3.1.

In general, the optimum design of a yokeless magnet may involve a multiplicity of single layer magnetic structures, each of them having a different geometry and a different value of parameter K. The optimum combination of number of layers and geometries of the individual single layer structures is a rather cumbersome computational problem, particularly in the case of an arbitrarily assigned geometry of the magnet cavity.

The computation of the optimum structure around each particular geometry of the magnet cavity involves the calculation of the figure of merit for each combination of the independent design parameters: number of layers and the value of K of the individual layers. As previously stated, the effect of each element of magnetized material on the field within the cavity decreases rather rapidly with its distance, and as a consequence, the external layers of a multilayered structure may affect adversely the figure of merit as the number of layers increases. To discuss in some detail how the addition of an extra layer may affect the performance of a multilayered structure, assume a magnet designed for a value K_1 of parameter K and let M_1 be the value of its figure of merit at $K = K_1$. Assume that the magnet is enclosed by an additional single layer magnetic structure. The figure of merit of the combined structure is

$$M = (K_1 + K)^2 \frac{A_c}{A_{m,1} + A_{m,2}}, \qquad (4.4.1)$$

where K is the design parameter of the second layer and $A_{m,1}$, $A_{m,2}$ are the areas of the cross-sections of the internal and external layers, respectively. The area of the cross-section of the cavity of the external layer is $A_c + A_{m,1}$. Thus the figure of merit of the external layer is

$$M_2 = K^2 \frac{A_c + A_{m,1}}{A_{m,2}}. \qquad (4.4.2)$$

Assume a thin external layer such that

$$K \ll K_1 < 1. \tag{4.4.3}$$

In this case the area $A_{m,2}$ of the external layer can be written in the form

$$A_{m,2} = \left[\frac{dA_{m,2}}{dK}\right]_0 K + \cdots, \tag{4.4.4}$$

and by virtue of Eq. 4.4.2, the figure of merit M_2 increases linearly with K at a rate

$$\left[\frac{dM_2}{dK}\right]_0 = (A_c + A_{m,1}) \left[\frac{dA_{m,2}}{dK}\right]_0^{-1}. \tag{4.4.5}$$

Thus the addition of the second thin layer results in a change of the figure of merit 4.4.1 given by

$$\left[\frac{dM}{dK}\right]_0 = M_1 \left[\frac{2}{K_1} - \left(1 + \frac{M_1}{K_1^2}\right) \left[\frac{dM_2}{dK}\right]_0^{-1}\right]. \tag{4.4.6}$$

Assume that the internal magnet is a multilayer structure of magnets with similar regular polygonal interfaces as defined, for instance, by Eq. 4.2.7 and consider the limit $m \to \infty$, $n \to \infty$ when the interfaces are concentric circles. In this limit, by virtue of Eq. 4.3.4, Eq. 4.4.5 reduces to

$$\left[\frac{dM_2}{dK}\right]_0 = \frac{1}{2}. \tag{4.4.7}$$

Assume optimal dimensions of the internal magnet that correspond to the maximum value of its figure of merit for the value K_0 given by Eq. 4.3.6. By virtue of Eqs. 4.3.4 and 4.3.5 one has

$$1 + \frac{M_0}{K_0^2} = \frac{1}{K_0}. \tag{4.4.8}$$

Thus, for $K_1 = K_0$ and $M_1 = M_0$, Eqs. 4.4.6 and 4.4.7 yield

$$\left[\frac{dM}{dK}\right]_0 = 0, \tag{4.4.9}$$

136 YOKELESS MAGNETS

i.e., as expected, the additional thin external layer does not modify the figure of merit of the internal magnetic structure designed for the optimal value K_0. This is not necessarily true in a structure designed around a polygonal contour with a finite number of sides, where the maximum of the figure of merit of a single layer magnet is achieved with a geometry of the magnetic structure whose internal and external boundaries are not similar polygons. A positive value in Eq. 4.4.6, i.e., an increase of the figure of merit, can be obtained with the addition of the external layer if

$$\left[\frac{dM_2}{dK}\right]_0 > \frac{K_1}{2}\left[1 + \frac{M_1}{K_1^2}\right]. \tag{4.4.10}$$

Equation 4.4.6 is valid as long as no air gap is left between the internal structure and the external layer. Having different geometries, the internal structure and the external layer exhibit different values of the derivative of the figure of merit at $K = 0$.

Consider, for instance, the particular case of the optimal design of a single layer structure around a square cross-sectional prismatic cavity. According to Table 4.3.3 the optimum values M_1, K_1 are

$$M_1 \approx 0.118, \qquad K_1 \approx 0.56. \tag{4.4.11}$$

On the other hand for the particular case of $n = 4$, the value of K_1 given by Eq. 4.4.11 is a solution of Eq. 4.1.15, i.e., the optimal magnetic structure around the square cross-sectional cavity has a regular octagonal external boundary inscribed in a circle of radius r_e whose value is listed in Table 4.1.1.

The structure defined by Eq. 4.4.11 results in the value of the right hand side of inequality 4.4.10

$$\frac{K_1}{2}\left[1 + \frac{M_1}{K_1^2}\right] \approx 0.39. \tag{4.4.12}$$

The value of the left hand side of the same inequality is provided by the plotting of M versus K for an octagonal cavity. One has

$$\left[\frac{dM_2}{dK}\right]_{\substack{K=0 \\ 2n=8}} \approx 0.485. \tag{4.4.13}$$

Table 4.4.1. Optimal values of K, M for two-layer magnets designed around regular polygons.

n	3	4	5	6	7
K_1	0.66	0.51	0.43	0.38	0.36
K_2	0.09	0.27	0.29	0.29	0.30
K	0.77	0.78	0.72	0.67	0.66
M	0.0966	0.1299	0.1418	0.1456	0.1463

Because value 4.4.13 is significantly larger than value 4.4.12, the addition of an external layer to the optimal magnetic structure designed around the square cross-sectional cavity leads to an improvement of the figure of merit in addition to the increase of the value of K.

As the thickness of the external layer increases, one can expect the figure of merit to reach a maximum for a value K_2 of parameter K. In an optimal combination of internal and external magnets, K_1 does not necessarily coincide with the optimal value of the internal magnet. For regular polygon boundaries of the magnet cavity the results of the optimization of two-layer structures are summarized in Table 4.4.1. The effect of the optimization is apparent if Table 4.4.1 is compared with Table 4.3.3, which lists the optimal values of M_1 and K_1 for single layer structures. Though the improvement is marginal for a triangular cavity ($n = 3$), the beneficial effect of the external layer becomes significant for increasing values of n.

In Table 4.4.1, it is of interest to observe that, as n increases, the internal and external layers contribute essentially in equal measure to the field inside the cavity.

It is also of interest to compare Table 4.4.1 with Tables 4.3.1 and 4.3.2, which list the values of the figure of merit of multilayer structures with similar polygonal interfaces between layers. As Eqs. 4.2.7 and 4.2.15 have shown, as n increases, an increasingly larger number of layers is required in the multilayer structures with similar geometries to attain the optimal value of K. This is particularly apparent for the structures listed in Table 4.3.1. For the same values of n of the polygonal cavity, the two-layer structures listed in Table 4.4.1 exhibit a higher value of the figure of merit compared to both multilayered structures listed in Tables 4.3.1 and 4.3.2.

Table 4.4.2 shows the values of the figure of merit resulting from the combinations of K_1 and K_2 in the range

$$0.44 \leq K_1 \leq 0.60, \qquad 0.18 \leq K_2 \leq 0.34. \qquad (4.4.14)$$

One observes that in the range 4.4.14 of values of parameters K_1, K_2, the figure of merit is larger than the value M_1 given by Eq. 4.4.11, i.e., the addition of the external layer improves the optimum figure of merit of the single layer design.

Table 4.4.2. Figure of merit M of two-layer magnets designed around a square cross-sectional cavity.

K_1 \ K_2	.18	.20	.22	.24	.26	.28	.30	.32	.34
.44	.1244	.1252	.1258	.1264	.1268	.1270	.1271	.1271	.1269
.46	.1260	.1267	.1274	.1278	.1282	.1283	.1284	.1282	.1279
.48	.1273	.1280	.1285	.1289	.1292	.1293	.1292	.1290	.1286
.50	.1281	.1288	.1292	.1296	.1298	.1298	.1297	.1294	.1289
.52	.1285	.1290	.1294	.1297	.1299	.1298	.1296	.1292	.1286
.54	.1282	.1287	.1291	.1293	.1293	.1292	.1289	.1284	.1278
.56	.1274	.1277	.1280	.1281	.1281	.1279	.1275	.1270	.1262
.58	.1258	.1261	.1262	.1262	.1261	.1258	.1253	.1247	.1239
.60	.1235	.1236	.1236	.1235	.1233	.1229	.1224	.1217	.1208

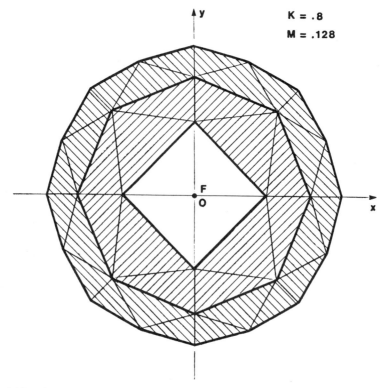

Fig. 4.4.1. Two-layer magnet around a square cross-sectional cavity ($K_1 = 0.56$, $K_2 = 0.24$).

Figure 4.4.1 shows the two-layer structure with the regular octagonal interface between internal and external layers. The external layer is designed for

$$K_2 = 0.24 , \qquad (4.4.15)$$

and as a consequence, the value of K of the two-layer structure of Fig. 4.4.1 is $K = 0.8$. As shown by Table 4.4.2, the values of K_2 given by Eq. 4.4.15 correspond to the optimal geometry of the external layer around the internal magnet designed for $K_1 = 0.56$.

4.5 THREE-DIMENSIONAL YOKELESS MAGNETS

The analysis of the spherical magnet presented in Section 5 of Chapter 2 has shown that, compared to a two-dimensional geometry, a three-dimensional yokeless magnet requires a more complex distribution of the remanence in order to achieve the field confinement. An identical situation is found in the development of a three-dimensional structure composed of uniformly magnetized polyhedrons. The design procedure of two-dimensional structures cannot be extended to a three-dimensional geometry without the introduction of additional components of magnetized material designed to satisfy the boundary conditions at the external surface of the magnet [4].

The approach discussed in this section is based on the separation of the components of the magnet in two independent groups. One group consists of magnetized components designed to generate the required field intensity $\vec{H}_0$ within the three-dimensional closed cavity and to close the flux of the magnetic induction. The geometry and the orientation of the remanences of the components of this first group are computed by means of the same two-dimensional method defined in Chapter 3 and in the Section 1 of this chapter. The second group of magnetized components is designed not to channel the flux of the magnetic induction but rather as a transition layer between the structure of the first group and the external nonmagnetic medium. The geometry and the orientation of the remanence of each component of the second group are computed to satisfy the condition

$$\mu_0 \vec{H} = -\vec{J}. \tag{4.5.1}$$

Consider, for instance, a magnet with a cubic cavity and assume that the intensity $\vec{H}_0$ within the cavity is perpendicular to a face of the cube and parallel to the axis y, as shown in Fig. 4.5.1. To apply the two-dimensional method to the structure of magnetic material, divide first the geometry of Fig. 4.5.1 in four identical wedges with the axis y as the common edge. The interfaces between wedges are the planes that contain the axis y and the edges of the cube parallel to the axis y, as shown in Fig. 4.5.1. Consider each wedge as part of a two-dimensional cavity, and let us compute the structure of magnetic material by means of the two-dimensional method in the plane per-

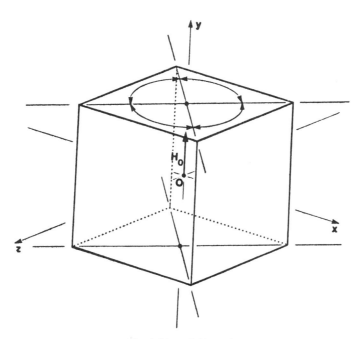

Fig. 4.5.1. Cubic cavity.

140 YOKELESS MAGNETS

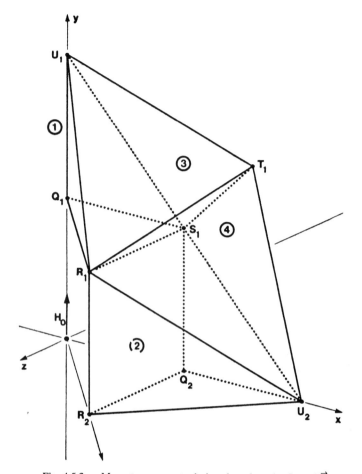

Fig. 4.5.2. Magnet components designed to close the flux of $\vec{B}$.

pendicular to the z axis and in the plane perpendicular to the x axis, with the assumption that in both computations the center F of the field configuration is located on the edge common to the four wedges, at the center of the cube.

Because of symmetry, the computation of geometry and distribution of remanence can be limited to the space confined between the three planes

$$x = z, \quad y = 0, \quad z = 0. \tag{4.5.2}$$

The four components shown in Fig. 4.5.2 are designed to close the lines of flux of the magnetic induction in planes parallel to the $z = 0$ plane. Their geometry is computed on the basis of the two-dimensional calculation of a structure designed around a square cross-sectional cavity. Assuming, for instance, a value of parameter K

$$K = \frac{1}{2}, \tag{4.5.3}$$

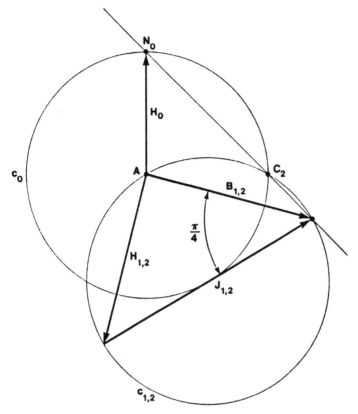

Fig. 4.5.3. Computation of the field in the components of Fig. 4.5.2.

the field in components $(R_1S_1T_1U_1)$ and $(R_1S_1T_1U_2)$ of Fig. 4.5.2 is computed by means of the vector diagram of Fig. 4.5.3, as if these components were part of the single layer two-dimensional magnet shown in Fig. 4.5.4. The coordinates of points U_1, T_1, U_2 are

$$
\begin{aligned}
x = 0, \qquad y = 2x_0, \qquad z = 0 & \qquad (U_1) \\
x = y = \left[1 + \tan\frac{\pi}{6}\right]x_0, \qquad z = 0 & \qquad (T_1) \\
x = 2x_0, \qquad y = z = 0, & \qquad (U_2)
\end{aligned}
\qquad (4.5.4)
$$

where $2x_0$ is the side of the cubic cavity.

In Fig. 4.5.2 the two interfaces $(U_1R_1S_1)$ and $(U_2R_1S_1)$ have segment (R_1S_1) of the edge of the cavity in common. The flux of the magnetic induction within the $\pi/4$ wedge of the cavity is channeled through tetrahedron $(U_1T_1R_1S_1)$ in the direction parallel to edge (U_1T_1), through tetrahedron $(U_2T_1R_1S_1)$ in the direction parallel to edge (U_2T_1), and through pentahedron $(U_2R_1S_1R_2Q_2)$ in the opposite direction of the axis y. Thus, the structure of Fig. 4.5.2 closes the flux of $\vec{B}$ flowing within the $\pi/4$ wedge of the cavity.

142 YOKELESS MAGNETS

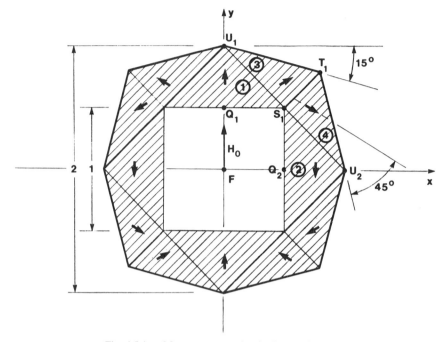

Fig. 4.5.4. Magnet cross-section in the $z = 0$ plane.

By definition, lines $(T_1 U_1)$ and $(T_1 U_2)$ in Fig. 4.5.2 are $\Phi = 0$ equipotential lines. However, the potential of point R_1 is different from zero. Thus surfaces $(T_1 R_1 U_1)$ and $(T_1 R_1 U_2)$ are not equipotential surfaces and, as a consequence, they cannot be the interfaces between the magnetic structure and an external, nonmagnetic equipotential medium. The structure of Fig. 4.5.2 must be enclosed in a second structure designed as a transition layer that satisfies condition 4.5.1 in order not to generate an additional flux of $\vec{B}$.

The transition structure, within the same $\pi/4$ wedge of the cavity, is shown in Fig. 4.5.5. It consists of two tetrahedrons $(R_1 T_1 U_1 V_1)$, $(R_1 R_2 U_2 V_2)$ and a pentahedron $(R_1 T_1 V_1 V_2 U_2)$. By virtue of Eqs. 4.5.1 and 4.5.3, if the same remanence $\vec{J_i}$ is used in these three components, their external surfaces $(T_1 U_1 V_1)$, $(T_1 V_1 V_2 U_2)$ $(U_2 V_2 R_2)$ are $\Phi = 0$ equipotential surfaces if they are tangent to a sphere whose center is located at point R_1 and whose radius is

$$r_i = \frac{x_0}{2} \frac{J_0}{J_i}. \tag{4.5.5}$$

Obviously, in each component of the transition structure vector $\vec{J_i}$ is perpendicular to the external surface and oriented toward the nonmagnetic medium that surrounds the magnet.

The value of the radius r_i given by Eq. 4.5.5 must be smaller than the distance of point R_1 from edge $(R_2 U_2)$, i.e.,

$$r_i < x_0. \tag{4.5.6}$$

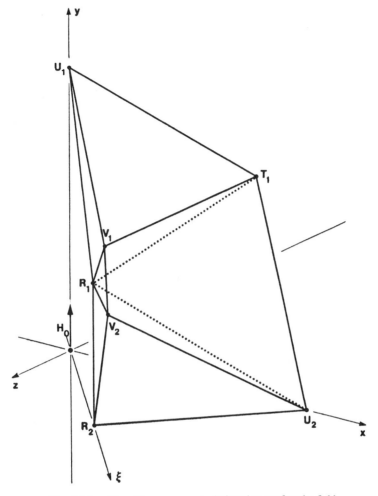

Fig. 4.5.5. Transition components designed to confine the field.

Thus in the transition structure one can choose any value of remanence J_i as long as its magnitude satisfies the condition

$$J_i \geq \frac{1}{2} J_0. \qquad (4.5.7)$$

In particular if

$$J_i = \frac{J_0}{\sqrt{2}}, \qquad (4.5.8)$$

r_i becomes equal to the distance of point S_1 from either $(T_1 U_1)$ or $(T_1 U_2)$, and edge

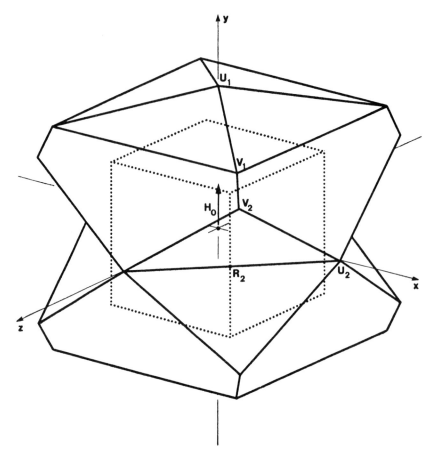

Fig. 4.5.6. View of the yokeless magnet with the cubic cavity.

$(T_1 V_1)$ in Fig. 4.5.5 becomes parallel to side $(R_1 S_1)$ of the cavity. Surface $(U_2 V_2 R_2)$ is inclined at an angle $\pi/4$ with respect to the $y = 0$ plane and the external surface of the yokeless magnet has the geometry shown in Fig. 4.5.6. Point R_2 is one of the four points common to the cubic cavity and the external surface.

REFERENCES

[1] M. G. Abele, Yokeless permanent magnets, Technical Report 14, New York University, Nov 1, 1986.
[2] M. G. Abele, H. A. Leupold, A general method for flux confinement in permanent magnet structures, *Journal of Applied Physics*, 64 (10), p. 5988-5990, 1988.
[3] M. G. Abele, H. Rusinek, Optimum design of yokeless permanent magnets, *Journal of Applied Physics*, 67 (9), p. 4644-4646, 1990.
[4] M. G. Abele, Three-dimensional design of a permanent magnet, Technical Report 16, New York University, June 1, 1987.

CHAPTER 5

Yoked Magnets

INTRODUCTION

The yoke of the magnetic structures presented in this chapter is defined as a medium of infinite magnetic permeability. In such an idealized model of a permanent magnet, a yoke reduces to an equipotential surface that acts like a perfect sink or source of the flux of the magnetic induction generated by the structure of magnetized material.

Furthermore, the ideal yoke is a closed surface totally enclosing the magnetic structure, thereby making it possible to maintain the field configuration developed in Section 3.3.

The design of two-dimensional yoked structures is developed in the following sections and the procedure is extended to the three-dimensional problem of a cubic cavity.

5.1 TWO-DIMENSIONAL YOKED MAGNETIC STRUCTURES

The method for computing the geometry and distribution of magnetization of a yoked two-dimensional structure has been introduced in Section 3 of Chapter 3. The vector diagram of Fig. 3.3.1 will be used in this chapter to design and analyze the properties of magnetic structures around arbitrarily assigned prismatic cavities.

Similarly to the design of a yokeless structure, the computation of the geometry of a yoked magnet starts with the selection of parameter K and orientation of intensity $\vec{H}_0$ within the cavity. The design of a yokeless structure requires also the selection of the position of the characteristic point F. The formulation of the design of a yoked structure requires only the selection of the position of the reference line $\Phi = 0$ within the cavity, according to the conclusion of Section 3.3.

Assume again a prismatic cavity and let s_h be one side of the polygonal contour of the cavity, as shown in Fig. 5.1.1, where the x axis is selected to coincide with the equipotential line $\Phi = 0$ within the cavity. Let $0, y_{h-1}$ be the coordinates of vertex S_{h-1} common to sides s_{h-1}, s_h. Assume that $\vec{H}_0$ is oriented in the positive direction of the y axis and S_{h-1} is the vertex where the scalar potential within the cavity attains its maximum volume.

Once the value of parameter K has been assigned, the vector diagram of Fig. 5.1.2 provides the values of vectors $\vec{J}_{h-1}, \vec{J}_h, \vec{J}_{h+1}$ in the components of magnetized material that have sides s_{h-1}, s_h, s_{h+1} in common with the magnet cavity, respectively.

The straight line that contains side s_h intersects axis x at point P_h as shown in Fig. 5.1.1. Thus, as in Fig. 3.2.2 and by virtue of the geometric invariance theorem of

146 YOKED MAGNETS

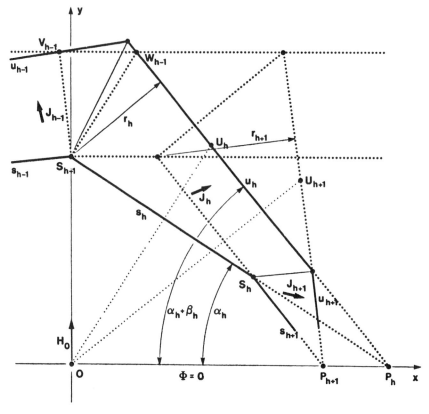

Fig. 5.1.1. Yoked two-dimensional structure.

yokeless structures, the $\Phi = 0$ line in the region of remanence $\vec{J}_h$ is line u_h which must contain point P_h and the same point U_h defined in the schematic of Fig. 4.1.2. The distance $\xi_{1,h}$ of point U_h from line s_h is given by Eq. 4.1.7, where $\xi_{0,h}$ is the distance of line s_h from the origin O of the frame of reference.

As shown in Fig. 5.1.1, line u_h forms an angle β_h with respect to s_h given by

$$\tan \beta_h = \frac{K}{1-K} \tan \alpha_h , \qquad (5.1.1)$$

where α_h is the angle of side s_h with respect to the x axis. Thus line u_h is at a distance r_h from vertex S_{h-1}

$$r_h = y_{h-1} \frac{1}{\sqrt{1 - \frac{1}{K^2}(2K-1)\cos^2 \alpha_h}} , \qquad (5.1.2)$$

and by virtue of the vector diagram of Fig. 3.2.16, the intensity $\vec{H}_h$ in the magnetic material contained between boundaries s_h, u_h is oriented perpendicular to line u_h, and its magnitude is given by

$$\left[\frac{H_h}{H_0}\right]^2 = 1 - \frac{1}{K^2}(2K - 1)\cos^2\alpha_h . \tag{5.1.3}$$

In the particular case $K = \frac{1}{2}$, one has

$$H_h = H_0 = \frac{1}{2}J_0, \qquad \beta_h = \alpha_h . \tag{5.1.4}$$

The external boundary of the region of remanence $\vec{J}_h$ is a segment of the $\Phi = 0$ line u_h. In a similar manner one determines lines u_{h-1} and u_{h+1}. One notices in Fig. 5.1.1 that r_{h+1} is the distance of line u_{h+1} from the point of intersection of line s_{h+1} with line $y = y_{h-1}$.

To fully determine the geometry of the prismatic components of the magnetic structure and the interface between structure and magnetic yoke, one must compute the geometry of the air wedges between magnetized components. Following the procedure defined in Section 3.3, the orientations of the boundaries of the air wedge are defined by the position of points D_{h-1}, D_h on the circular diagram of Fig. 5.1.2. Assume, for instance, that the desired intensity is oriented in the direction opposite of $\vec{H}_0$ in the wedge with vertex at point S_{h-1}. Then point D_{h-1} in Fig. 5.1.2 is the intersection of

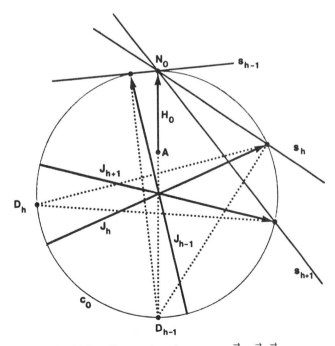

Fig. 5.1.2. Computation of remanences $\vec{J}_{h-1}, \vec{J}_h, \vec{J}_{h+1}$.

148 YOKED MAGNETS

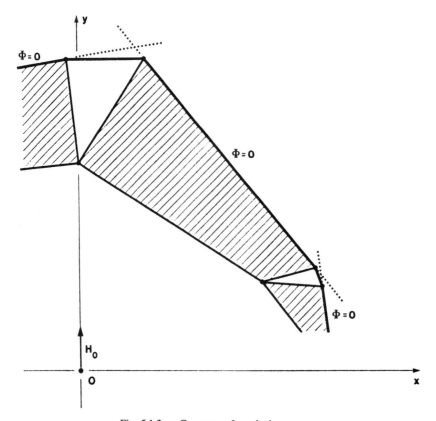

Fig. 5.1.3. Geometry of a yoked structure.

the circle with the line that contains $\vec{H}_0$, and the external boundary of the air wedge is the dotted line parallel to the x axis, which passes through points V_{h-1}, W_{h-1} of Fig. 5.1.1. Similarly the boundaries of the wedge with vertex at point S_h are defined by point D_h as shown in Fig. 5.1.2. The resulting geometry of the cross-section of the magnetic structure is shown in Fig. 5.1.3, where the heavy line is the $\Phi = 0$ equipotential interface between the structure of magnetized material and the external yoke of infinite magnetic permeability.

As an example, consider the design of a yoked magnetic structure around the same regular hexagonal cavity of the yokeless magnet of Fig. 4.1.5 for the same value of parameter K,

$$K = \frac{1}{4}.$$ (5.1.5)

Assume an intensity $\vec{H}_0$ oriented in the direction perpendicular to a side of the hexagon and assume that the equipotential line $\Phi = 0$ within the cavity passes through the center of the hexagon.

The geometry of the magnetic structure of Fig. 5.1.4 around the hexagonal cavity is provided by the vector diagram of Fig. 5.1.5. The heavy arrows in Fig. 5.1.4 show

TWO-DIMENSIONAL YOKED MAGNETIC STRUCTURES **149**

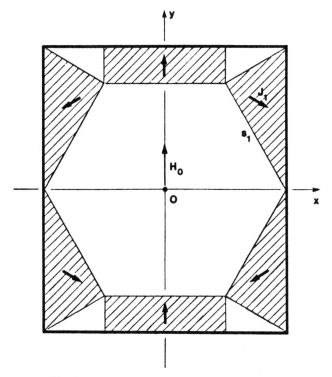

Fig. 5.1.4. Yoked magnet with hexagonal cavity.

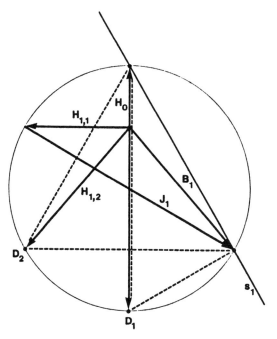

Fig. 5.1.5. Geometry of the air wedges of a yoked magnet with hexagonal cavity.

150 YOKED MAGNETS

the orientation of the remanences in the six components of the magnetic structure. The interfaces between the external yoke and the components with triangular cross-section are parallel to the y axis. If point D_1 in the vector diagram of Fig. 5.1.5 is selected in such a way that the external boundaries of the air wedges are parallel to the x axis, the cross-section of the $\mu = \infty$ external yoke is a rectangle shown in Fig. 5.1.4. The side $2x_1$ of the rectangle is equal to the diameter of the circle which inscribes the hexagonal cavity. The side $2y_1$ is equal to

$$2y_1 = \frac{2x_1}{1-K} \cos\frac{\pi}{6} = \frac{8}{3} x_1 \cos\frac{\pi}{6}. \qquad (5.1.6)$$

A modification of the magnet geometry of Fig. 5.1.4 is shown in Fig. 5.1.6. The new geometry is obtained from the vector diagram of Fig. 5.1.4 by selecting point D_2 in such a way that the interfaces between the air wedges and the magnetic components of triangular cross-section are parallel to the x axis.

Equations 5.1.1 and 5.1.2 show that for $K = 1$, line u_h is perpendicular to s_h and its distance from S_{h-1} is equal to the distance of point P_h from S_{h-1}. For $K = 1$ and $\alpha_h = 0$ one has

$$r_h = \infty. \qquad (5.1.7)$$

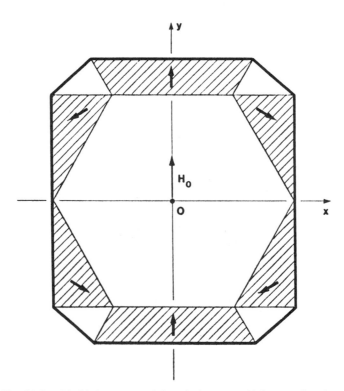

Fig. 5.1.6. Modified geometry of the yoked magnet with hexagonal cavity.

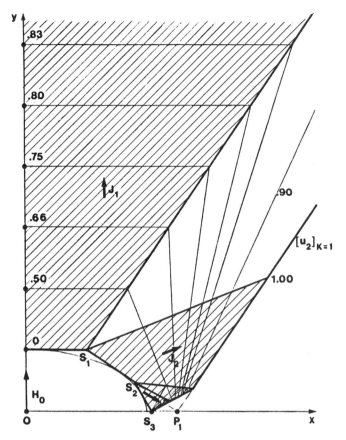

Fig. 5.1.7. Effect of parameter K on the external boundary of a yoked magnet.

Consequently $K = 1$ is the upper limit of parameter K in both yoked and single layer yokeless magnets.

The dependence of the geometry of a two-dimensional yoked magnet on the value of parameter K is illustrated in Fig. 5.1.7 which shows the family of external boundaries of a magnetic structure designed around a cavity with a decagonal cross-section with the intensity $\vec{H}_0$ oriented along the axis y. The three regions of magnetized material in the first quadrant of the magnet have remanences $\vec{J}_1, \vec{J}_2, \vec{J}_3$ which are independent of K. The y dimension of the region with remanence $\vec{J}_1$ increases as K increases, and in agreement with Eq. 5.1.7, it diverges for $K = 1$. The dimensions of the other two regions of remanences $\vec{J}_2$ and $\vec{J}_3$ are always finite. For $K = 1$, the external boundary of the region of remanence $\vec{J}_2$ becomes perpendicular to side $(S_1 S_2)$ of the cavity, and the external boundary of the region of remanence $\vec{J}_3$ becomes perpendicular to side $(S_2 S_3)$ of the cavity.

Chapter 4 has shown that multilayered structures of yokeless magnets can be designed if the magnetic material is transparent to the external field. Obviously a yoked magnet can always be designed to enclose a single or multilayered yokeless structure. Thus a value of the intensity of the order or larger than the remanence can

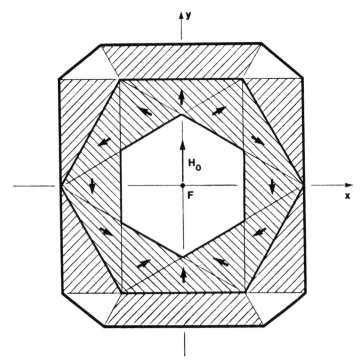

Fig. 5.1.8. Two-layer yoked magnet.

always be attained with a multilayer design where the external layer is enclosed by a yoke.

An example of a multilayered yoked magnet is the two-layer magnet shown in Fig. 5.1.8. The internal structure is a yokeless magnet designed around a regular hexagonal cross-sectional cavity for

$$K_i = \frac{1}{2} \tag{5.1.8}$$

and point F selected at the center of the hexagon. Figure 5.1.8 shows the orientation of the remanence in the components of the internal structure that generate an intensity oriented along axis y. By virtue of Eq. 4.2.15, the value of parameter K given by Eq. 5.1.8 yields a regular hexagonal cross-section of the external boundary. By virtue of Eq. 4.2.16, the two hexagons are inscribed in two circles of radii r_u, r_0:

$$r_u = \sqrt{3} r_0 . \tag{5.1.9}$$

The external structure in Fig. 5.1.8 is the yoked magnet of Fig. 5.1.6 designed around the hexagonal cavity inscribed in the circle of radius r_u, for the same value of K given by Eq. 5.1.5, for the field oriented along the y axis, and for the zero equipotential lines passing through the center of the cavity.

The two-layer structure of Fig. 5.1.8 which combines yokeless and yoked structures is an example of the hybrid magnets that will be discussed in more detail in the following chapter. It is of interest to point out that a multilayered structure of the type shown in Fig. 5.1.8 may yield higher values of the figure of merit compared to single or multilayered yokeless magnets. In the example of Fig. 5.1.8 the area of the cross-section of the internal layer is

$$A_i = 6r_0^2 \cos\frac{\pi}{6}. \tag{5.1.10}$$

The area of the cross-section of the external yoked layer is

$$A_e = \frac{16}{3}r_0^2 \cos\frac{\pi}{6}, \tag{5.1.11}$$

and the area of the cross-section of the hexagonal cavity is

$$A_c = 3r_0^2 \cos\frac{\pi}{6}. \tag{5.1.12}$$

Thus the value of the figure of merit of the structure of Fig. 5.1.8 at

$$K = \frac{3}{4} \tag{5.1.13}$$

is

$$M = K^2 \frac{A_c}{A_i + A_e} \approx 0.149. \tag{5.1.14}$$

As shown in Table 4.3.1, the maximum value of the figure of merit of a multilayered yokeless structure designed around a hexagonal cavity is achieved at approximately the same value of K as given by Eq. 5.1.13 with five layers of similar geometries, and it is lower than the value 5.1.14.

Figure 5.1.9 shows the first quadrant of the cross-section of a yoked magnet designed around the regular hexagonal cavity for the same value of K as given by Eq. 5.1.13. The air wedge between the two magnetized components corresponds to the same position of point D_2 as shown in the vector diagram of Fig. 5.1.4. For the value of K given by Eq. 5.1.13, the intensity $\vec{H}_i$ within the air wedge is parallel to the x axis; i.e., the external boundary of the wedge is parallel to the y axis as shown in Fig. 5.1.9. The ratio of the area of the cross-section of the magnetic material to the area of the cavity is

$$\frac{A_m}{A_c} = 6. \tag{5.1.15}$$

154 YOKED MAGNETS

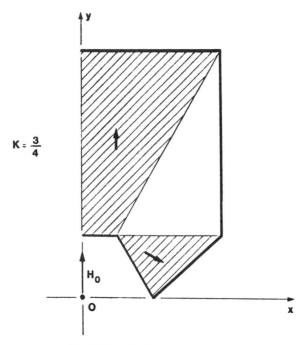

Fig. 5.1.9. Single layer yoked magnet.

Thus the figure of merit of the structure of Fig. 5.1.9 is

$$M \approx 0.094 , \tag{5.1.16}$$

which is substantially lower than the value given by Eq. 5.1.14 of the two-layer magnet.

5.2 SQUARE CROSS-SECTIONAL CAVITY

The theorem of geometric invariance, which states that the geometry of a yokeless magnet is independent of the orientation of $\vec{H}_0$, does not apply to yoked two-dimensional structures. Once the geometry of the magnet cavity has been selected, the orientation of the intensity within the cavity of a yoked magnet becomes a design parameter. Thus the orientation of $\vec{H}_0$ is an additional degree of freedom which can be used to optimize the magnet design.

To illustrate the effect of a change of the orientation of $\vec{H}_0$, consider again the two-dimensional square cross-sectional cavity discussed in the preceding chapter, and assume first an intensity $\vec{H}_0$ oriented parallel to a side of the square as shown in Fig. 5.2.1. Assume that the center of the square coincides with the origin of the cartesian frame of reference x, y, whose axes are parallel to the sides of the square. Furthermore assume that the equipotential line $\Phi = 0$ within the cavity coincides with the $y = 0$ axis. Thus, the geometry of the magnet is symmetric with respect to the x and y axes.

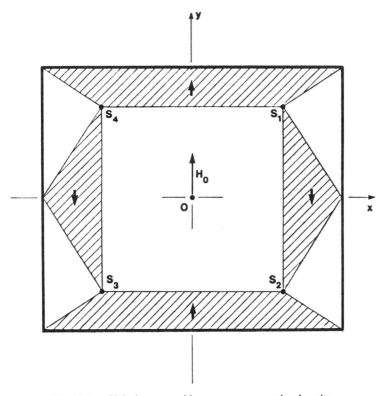

Fig. 5.2.1. Yoked magnet with a square cross-sectional cavity.

The magnetic structure in the first quadrant of Fig. 5.2.1 is determined by the vector diagram of Fig. 5.2.2. By selecting a position of point D such that the intensity $\vec{H}_i$ within the air wedge is parallel to the x axis, the interface between magnetic structure and yoke is the rectangle shown in Fig. 5.2.1. The configuration of the lines of flux of the magnetic induction in the first quadrant of the magnet cross-section is shown in Fig. 5.2.3. If a different point D_1' is selected on the circle of Fig. 5.2.2 such that the intensity within the wedge is vector $\vec{H}_i'$, one obtains the geometry of Fig. 5.2.4.

If $2x_0$ is the side of the square cross-sectional cavity, the dimensions y_1 of both structures of Figs. 5.2.1 and 5.2.4 are given by

$$2y_1 = \frac{2x_0}{1-K}. \qquad (5.2.1)$$

In Fig. 5.2.4 interfaces $(S_1 V_1)$ and $(S_1 W_1)$ are perpendicular to each other, and the abscissas x_2 of point V_1 and x_1 of point W_1 are related to each other by the equation

$$(x_1 - x_0)(x_2 - x_0) = \frac{K}{1-K} x_0^2. \qquad (5.2.2)$$

156 YOKED MAGNETS

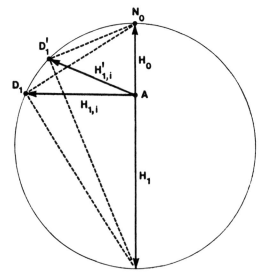

Fig. 5.2.2. Computation of the geometry of the nonmagnetic wedges.

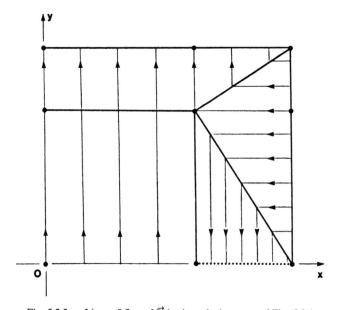

Fig. 5.2.3. Lines of flux of $\vec{B}$ in the yoked magnet of Fig. 5.2.1.

Thus the area of the cross-section of the magnetic material in Fig. 5.2.4 is

$$A_m = \frac{x_0^2}{1-K}\left\{K + \frac{1}{2}(1-K)\left[\frac{K^2}{(1-K)^2}\frac{x_0}{x_1-x_0} + \frac{x_1-x_0}{x_0}\right]\right\}. \qquad (5.2.3)$$

At constant K, A_m is a function of x_1, and it attains a minimum at

$$x_1 - x_0 = \frac{K}{1-K}x_0, \qquad (5.2.4)$$

i.e.,

$$x_1 = \frac{x_0}{1-K} = y_1. \qquad (5.2.5)$$

By virtue of Eqs. 5.2.1 and 5.2.5, the minimum value of A_m is

$$A_{min} = 2\frac{K}{1-K}x_0^2. \qquad (5.2.6)$$

It is of interest to point out that A_{min} is twice the area of the rectangular region of magnetic material. At $A = A_{min}$ the figure of merit is

$$M = \frac{1}{2}K(1-K). \qquad (5.2.7)$$

Thus the maximum value of the figure of merit of the structure of Fig. 5.2.4 is

$$M_{max} = \frac{1}{8}, \qquad (5.2.8)$$

which is achieved at

$$K = \frac{1}{2}. \qquad (5.2.9)$$

At the optimum value of K given by Eq. 5.2.9, Eq. 5.2.5 yields

$$x_1 = y_1 = 2x_0, \qquad (5.2.10)$$

i.e., the interface between magnetic structure and yoke is a square whose side is twice the side of the cavity. In the vector diagram of Fig. 5.2.2, point A coincides with the

158 YOKED MAGNETS

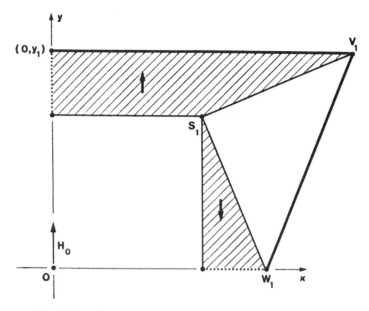

Fig. 5.2.4. Optimization of the geometry of the magnetic structure.

center of circle c_0, and the magnitudes of the intensities are

$$H_0 = H_1 = H_{h,i} = \frac{J_0}{2\mu_0}. \qquad (5.2.11)$$

The intensity $\vec{H}_2$ in the triangular components of magnetized material of the structure of Fig. 5.2.1 is

$$\vec{H}_2 = \vec{H}_0. \qquad (5.2.12)$$

Thus at $K = 0.5$, the magnitude of the intensity is uniform throughout the magnetic structure, as in the case of the one-dimensional magnet discussed in Section 2.1, and the magnetic material operates at the peak of the energy product curve. At $K = 0.5$, Eq. 5.2.6 shows that the area of the magnetized material is twice the area of the cavity cross-section. Thus, by virtue of Eq. 5.2.10, the area of the magnetized material is equal to the total area of the cavity and the nonmagnetic wedges of the magnet.

The structure shown in Fig. 5.2.4 corresponds to $K = 0.3$. Its geometry is the result of the assumption that a single air wedge separates the two components of magnetized material. A modification of the geometry of Fig. 5.2.4, suggested by the analysis presented in Section 3.3, is the structure of Fig. 5.2.5 designed for the same value of K, with two air wedges separated from each other by the wedge of magnetized material whose remanence $\vec{J}_3$ is parallel to the axis x.

The values of intensity $\vec{H}_3$ and magnetic induction $\vec{B}_3$ within the wedge of magnetized material of Fig. 5.2.5 are provided by the vector diagram of Fig. 5.2.6. Point D_1

SQUARE CROSS-SECTIONAL CAVITY   **159**

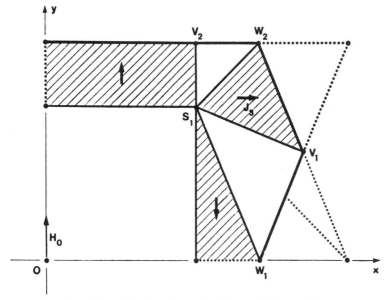

Fig. 5.2.5.   Insertion of a wedge of magnetized material.

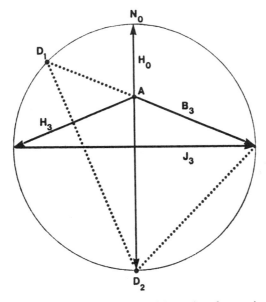

Fig. 5.2.6.   Computation of the geometry of the wedge of magnetized material.

160 YOKED MAGNETS

provides the interfaces of the wedge $(S_1 V_1 W_1)$ of nonmagnetic material that separates the two media of remanences $\vec{J}_2$ and $\vec{J}_3$, and point D_2 provides the interfaces of the wedge $(S_1 V_2 W_2)$ that separates the two media of remanences $\vec{J}_1$ and $\vec{J}_3$. Interface $(S_1 V_2)$ is parallel to the y axis and interface $(S_1 W_2)$ is oriented at the angle $\pi/4$ with respect to the x axis. The orientation of interface $(S_1 W_1)$ and the abscissa of point W_1 in Fig. 5.2.5 are the same as in Fig. 5.2.4.

Assume now that $\vec{H}_0$ is oriented parallel to a diagonal of the square. The yoked structure of Fig. 5.2.7 consists of four right angle triangular magnetized components, whose remanences are oriented parallel to the axis x by virtue of theorem 2 of Section 3.2.

The remanence $\vec{J}_1$ of the magnetized components in the first quadrant of Fig. 5.2.7 is given by the vector diagram of Fig. 5.2.8. The magnetized components are separated from each other by two nonmagnetic wedges, and the interfaces are oriented at an angle $\pi/2$ with respect to each other, regardless of the selection of point D_1 on

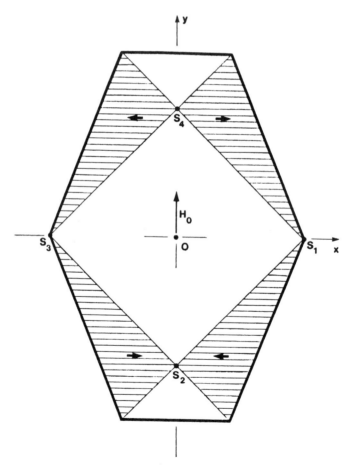

Fig. 5.2.7. Yoked magnet with $\vec{H}_0$ oriented along a diagonal of the cavity.

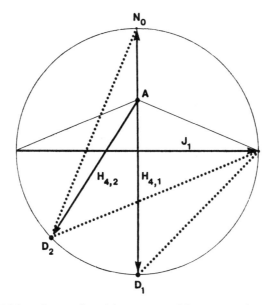

Fig. 5.2.8. Computation of the geometry of the nonmagnetic wedges.

the circle of the vector diagram of Fig. 5.2.8. By selecting D_1 as the point diametrically opposite to N_0, the magnetic structure is symmetric with respect to the axis y and the interface in the first quadrant is oriented at an angle $\pi/4$ with respect to the axis y. The intensity $H_{4,1}$ in the wedges is given by the vector diagram of Fig. 5.2.8. One has

$$\vec{H}_{4,1} = -\frac{1-K}{K}\vec{H}_0 . \qquad (5.2.13)$$

The cross-sectional area of each of the four magnetized components of the structure of Fig. 5.2.7 is

$$A_m = \frac{2K}{1-K} x_0^2 , \qquad (5.2.14)$$

where $2x_0$ is the side of the square. The value of A_m given by Eq. 5.2.14 coincides with the minimum value 5.2.6 obtained in the structure of Fig. 5.2.4, and the figure of merit of the magnet of Fig. 5.2.7 is given by Eq. 5.2.7. Hence the optimum figure of merit is independent of the orientation of the intensity within the cavity.

The structure of Fig. 5.2.7 is designed for $K = 0.3$. Its geometry can be modified as shown in Fig. 5.2.9, by introducing two wedges of magnetized material, whose geometry is determined by the position of point D_2 on the circle of the vector diagram of Fig. 5.2.8. To maintain the symmetry of the magnetic structure with respect to the axis y, the remanence $\vec{J}_3$ of the wedges of magnetized material must be parallel to $\vec{H}_0$. Assume that the angular position of point D_2 relative to D_1 on the circle of Fig. 5.2.8

162 YOKED MAGNETS

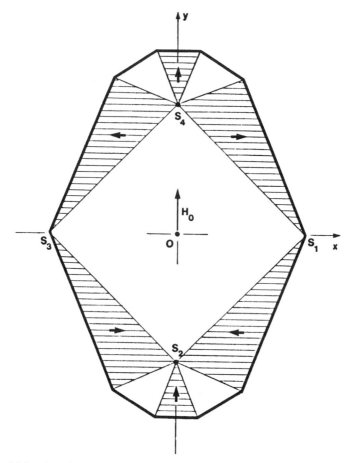

Fig. 5.2.9. Insertion of wedges of magnetized material in the structure of Fig. 5.2.7.

is equal to $\pi/4$. The boundaries of the nonmagnetic wedge in the first quadrant of Fig. 5.2.7 are oriented at the angles $\pi/8$ and $3\pi/8$ with respect to the axis x and the interface between the nonmagnetic wedge and the yoke is perpendicular to the intensity $\vec{H}_{4,2}$ within the wedge, as shown by the vector diagram of Fig. 5.2.8. Thus the angles of the wedges of magnetized and nonmagnetic materials are all equal to $\pi/4$. The lines of flux of the magnetic induction in the first quadrant of the structure of Fig. 5.2.9 are shown in Fig. 5.2.10.

It is of interest to compare the figure of merit of the yoked structure of Figs. 5.2.4 with the figure of merit of the single layer yokeless magnet designed around the square cross-sectional cavity, which has been plotted in Fig. 4.3.5. As shown in Table 4.3.3, the maximum value of M of the yokeless magnet is 0.118 and it is achieved at $K = 0.56$. The maximum of the figure of merit of the yoked magnet is slightly higher and it is achieved at a somewhat lower value of K, as shown by Eqs. 5.2.8 and 5.2.9.

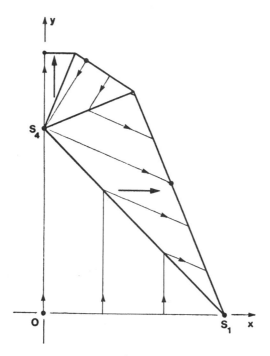

Fig. 5.2.10. Lines of flux of $\vec{B}$ in the yoked magnet of Fig. 5.2.9.

5.3 POSITION OF Φ = 0 LINE WITHIN THE MAGNET CAVITY

In a yoked magnet, the flux of the magnetic induction within the cavity closes through the yoke. The selection of the path of the flux within the yoke is determined by the way the yoke is designed rather than by the magnetic structure itself. In the limit of an ideal yoke of infinite magnetic permeability, the direction of the flux within the yoke is simply a matter of definition which has no bearing on the computation of the geometry of the magnetic structure. Consider, for instance, the yoked structure designed around the square cross-sectional cavity of Fig. 5.2.9. Assume two imaginary cuts at points G_1 and G_2 in the cross-section of the $\mu = \infty$ yoke as shown in Fig. 5.3.1(a). Points G_1, G_2 are located on the y axis that contains the center of the cavity. Thus the flux of $\vec{B}$ is divided in two equal parts in the direction of the arrows in the two sections of the yoke separated by the imaginary cuts. If one assumes a single imaginary cut on the axis x as shown in Fig. 5.3.1(b), the entire flux of $\vec{B}$ flows in the same direction as indicated by the arrows of Fig. 5.3.1(b).

Thus because the yoked magnetic structure is independent of the distribution of the flux within the yoke, the computation of a yoked structure is based on a selection of design parameters different from the ones that control a yokeless magnet design.

Chapter 4 has shown that the geometry of a two-dimensional yokeless magnet is determined by three parameters: the value of K and the two coordinates of the characteristic point F. The position of point F is not a parameter of the design of a yoked magnet. On the other hand, because the theorem of geometric invariance does not

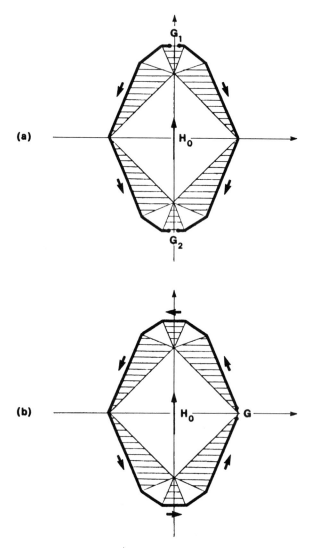

Fig. 5.3.1. Flux of $\vec{B}$ in the yoke of a two-dimensional magnet.

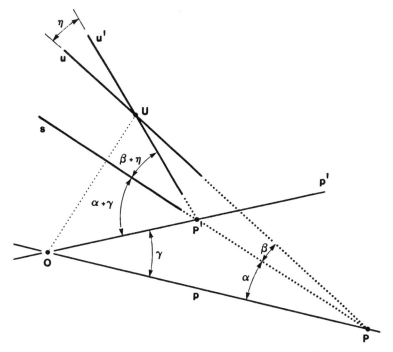

Fig. 5.3.2. Effect of the change of orientation of the $\Phi = 0$ reference line on the geometry of a yoked magnet.

apply, the geometry of the yoked magnet depends upon the selection of the reference $\Phi = 0$ equipotential surface. Thus the geometry of a two-dimensional yoked structure is determined by three parameters: the value of K and the position and orientation of the equipotential line $\Phi = 0$ within the cross-section of the cavity.

As a consequence, while the design of the yokeless magnet enjoys the advantage that the magnetic structure can be computed independently for each face of the prismatic cavity, the design of the yoked structure must be developed for the entire structure on the basis of the selected $\Phi = 0$ line. The effect of the orientation of the $\Phi = 0$ reference line on the geometry of the magnetic structure can be analyzed on the basis of the schematic of Fig. 5.3.2. Assume that line p in Fig. 5.3.2 is the selected $\Phi = 0$ equipotential line within the cavity, oriented at an angle α with respect to side s of the polygonal contour of the cavity.

Assume that the $\Phi = 0$ equipotential line is rotated in the counterclockwise direction by an angle γ about point O to the new position p' shown in Fig. 5.3.2. The line that contains side s of the cavity intersects the $\Phi = 0$ equipotential line u' at point P'.

As shown in Section 3.5, if point O were the characteristic point F of a yokeless magnet, the region of magnetized material that interfaces with the cavity through side s would be a triangle whose vertex U opposite to side s would be located on the line perpendicular to s drawn from point O. Thus in Fig. 5.3.2 the distance ξ_1 of point U from s is related to the distance ξ_0 of point O from s by Eq. 3.5.3.

166 YOKED MAGNETS

The field within the region of magnetized material that interfaces with the cavity is the same regardless of whether this region is part of a yokeless or a yoked structure. Thus point U is always a point of zero potential, independent of the geometry of the yoked structure.

By virtue of the geometric invariance theorem of yokeless structures, the position of point U is independent of the orientation of the $\Phi = 0$ equipotential lines that contain point O. In general, the rotation of the $\Phi = 0$ equipotential line about an arbitrary point O of the cavity results in a different field configuration. However, each point O in a polygonal cavity of n sides results in n conjugate points U_h whose position is invariant to a rotation of the $\Phi = 0$ equipotential line about O.

In Fig. 5.3.2 the rotation of line p about O by an angle γ results in the rotation of line u about U by an angle η. With the notation used in Fig. 5.3.2, angles η, γ satisfy the equation

$$\tan(\beta + \eta) = \frac{K}{1 - K} \tan(\alpha + \gamma). \qquad (5.3.1)$$

One observes in Fig. 5.3.2 that if the $\Phi = 0$ equipotential line p within the cavity rotates in the clockwise direction, the $\Phi = 0$ line within the magnetized medium rotates in the counterclockwise direction.

In the particular case $K = 0.5$, angles α, β are equal to each other and a rotation of line p by an angle γ about point O generates a rotation of the line u by the same angle γ about point U in the opposite direction.

It is of importance to recognize that point U is not necessarily a point of the external boundary of the magnetic structure of a yoked magnet. Consider, for instance, the trapezoidal cavity cross-section shown in Fig. 5.3.3 with the two sides $(S_1 S_2)$ and $(S_3 S_4)$ oriented at an angle $\pi/3$ relative to each other, and assume that $\vec{H}_0$ is perpendicular to the two sides $(S_1 S_4)$ and $(S_2 S_3)$. Select the line p indicated in Fig. 5.3.3. The four points U_1, U_2, U_3, U_4 correspond to the value $K = 0.5$ and the position of point O at the center of the segment of line p within the cavity. Points U_1 and U_3 belong to the external boundaries of the trapezoidal components of magnetized material. By virtue of Eq. 5.1.1, these two external boundaries are parallel to line p.

By definition points U_2 and U_4 belong to the $\Phi = 0$ equipotential lines that emerge from the points where line p intersects the interfaces between the cavity and the triangular components of magnetized material. The points where the two $\Phi = 0$ lines intersect the external boundary of the magnetic structure are the result of the selection of the points D_h on the circle of the vector diagram of Fig. 5.3.4 that determine the geometry of the air wedges between magnetized components. Assume, for instance, that point D_1 in Fig. 5.3.4 is selected so that one side of the nonmagnetic wedge between the components of remanences $\vec{J}_1$ and $\vec{J}_2$ is parallel to interface $(S_1 S_4)$ as shown in Fig. 5.3.3. The geometry of the air wedge is fully determined, and in particular interface $(V_1 W_1)$ between the wedge and the $\mu = \infty$ yoke is perpendicular to vector $\vec{H}_i$ in the vector diagram of Fig. 5.3.4. In a similar manner one determines the geometry of the air wedge between the components of remanences $\vec{J}_1$ and $\vec{J}_4$. As shown in Fig. 5.3.3, the selection of point D_1 in the diagram of Fig. 5.3.4 leaves points U_2, U_4 outside the external boundaries $(V_1 W_1)$ and $(W_3 V_4)$.

POSITION OF Φ = 0 LINE WITHIN THE MAGNET CAVITY **167**

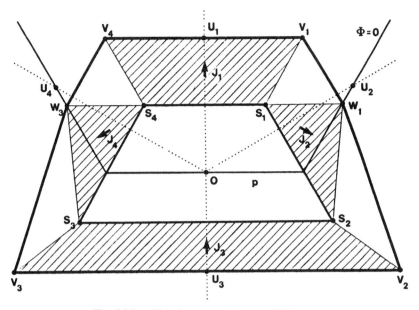

Fig. 5.3.3. Yoked magnet with trapezoidal cavity.

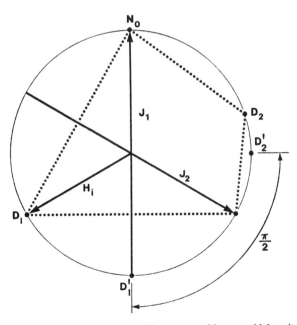

Fig. 5.3.4. Computation of the geometry of the magnet with trapezoidal cavity for $K = 0.5$.

168 YOKED MAGNETS

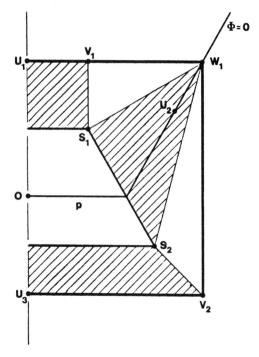

Fig. 5.3.5. Magnetic structure with rectangular external boundary.

If point D_2 in the vector diagram of Fig. 5.3.4 is selected so that $(W_1 V_2)$ is the external boundary of the nonmagnetic wedge with vertex at S_2, point U_2 is outside the interface between yoke and magnetic structure, as shown by Fig. 5.3.3.

Assume that point D_1 is moved to D_1' and point D_2 is moved to D_2' at an angle $\pi/2$ relative to D_1' as indicated in Fig. 5.3.4. Figure 5.3.5 shows that the resulting magnetic structure has a rectangular external boundary, and point U_2 is found inside the magnetic structure.

It is worthwhile pointing out that the $\Phi = 0$ equipotential line inside the magnet divides the structure in two parts that can be designed independently from each other. The computation of the geometries of the two parts corresponds to an arbitrary selection of D_1 and D_2, independent from each other. In this case the external boundaries of the two parts of the magnetic structure intersect the equipotential line $\Phi = 0$ at different points, and the boundary of the yoked magnet is closed by segments of the $\Phi = 0$ equipotential lines that pass through points U_2 and U_4.

To illustrate the dependence of the geometry of a yoked magnet on the position of the $\Phi = 0$ equipotential line, consider again the example of the equilateral triangular cross-section of the magnet cavity analyzed in Section 4.3.

The triangular cavity of Fig. 5.3.6 is the limit of the cavity of Fig. 5.3.3 for zero length of side $(S_1 S_4)$. Thus, for $K = 0.5$ and $\vec{H}_0$ oriented along the axis y, the geometry of the structure of Fig. 5.3.6 is obtained from the vector diagram of Fig. 5.3.4. The selection of the same points D_1 and D_2 in Fig. 5.3.6 results in identical

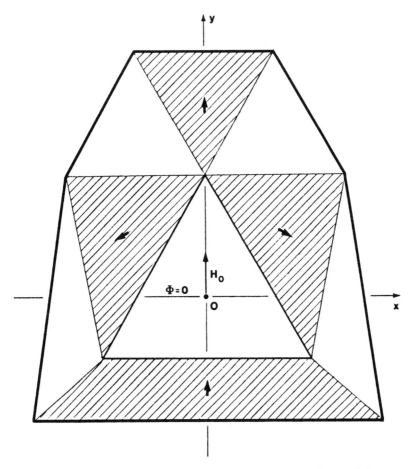

Fig. 5.3.6. Yoked magnet with equilateral triangular cavity for $K = 0.5$.

geometries of the nonmagnetic wedges in both structures of Figs. 5.3.3 and 5.3.6 for the same position of the $\Phi = 0$ equipotential line within the cavity. The line p selected in Fig. 5.3.6 contains the center O of the triangle. One observes that, as a result of the change of the geometry of the cavity, points U_2 and U_4 are found inside the magnetic structure of Fig. 5.3.6. The configuration of equipotential lines is shown in Fig. 5.3.7.

Assume now that the $\Phi = 0$ equipotential line p is translated to a different position inside the cavity, and compute the new geometry of the magnetic structure without changing points D_1 and D_2 in the vector diagram. The orientation of the remanence and the interfaces between magnetized and nonmagnetic materials are not affected by the translation of line p.

Figure 5.3.8 shows the magnetic structures that result from the two extreme positions of line p inside the triangular cavity. If the $\Phi = 0$ line is selected to coincide with one of the sides of the triangle, one obtains the perfectly symmetric hexagonal structure of Fig. 5.3.8(a) whose center coincides with the vertex of the triangle opposite to the $\Phi = 0$ side. In this limit, two of the four air wedges of the structure of Fig.

170 YOKED MAGNETS

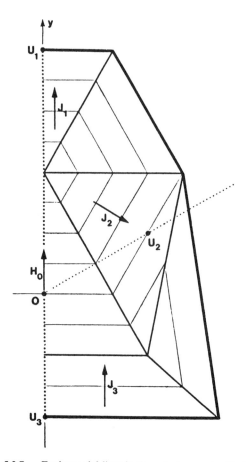

Fig. 5.3.7. Equipotential lines in the yoked magnet with triangular cavity.

5.3.6 are eliminated, and the six regions of the magnet that include the cavity are identical equilateral triangles shown in Fig. 5.3.8(a).

The equipotential lines are regular hexagons concentric with the external $\Phi = 0$ hexagonal boundary. One observes that points U_1, U_2, U_4 belong to the external hexagonal boundary and point U_3 coincides with the center O of line p.

If the $\Phi = 0$ line is selected to pass through a vertex of the triangular cavity, the external boundary of the magnetic structure cannot be closed at a finite distance, and the magnet reduces to the one-dimensional structure of Fig. 5.3.8(b). The interface between the magnetized material and the nonmagnetized medium is the line that contains the side of the triangle opposite to the vertex where $\Phi = 0$. Because $K = 0.5$, both regions have the same thickness in the direction of the axis y. The dotted line in Fig. 5.3.8(b) is the boundary of the triangular cavity. The elimination of the physical boundary of the cavity is the consequence of the elimination of the region of the remanence $\vec{J}_2$ caused by the $\Phi = 0$ line passing through the vertex of the triangle. In this limit, the position of points U_1, U_2, and U_4 coincides with the position of the vertex where $\Phi = 0$.

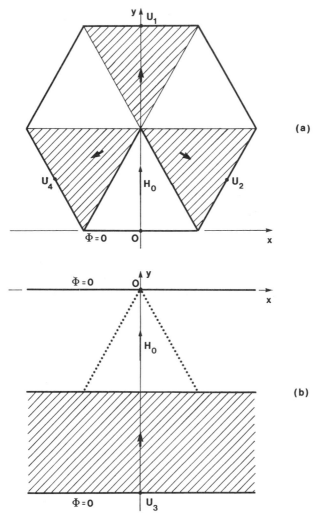

Fig. 5.3.8. Effect of position of the $\Phi = 0$ line on the magnet geometry: (a) point O selected on a side of the triangle; (b) point O coincident with a vertex of the triangle.

5.4 CLOSED CURVILINEAR BOUNDARY OF THE MAGNET CAVITY

Assume that the polygonal contour of n sides of the magnet cavity is formed by n tangents to a closed line s. In the limit $n \to \infty$, the polygonal contour approaches line s. Assume a curve s defined by the equation

$$y = y(x). \tag{5.4.1}$$

Figure 5.4.1 shows the first quadrant of line s in the particular case of symmetry with respect to the x and y axes. As indicated in this figure, $\vec{n}_1$ is the unit vector perpendicular to s at a point S and oriented outwards with respect to the cavity of the magnet. The angle α between $\vec{n}_1$ and the y axis is given by

$$\tan \alpha = \left[\frac{dy}{dx}\right]_S. \tag{5.4.2}$$

Assume that the intensity $\vec{H}_0$ within the cavity is oriented along the axis y, and assume that the axis x coincides with the equipotential line $\Phi = 0$. To be consistent with the formulation of the vector diagram of Fig. 3.2.14, the distribution of the remanence $\vec{J}$ within the magnetized material must satisfy the condition that the line of force of $\vec{J}$ is oriented at an angle 2α with respect to the y axis, and the magnitude J_0 of $\vec{J}$ is independent of position [1].

The equation of the external boundary u of the magnetized material is obtained by computing the position of the point U where line u intersects the line of force of vector $\vec{J}$ that emerges from S. In Fig. 5.4.1, $\vec{n}_2$ is the unit vector perpendicular to boundary u and oriented inwards with respect to the magnetized material. If u is the boundary of the magnetic structure of a yoked magnet, by definition u is an equipotential line $\Phi = 0$, and the intensity $\vec{H}$ at point U is oriented in the direction parallel to $\vec{n}_2$.

In the schematic of Fig. 5.4.2, t_1 and t_2 are the tangents to lines s, u at points S and U, respectively. The distance r of point S from t_2 is

$$r = \frac{H_0}{H} y_1, \tag{5.4.3}$$

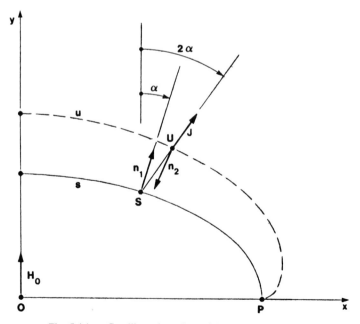

Fig. 5.4.1. Curvilinear boundary of the cavity cross-section.

CLOSED CURVILINEAR BOUNDARY OF THE MAGNET CAVITY 173

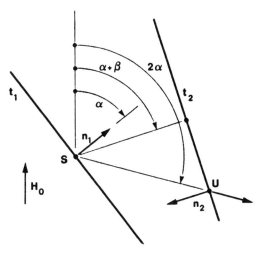

Fig. 5.4.2. Derivation of the differential equation of the external boundary.

where y_1 is the ordinate of point S. By virtue of Eq. 5.1.2,

$$r = \frac{y_1}{\sqrt{1 - \frac{1}{K^2}(2K - 1)\cos^2\alpha}}, \tag{5.4.4}$$

and the distance $r_{1,2}$ between points S and U is

$$r_{1,2} = \frac{r}{\cos(\alpha - \beta)} = \frac{y_1}{1 - \frac{2K-1}{K}\cos^2\alpha}, \tag{5.4.5}$$

where β is the angle between t_1 and t_2. For

$$K = \frac{1}{2}, \tag{5.4.6}$$

Eqs. 5.4.4 and 5.4.5 yield

$$r = r_{1,2} = y_1, \qquad \beta = \alpha \tag{5.4.7}$$

independent of the orientation of t_1. The limits 5.4.7 reflect the fact that at $K = 1/2$, the intensity within the magnetic material is oriented in the direction of the remanence and its magnitude is equal to $J_0/2$ everywhere.

The coordinates x_2 and y_2 of point U of the external boundary of the magnetic

174 YOKED MAGNETS

structure are related to the coordinates x_1 and y_1 of the internal boundary s by the equations

$$x_2 = x_1 + y_1 \frac{\sin 2\alpha}{1 - \frac{2K-1}{K}\cos^2\alpha} \qquad (5.4.8)$$

$$y_2 = \frac{1}{K}y_1 \frac{\cos^2\alpha}{1 - \frac{2K-1}{K}\cos^2\alpha} . \qquad (5.4.9)$$

Let x_0, 0 be the coordinates of point P_0 in Fig. 5.4.1 where line s intersects the x axis and assume that $\alpha = \pi/2$ at P_0. Equations 5.4.8 and 5.4.9 show that the external boundary intersects the x axis at

$$x_2 = x_0 . \qquad (5.4.10)$$

Thus point P_0 is common to both internal and external boundaries of the magnetic structure, as indicated in Fig. 5.4.1. As a consequence, the lines of flux of the magnetic induction cannot close upon themselves within the region of the magnetized material.

In the limit $\alpha = 0$, Eqs. 5.4.8 and 5.4.9 yield

$$x_2 = x_1 , \qquad y_2 = \frac{y_1}{1-K} \qquad (5.4.11)$$

and because of Eq. 5.1.1,

$$\beta = 0 . \qquad (5.4.12)$$

Consider the example of a circular cross-section of radius r_0 and center at O. In polar coordinates r, ψ one has

$$x_1 = r_0 \sin\psi , \qquad y_1 = r_0 \cos\psi , \qquad (5.4.13)$$

where ψ is the angular coordinate relative to the y axis. Angle α is equal to ψ. Thus the equations of the external boundary u are

$$x_2 = r_0 \sin\alpha \frac{2K + 1 + \cos 2\alpha}{1 - (2K - 1)\cos 2\alpha}$$

$$y_2 = r_0 \cos\alpha \frac{1 + \cos 2\alpha}{1 - (2K - 1)\cos 2\alpha} . \qquad (5.4.14)$$

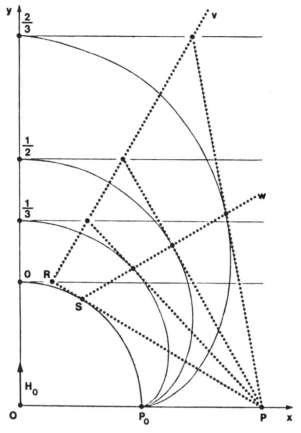

Fig. 5.4.3. External boundaries of magnets with a cylindrical cavity of circular cross-section for several values of K.

The external boundary given by Eq. 5.4.14 is shown in Fig. 5.4.3 for several values of parameter K. It is of interest to compare the geometries of Fig. 5.4.3 with the geometries of the yoked structures designed around the polygonal cavity of Fig. 5.1.7. Point R in Fig. 5.4.3 is the point of intersection of the line

$$y = r_0 \tag{5.4.15}$$

with the line tangent to the circular boundary of cavity, drawn from a point P of the axis x. Line v is the line perpendicular to (RP) drawn from point R. As in the case of the magnet with the prismatic cavity of Fig. 5.1.7, line v is the locus of the points where the lines

$$y = \frac{r_0}{1-K} \tag{5.4.16}$$

176 YOKED MAGNETS

intersect the lines drawn from P tangent to the external boundaries as shown in Fig. 5.4.3. Again the value of y given by Eq. 5.4.16 diverges in the limit $K = 1$. Also all points where the lines drawn from P are tangent to the external boundaries belong to the same line w drawn from point S parallel to $\vec{J}$, as shown in Fig. 5.4.3.

Figure 5.4.4 shows the cross-section of the yoked magnet with circular cavity computed for $K = 1/2$, superimposed to the coaxial geometry of the yokeless magnet with the distribution of remanence given by Eq. 2.4.1. By virtue of Eq. 2.4.5, the radius r_e of the external boundary of the yokeless magnet is

$$\frac{r_e}{r_0} = (e^K)_{K=1/2} \approx 1.65 . \tag{5.4.17}$$

The external boundary of the yoked magnet in Fig. 5.4.4 is the epicycloid

$$\begin{aligned} x &= r_0 \sin \psi \, (2 + \cos 2\psi) \\ y &= r_0 \cos \psi \, (1 + \cos 2\psi) . \end{aligned} \tag{5.4.18}$$

As K approaches unity, the value of y_2 in Eqs. 5.4.14 diverges at $\alpha = 0$, and the area of the magnetized material of the yoked magnet diverges. By contrast, at $K = 1$, the area A_m of the yokeless magnet discussed in Section 2.4 has the finite value

$$(A_m)_{K=1} = \pi r_0^2 \, (e^2 - 1) . \tag{5.4.19}$$

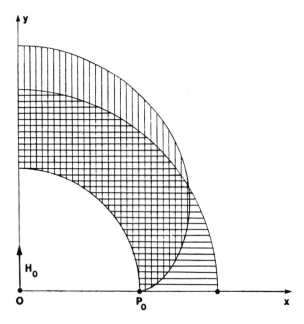

Fig. 5.4.4. Comparison of geometries of yoked and yokeless magnets for $K = 0.5$.

5.5 GENERATION OF A UNIDIRECTIONAL FLUX OF $\vec{B}$

Consider again a prismatic cavity with a rectangular cross-section of dimensions $2x_0$, $2y_0$ and assume that intensity $\vec{H}_0$ is oriented parallel to a side of the rectangle as indicated in Fig. 5.5.1, which shows one quadrant of the cavity cross-section.

Section 5.2 and the vector diagram of Fig. 5.2.2 in particular have shown how to generate a yoked structure around a rectangular cavity. The vector diagram of Fig. 5.2.2 provides the remanence $\vec{J}_1$ in the rectangular region of magnetized material of Fig. 5.5.1, limited by interface s_1 and the $\Phi = 0$ equipotential line u_1. As shown by Fig. 5.2.3, the flux of $\vec{B}$ within the region ($O\ V_0 U_1 F$) is uniform and oriented in the positive direction of the axis y.

Assume that u_1 is part of a surface of $\mu = \infty$ material. In the yoked structure of Fig. 5.2.1, as shown by Fig. 5.2.3, side s_2 is the interface between the cavity and a triangular region of magnetized material where both $\vec{J}_2$ and $\vec{B}_2$ are oriented in the negative direction of the axis y and the magnitude of $\vec{B}_2$ is different from the magnitude of the induction within the region of remanence $\vec{J}_1$. In the triangular regions of nonmagnetic material of the yoked structure, the magnetic induction is parallel to the axis x. Thus the distribution of the flux of $\vec{B}$ within the yoked magnet of Fig. 5.2.3 is not uniform.

Assume now that the method defined in Chapter 4 for the design of yokeless magnets is used to determine the magnetic structure in the region $x > x_0$ in Fig. 5.5.1.

The design of a yokeless magnet requires the selection of the characteristic point F. Assume that F is selected to coincide with point $(x_0, 0)$ of interface s_2, as shown is Fig. 5.5.1. Following the nomenclature of Figs 4.1.2 and 4.1.4, by virtue of 4.1.7,

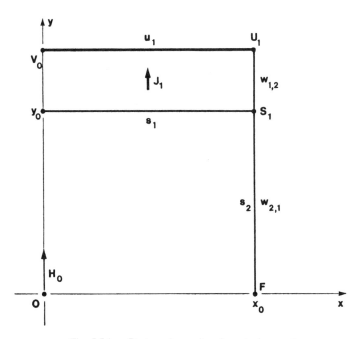

Fig. 5.5.1. Rectangular cavity of a yoked magnet.

178 YOKED MAGNETS

the vertex U_2 of the triangle of remanence $\vec{J}_{2,1}$ coincides with the characteristic point F. The region of remanence $\vec{J}_2$ is eliminated and interface $w_{2,1}$ coincides with interface s_2. Vertex U_1 is located on the line $x = x_0$, and its distance from interface s_1 is

$$w_{1,2} = \frac{K}{1-K} y_0 ; \qquad (5.5.1)$$

i.e., U_1 is located on the $\Phi = 0$ equipotential line u_1 of the yoked region of remanence $\vec{J}_1$. The calculation of the remanences $\vec{J}_{1,2}$ and $\vec{J}_{2,1}$ of the regions that interface with any external nonmagnetic medium and the magnetic structure of Fig. 5.5.1 is provided by the vector diagram of Fig. 5.5.2. Vectors $\vec{J}_{1,2}$ and $\vec{J}_{2,1}$ are inscribed in circles $c_{1,2}, c_{2,1}$ of diameter J_0 that pass through points A, N and A, N_0, respectively. Because of the selection of point F, A is the tip of both vectors $\vec{J}_{1,2}$ and $\vec{J}_{2,1}$, and their origin is at the points of intersection of $c_{1,2}$ and $c_{2,1}$ with the lines perpendicular to $\vec{H}_0$, drawn from points N and N_0, respectively, as shown in Fig. 5.5.2. Thus in the regions of remanences $\vec{J}_{1,2}$ and $\vec{J}_{2,1}$ one has

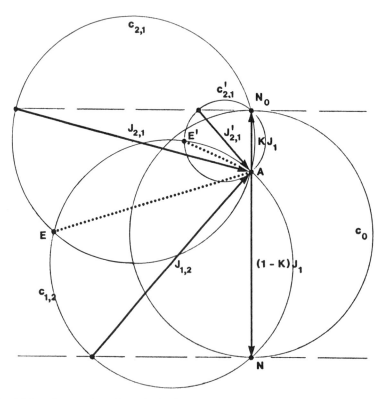

Fig. 5.5.2. Computation of the distribution of remanences designed to generate a unidirectional flux of $\vec{B}$.

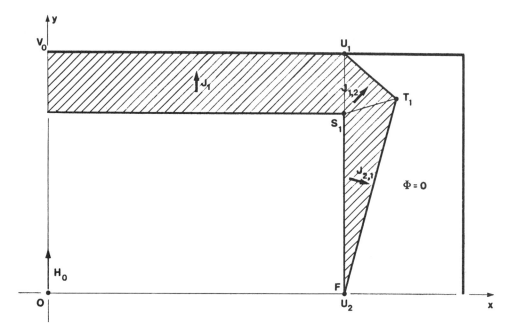

Fig. 5.5.3. Magnetic structure designed to generate a unidirectional flux of $\vec{B}$.

$$\vec{B}_{1,2} = \vec{B}_{2,1} = 0 , \qquad (5.5.2)$$

and intensities $\vec{H}_{1,2}$ and $\vec{H}_{2,1}$ are equal and opposite to $\vec{J}_{1,2}, \vec{J}_{2,1}$, respectively.

The structure resulting from the selection of F at point $(x_0, 0)$ is shown in Fig. 5.5.3. Interfaces $(U_1 T_1)$ and $(T_1 U_2)$ are perpendicular to $\vec{J}_{1,2}$ and $\vec{J}_{2,1}$, respectively, and interface $(S_1 T_1)$ is parallel to the dotted line in Fig. 5.5.2 that contains point A and the point E where circles $c_{1,2}$, $c_{2,1}$ intersect each other.

The region outside boundary $(U_1 T_1 U_2)$ in Fig. 5.5.3 is a $\Phi = 0$ equipotential region. Thus by virtue of Eq. 5.5.2, the two-dimensional structure of Fig. 5.5.3 generates a uniform unidirectional flux of B. However, a closed boundary of the external yoke is necessary to close the flux of the magnetic induction which crosses line u_1 in Fig. 5.5.1 In the ideal limit of an infinite permeability yoke, any arbitrary closed boundary can be chosen outside boundary $(U_1 T_1 U_2)$, as, for instance, the rectangular geometry shown in Fig. 5.5.3. Obviously the same boundary $(V_0 U_1 T_1 U_2)$ of the magnetic structure could be the interface between structure and yoke.

The figure of merit of the two-dimensional structure of Fig. 5.5.3 is

$$M = \frac{K(1-K)}{1 + \frac{1}{2} \frac{y_0}{x_0} \frac{1}{\sqrt{(1-K)^3(1+K)} + \sqrt{K^3(2-K)}}} . \qquad (5.5.3)$$

The geometry of the regions of remanences $\vec{J}_{1,2}$ and $\vec{J}_{2,1}$ is independent of the dimension $2x_0$ of the rectangular cavity. Thus in the limit $x_0 \gg y_0$, the area of these two regions is negligible compared to the area of the rectangular region of remanence $\vec{J}_1$, and the value of M given by Eq. 5.5.3 approaches the limit of the figure of merit of the one-dimensional yoked magnet of Section 2.1.

In the limit $x_0 \to \infty$, Eq. 5.5.3 yields

$$\lim_{x_0 \to \infty} M = K(1-K) \tag{5.5.4}$$

whose maximum value is

$$M_{max} = \frac{1}{4} \tag{5.5.5}$$

at

$$K = \frac{1}{2}. \tag{5.5.6}$$

Because the two triangular regions of Fig. 5.5.3 with remanences $\vec{J}'_{1,2}, \vec{J}'_{2,1}$ do not contribute to the energy of the field inside the cavity, it may be of practical interest to modify the structure of Fig. 5.5.3 by using materials with lower remanences to design the two triangular regions. This is possible as long as the two regions have remanences $\vec{J}'_{1,2}, \vec{J}'_{1,1}$:

$$\vec{J}'_{1,2} \geq (1-K)\vec{J}_1, \qquad \vec{J}'_{2,1} \geq K\vec{J}_1. \tag{5.5.7}$$

For instance, the vector diagram of Fig. 5.5.2 provides the change of the geometry of the magnetic structure resulting from a selection of remanences:

$$J'_{1,2} = J_1, \qquad J'_{2,1} < J_1. \tag{5.5.8}$$

The geometry and the orientation of the remanences in the new $\vec{B} = 0$ region of magnetized material are provided by the vector diagrams inscribed in circles $c_{1,2}, c'_{2,1}$ of Fig. 5.5.2, where $c'_{2,1}$ is the circle of diameter $J'_{2,1}$. The resulting magnetic structure is shown in Fig. 5.5.4. The interface $S_1 T'_1$ is parallel to the dotted line that joins point A and the point E' of intersection of circles $c_{1,2}$ and $c'_{2,1}$ in Fig. 5.5.2.

As shown by Fig. 5.5.4, the selection of a remanence $J'_{2,1} < J_1$ results in an area of triangle $(U_1 U_2 T'_1)$ larger than the area of triangle $(V_1 U_1 T_1)$. Thus, in a general case, the selection of a lower remanence material does not necessarily result in a higher figure of merit and a lower cost of the magnetic material.

Let us compare the structure of Fig. 5.5.3 with the yoked structure of Fig. 5.2.4. If the cavity of the magnet of Fig. 5.2.4 becomes a rectangular cavity of dimensions

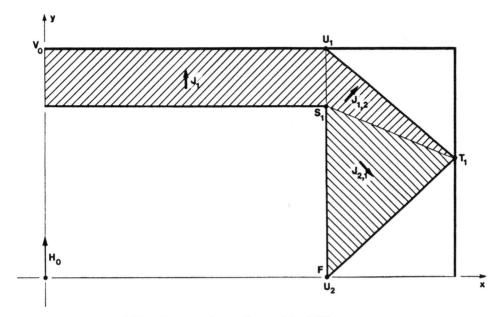

Fig. 5.5.4. Structure of magnetic materials of different remanences.

$2x_0$, $2y_0$, the minimum area A_1 of the magnetic material in the region $x > x_0$ of Fig. 5.2.4 becomes

$$A_1 = \frac{K}{1-K} y_0^2 \qquad (5.5.9)$$

independent of x_0. The area of the cross-section of magnetic material in the two triangular components of remanences $\vec{J}_{1,2}$ and $\vec{J}_{2,1}$ of Fig. 5.5.3 is

$$A_2 = \frac{1}{2} \frac{K}{1-K} \frac{y_0^2}{\sqrt{(1-K)^3(1+K)} + \sqrt{K^3(2-K)}} . \qquad (5.5.10)$$

Thus

$$\frac{A_1}{A_2} = 2\left[\sqrt{(1-K)^3(1+K)} + \sqrt{K^3(2-K)}\right]. \qquad (5.5.11)$$

The area ratio 5.5.11 has a minimum value at $K = 0.5$ which results in the two terms within the bracket of Eq. 5.5.11 being equal to each other, and one has

$$\left[\frac{A_1}{A_2}\right]_{min} = 1.732 . \qquad (5.5.12)$$

182 YOKED MAGNETS

Thus A_1/A_2 is always larger than unity. As a consequence, the structure of Fig. 5.5.3 has a higher figure of merit than the yoked structure of Fig. 5.2.4, regardless of the value of K and the ratio y_0/x_0. The value A_1 is substantially larger than A_2, because in the structure of Fig. 5.2.4 a large area of magnetized material is used to circulate the flux of $\vec{B}$ outside the magnet cavity, as shown in the schematic of Fig. 5.2.3 [2].

5.6 THREE-DIMENSIONAL YOKED MAGNETS

The basic difference between the three-dimensional yokeless magnet developed in Section 4.5 and a two-dimensional yokeless magnet is the transition structure of Fig. 4.5.5 required to close and confine the three-dimensional flux of the magnetic induction. This problem does not arise in a fully yoked magnet where the perfect yoke automatically ensures the closing of the flux regardless of the magnet geometry.

In a way similar to the two-dimensional yoked structures developed in Section 1 of this chapter, the design of a yoked magnet totally enclosing a polyhedral cavity can be approached as a structure of uniformly magnetized polyhedrons as long as transition regions between polyhedrons can be developed to satisfy the boundary conditions. As in the case of two-dimensional magnets, one can expect the transition between magnetized polyhedrons to involve regions of nonmagnetic materials.

To illustrate the difference between yoked and yokeless three-dimensional magnets, let us develop a yoked structure around the same cubic cavity of Section 4.6 for the same orientation of $\vec{H}_0$.

Consider first a two-dimensional yoked magnet whose prismatic cavity has a rectangular cross-section in the x, y plane. Assume $\vec{H}_0$ oriented in the direction along the axis y parallel to a side of the rectangle. The geometry and the distribution of magnetization are given by the vector diagram of Fig. 5.6.1 for a given value of K. The

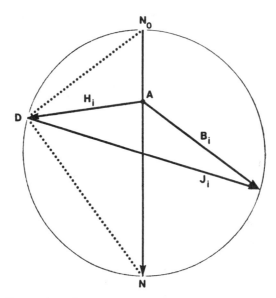

Fig. 5.6.1. Computation of geometry and distribution of magnetization in the three-dimensional magnet with a cubic cavity.

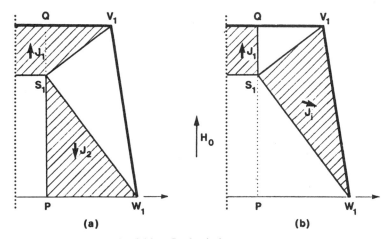

Fig. 5.6.2. Dual yoked structures.

schematic of Fig. 5.6.2(a) shows the detail of the first quadrant of the cross-section of the magnetic structure resulting from an arbitrary selection of point D on the circle of the vector diagram and the assumption that the line $\Phi = 0$ inside the cavity contains the center of the rectangle.

The dotted lines $(D\,N_0)$ and $(D\,N)$ in Fig. 5.6.1 determine the interfaces of the nonmagnetic wedge $(S_1 V_1 W_1)$ that separates the two components of magnetized material of remanences $\vec{J}_1$ and $\vec{J}_2$ in Fig. 5.6.2(a). The external boundary $(V_1 W_1)$ of the nonmagnetic wedge is perpendicular to intensity $\vec{H}_i$ in Fig. 5.6.1. The configuration of the equipotential lines in the magnet cross-section does not change if its magnetic structure is replaced by its dual structure shown in Fig. 5.6.2(b). The nonmagnetic material of the wedge $(S_1 V_1 W_1)$ in Fig. 5.6.2(a) is replaced by a magnetized material of remanences $\vec{J}_i$ provided by the vector diagram of Fig. 5.6.1. The origin of vector $\vec{J}_i$ coincides with the tip D of intensity $\vec{H}_i$. The material of remanence $\vec{J}_2$ in the triangle $(S_1 P W_1)$ is replaced with a nonmagnetic material. Thus the interface $(P S_1)$ between the rectangular cavity and the region of remanence $\vec{J}_2$ in Fig. 5.6.2(a) disappears in the dual structure of Fig. 5.6.2(b). The rectangular cross-sectional cavity is transformed into a hexagonal cross-sectional cavity.

The distribution of the magnetic induction is not the same in the dual structures of Fig. 5.6.2. In the nonmagnetic medium of triangle $(S_1 V_1 W_1)$ of Fig. 5.6.2(a) the induction is

$$\vec{B}_{i,a} = \mu_0 \vec{H}_i \tag{5.6.1}$$

while in the medium of remanence $\vec{J}_i$ in Fig. 5.6.2(b) the induction is

$$\vec{B}_{i,b} = \mu_0 \vec{H}_i + \vec{J}_i . \tag{5.6.2}$$

Within triangle $(S_1 Q V_1)$ of the dual structures one has

$$\vec{B}_{1,a} = K \vec{J}_1, \qquad \vec{B}_{1,b} = -(1-K)\vec{J}_1, \tag{5.6.3}$$

184 YOKED MAGNETS

and within triangle $(S_1 P W_1)$ one has

$$\vec{B}_{2,a} = -(1-K)\vec{J}_1, \qquad \vec{B}_{2,b} = K\vec{J}_1. \qquad (5.6.4)$$

Because the center point of vector $\vec{J}_i$ is the center of the circle of Fig. 5.6.1, the two components $\vec{B}_{i,a,n}$ and $\vec{B}_{i,b,n}$ of vectors $\vec{B}_{i,a}, \vec{B}_{i,b}$ perpendicular to $\vec{H}_0$ satisfy the condition

$$\vec{B}_{i,a,n} = -\vec{B}_{i,b,n}. \qquad (5.6.5)$$

The exchange of regions of magnetized material with regions of nonmagnetic material in the dual geometries of Fig. 5.6.2 provides the key for the design of the three-dimensional yoked structure.

Following the same approach of Section 4.5, the cubic cavity is divided in four wedges with the axis y as the common edge, and each wedge is considered as a part of the cavity of a two-dimensional yoked structure. Assume that in each wedge the magnetic structure is identical to the one defined by Fig. 5.6.2(a). Thus Fig. 5.6.2(a) is the cross-section of the three-dimensional yoked magnet in both planes

$$z = x = 0. \qquad (5.6.6)$$

On the two planes

$$z = \pm x \qquad (5.6.7)$$

the boundary conditions cannot be satisfied by simply joining together the four wedges of the magnetic structure of Fig. 5.6.2(a). Figure 5.6.3 shows the cross-section of the magnetic structure in the plane $y = 0$ in the proximity of the edge of the cube that contains point P of Fig. 5.6.2(a). Assume that the magnetic structure is cut in two planes with the edge of the cube in common, oriented at the angles $\pm\pi/8$ with respect to the plane

$$z = x, \qquad (5.6.8)$$

as indicated in Fig. 5.6.3. Assume that within the $\pi/4$ wedge the magnetic structure of Fig. 5.6.2(a) is replaced by the structure of Fig. 5.6.2(b), where axis ξ belongs to the plane defined by Eq. 5.6.8.

The boundary conditions of continuity of the tangential components of the intensity on the two interfaces of the $\pi/4$ wedge of Fig. 5.6.3 are automatically satisfied because the intensity is the same in the dual structures of Fig. 5.6.2. As shown by Eqs. 5.6.3 and 5.6.4, vectors $\vec{B}_{1,a}, \vec{B}_{1,b}, \vec{B}_{2,a}, \vec{B}_{2,b}$ are parallel to the axis y. Thus the boundary conditions of continuity of the normal components of the magnetic induction are automatically satisfied on the interfaces of the $\pi/4$ wedge between the regions of the

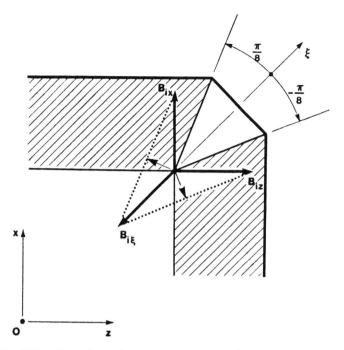

Fig. 5.6.3. Transition region between components of magnetized material.

structures of Figs. 5.6.2(a) and 5.6.2(b) corresponding to triangles $(S_1 Q V_1)$ and $(S_1 P W_1)$. The components $\vec{B}_{i,z}$ and $\vec{B}_{i,x}$ in the regions of nonmagnetic material of the structure of Fig. 5.6.3 that correspond to triangle $(S_1 V_1 W_1)$ of Fig. 5.6.2(a) are shown in Fig. 5.6.3, and their magnitude is

$$B_{i,z} = B_{i,x} = B_{i,a,n} , \qquad (5.6.9)$$

and the component $\vec{B}_{i,\xi}$ in the region of magnetized material of the $\pi/4$ wedge of Fig. 5.6.3 that corresponds to triangle $(S_1 V_1 W_1)$ of Fig. 5.6.2(b) is also shown in Fig. 5.6.3. By virtue of Eq. 5.6.5, its magnitude is

$$B_{i,\xi} = B_{i,b,n} = B_{i,a,n} . \qquad (5.6.10)$$

As shown by Eq. 5.6.5, $\vec{B}_{i,a,n}$ and $\vec{B}_{i,b,n}$ are oriented in opposite directions. Thus, the boundary condition of continuity of the normal components of the magnetic induction is satisfied on the two interfaces of the $\pi/4$ wedge of Fig. 5.6.3.

The resulting three-dimensional structure is shown without the external yoke in the region $y > 0$ in Fig. 5.6.4, and an exploded view is presented in Fig. 5.6.5 which shows one of the components of magnetized material of remanence $\vec{J}_2$ and the components of magnetized material of remanence $\vec{J}_i$ of one of the $\pi/4$ transition wedges. Finally the geometry of the interface between the magnetic structure and the external yoke is shown in Fig. 5.6.6.

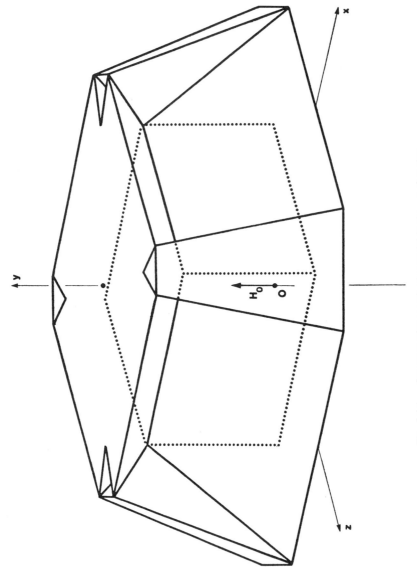

Fig. 5.6.4. Structure of magnetized material in the $y > 0$ region.

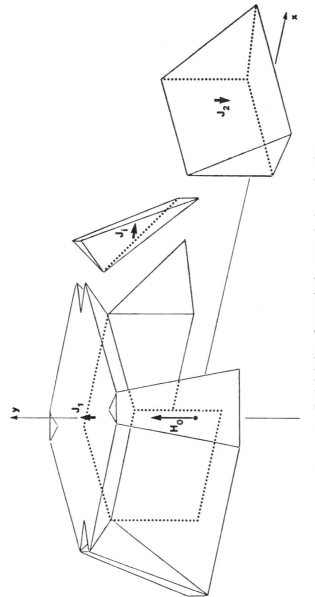

Fig. 5.6.5. Exploded view of magnetized structure in the $y > 0$ region.

188 YOKED MAGNETS

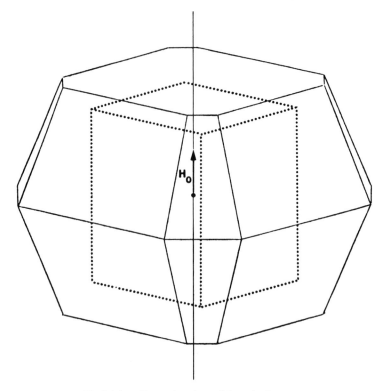

Fig 5.6.6. External surface of the yoked magnet.

REFERENCES

[1] M. G. Abele, Some considerations about permanent magnet design for NMR, Technical Report No. 13, New York University, Feb 1, 1986.
[2] M. G. Abele, A high efficiency yoked permanent magnet, Technical Report No. 23, New York University, Oct 1, 1990.

CHAPTER 6

Hybrid Magnets

INTRODUCTION

The presence of air spaces between the components of magnetized material in the yoked structures of Chapter 5 adversely affects the figure of merit. The magnetic material has to provide the energy of the field within the air spaces in addition to the energy of the field within the magnet cavity. This characteristic property of the magnets analyzed in Chapter 5 is equivalent to that of traditional magnets where part of the energy provided by the magnetic material is wasted in the fringe field outside the region of interest.

On the other hand, in the yokeless structures of Chapter 4, the magnetic material is used both to generate the field inside the cavity and to close the flux of the magnetic induction within the magnetic structure. In essence, the field confinement is achieved at the cost of replacing a high magnetic permeability yoke with additional magnetic material, which results in a lower figure of merit.

This chapter compares the properties of yokeless and yoked structures. In particular, it analyzes their figure of merit defined in the previous chapters as the fraction of the energy stored in the magnetic material which is used to generate the field in the magnet cavity. The following sections show how the figure of merit can be improved in magnets that combine yoked and yokeless structures. These hybrid magnets retain the basic property of confining the field within the region of the magnet, even if the yoke does not entirely enclose the magnetic structure. Furthermore, the introduction of high magnetic permeability components within the magnetic structure is also presented in this chapter as a means of filtering out or reducing the effects of potential sources of field distortion caused by a departure from the ideal conditions of geometry and magnetization.

6.1 COMPARISON OF YOKED AND YOKELESS STRUCTURES

To compare the properties of yoked and yokeless magnets, let us consider magnetic structures designed to generate a uniform field such that both the equipotential line $\Phi = 0$ and the dividing line of flux l_0 pass through two independent pairs of vertices of the polygonal cavity.

An example is the design around the quadrangular geometry of the cavity analyzed in Section 3 of Chapter 4 in the particular case of point F located at the intersection of

190 HYBRID MAGNETS

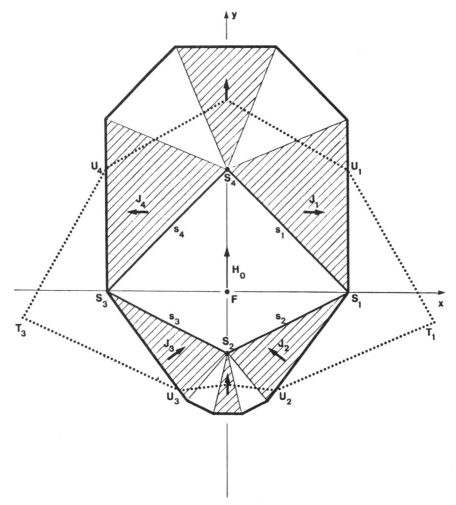

Fig. 6.1.1. Comparison of yoked and yokeless magnet geometries designed for $K = 0.5$ around a quadrangular cavity.

the two diagonals of the quadrangle. The cavity geometry is shown again in Fig. 6.1.1 with the two diagonals oriented along the x and y axes. As shown in Section 4.3, the position of F at the origin of the frame of coordinates does not yield the optimum design of the yokeless magnet around the cavity of Fig. 6.1.1.

The vector diagram of Fig. 6.1.2 provides the geometry and the distribution of magnetization of a yoked magnet designed around the same cavity of Fig. 6.1.1 for $K = 0.5$ with the condition that the $\Phi = 0$ equipotential line coincides with the $y = 0$ axis. Points D_1 and D_2 in Fig. 6.1.2 have been selected to compute the geometries of the air wedges at points S_4 and S_2. The total area of the air wedges is about 63% of the area of the cavity cross-section, and the field intensity within the air wedge is equal to the intensity within the cavity. Thus a large fraction of the energy stored in the magnetic material is spent to generate the field within the air wedges.

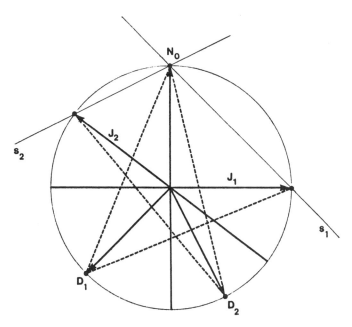

Fig. 6.1.2. Computation of geometry of air wedges in the yoked magnets with a quadrangular cavity.

The dotted line in Fig. 6.1.1 is the external boundary of the yokeless magnet designed around the same cavity for $K = 0.5$. The geometry and the distribution of magnetization of the yokeless structure are shown in Fig. 6.1.3, and their computation is provided by the vector diagram of Fig. 6.1.4. As shown by the vector diagram, because vertices S_1, S_3 of the cavity belong to the $\Phi = 0$ equipotential line, both regions $(S_1 U_1 T_1 U_2)$ and $(S_3 U_3 T_3 U_4)$ of the yokeless structure are also regions of constant potential $\Phi = 0$. The magnetic material in both regions is used only to close the flux of the magnetic induction without contributing to the field inside the cavity.

Obviously, in an ideal design the energy stored in the magnetic material would be used to generate the field within the cavity only. Even if this ideal limit cannot be achieved in practice, a magnet design should be aimed at minimizing the amount of magnetic material required to confine the field outside the region of interest. Thus, in general, for a given geometry of the magnet cavity and a given value of parameter K, neither a yokeless nor a yoked design approach may be the optimum choice. Instead, the optimum magnetic structure may result from a hybrid design, which combines yokeless and yoked structures to take advantage of both approaches [1].

In the example of Fig. 6.1.1, an optimum hybrid design may be a magnetic structure confined to the area common to both yokeless and yoked magnets, where both the air wedges and the magnetic material within the two regions $(S_1 U_1 T_1 U_2)$ and $(S_3 U_3 T_3 U_4)$ are eliminated. In a practical design two options are available to save the magnetic material within the two regions $(S_1 U_1 T_1 U_2)$ and $(S_3 U_3 T_3 U_4)$. The closing of the flux of the magnetic induction can be achieved either with a lower energy product material or with a passive ferromagnetic material of high magnetic permeability.

192 HYBRID MAGNETS

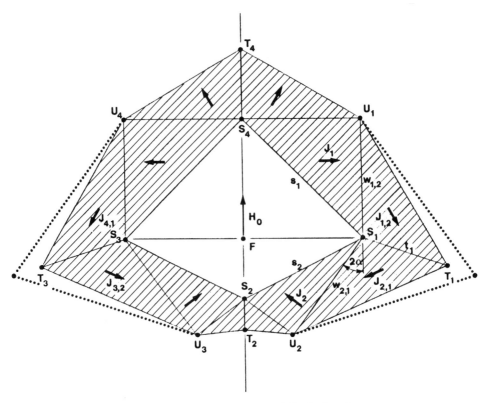

Fig. 6.1.3. Yokeless magnet geometry and distribution of remanences.

The remanences $\vec{J}_{1,2}$, $\vec{J}_{2,1}$ in the regions $(S_1 T_1 U_1)$ and $(S_1 T_1 U_2)$ of the yokeless magnet can be chosen arbitrarily as long as the condition $\vec{H} = 0$, and the continuity of the flux of $\vec{B}$ across sides $(S_1 U_1)$, $(S_1 U_2)$ are satisfied. The magnetic inductions $\vec{B}_{1,2}$ and $\vec{B}_{2,1}$ generated in the two regions are equal to $\vec{J}_{1,2}$ and $\vec{J}_{2,1}$, respectively. Thus the components of $\vec{J}_{1,2}$ and $\vec{J}_{2,1}$ perpendicular to interface $(S_1 U_1)$ and $(S_1 U_2)$ must satisfy the conditions

$$(J_{1,2})_n = \frac{J_1}{2}, \quad (J_{2,1})_n = \frac{J_2}{2}, \quad (6.1.1)$$

where J_1 and J_2 are the magnitudes of vectors $\vec{J}_1$ and $\vec{J}_2$. Both magnitudes J_1 and J_2 are equal to J_0, as shown in the diagram of Fig. 6.1.4. By virtue of Eq. 6.1.1, if t_1 is the chosen orientation of the interface between the media of remanence $\vec{J}_{1,2}$ and $\vec{J}_{2,1}$, as shown in the vector diagram of Fig. 6.1.4, vectors $\vec{J}_{1,2}$ and $\vec{J}_{2,1}$ can be chosen arbitrarily as long as

$$J_{1,2} = J_{2,1} \geq \frac{J_0}{2 \cos \alpha}, \quad (6.1.2)$$

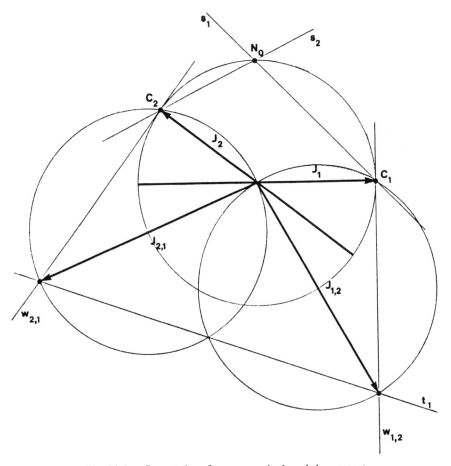

Fig. 6.1.4. Computation of remanences in the yokeless magnet.

where 2α is the angle between $(S_1 U_1)$ and $(S_1 U_2)$ in Fig. 6.1.3. If the magnitudes of $\vec{J}_{1,2}$ and $\vec{J}_{2,1}$ were selected within the range

$$\frac{J_0}{2} < J_{1,2} = J_{2,1} < \frac{J_0}{2 \cos \alpha}, \qquad (6.1.3)$$

the two new components of magnetic material would still be equipotential, but an additional flux of $\vec{B}$ closing at infinity would be generated by $\vec{J}_{1,2}$ and $\vec{J}_{2,1}$.

The dotted line in the schematic of Fig. 6.1.3 shows the geometry of the magnet cross-section corresponding to a magnitude of $\vec{J}_{1,2}$, $\vec{J}_{2,1}$,

$$J_{3,2} = J_{4,1} = J_{1,2} = J_{2,1} < J_0, \qquad (6.1.4)$$

where $\vec{J}_{3,2}$ and $\vec{J}_{4,1}$ are the remanences of the regions symmetric to the regions of remanences $\vec{J}_{1,2}$ and $\vec{J}_{2,1}$. The magnitude of the remanences of the rest of the magnetic

194 HYBRID MAGNETS

structure is equal to J_0. Obviously, the disadvantage of a larger volume resulting from the selection of remanences 6.1.4 may be compensated for by the lower cost of lower energy product magnetic materials.

The second option of replacing regions $(S_1 U_1 T_1 U_2)$ and $(S_3 U_3 T_3 U_4)$ with materials of high magnetic permeability is suggested by the consideration that, in the limit $\mu = \infty$, interfaces $(U_1 S_1 U_2)$ and $(U_3 S_3 U_4)$ would become equipotential lines and the lines of flux of $\vec{B}$ could close upon themselves in a zero thickness layer across both interfaces. In practice, a finite value of μ large compared to μ_0 introduces a perturbation of the potential distribution on the interfaces and, as a consequence, may generate a distortion of the field within the cavity, as will be discussed in Chapter 7. In principle, if μ is finite, the condition of field confinement cannot be satisfied any longer, and a small "leakage" of the field in the surrounding medium results from the replacement of the magnetic material with a passive ferromagnetic material. A minimum thickness of the ferromagnetic material may have to be used to minimize field distortion and field leakage. However, from a practical standpoint, even if the replacement may lead to a larger and heavier magnet, the passive ferromagnetic material may be much less expensive than the high energy product material used for the other components of the magnet.

To illustrate the design procedure for eliminating the air wedges of a yoked magnet, consider again the design of the yoked structure around a two-dimensional cavity of regular hexagonal cross-section, as shown in Fig. 6.1.5. The intensity of the magnetic field $\vec{H}_0$ is assumed to be perpendicular to one side of the hexagon. The first quadrant of the magnet cross-section is shown in Fig. 6.1.6. Remanence $\vec{J}_1$ is parallel to the y axis and remanence $\vec{J}_2$ is oriented at an angle $2\pi/3$ with respect to the y axis.

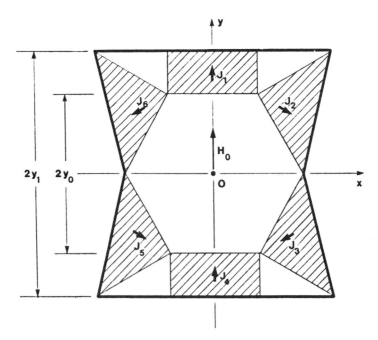

Fig. 6.1.5. Yoked magnet with a regular hexagonal cavity.

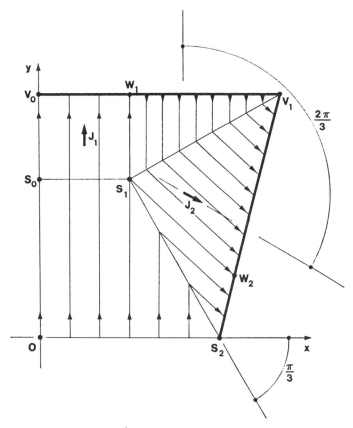

Fig. 6.1.6. Lines of flux of $\vec{B}$ in the yoked magnet with a regular hexagonal cavity.

Figure 6.1.6 shows the lines of flux of the magnetic induction in the first quadrant of the magnet cross-section. The flux of $\vec{B}$ within the cavity reaches the external yoke within the two segments $(V_0 W_1)$ and $(S_2 W_2)$ of the external yoke. The area $(S_1 W_2 V_1 W_1)$ of Fig. 6.1.6 contains the air wedge and it is the region where the flux of $\vec{B}$ circulates between the yoke and the magnetic structure outside the magnet cavity. The energy of the field within this region is provided by the area of the prism of remanence $\vec{J}_2$ contained within triangle $(S_1 W_2 V_1)$.

Assume that the area $(S_1 W_2 V_1)$ of the magnetic material is eliminated, and a new component of magnetic material of remanence $\vec{J}_7$ and cross-sectional area $(S_1 W_2 W_1)$ is inserted in the magnetic structure as shown in Fig. 6.1.7. Remanence $\vec{J}_7$ is assumed to be perpendicular to side $(W_1 W_2)$ and to satisfy the condition

$$\vec{J}_7 = -\mu_0 \vec{H}_7, \tag{6.1.5}$$

where $\vec{H}_7$ is the intensity of the field in the area $(S_1 W_1 W_2)$. Thus the magnetic induction is zero in this area. The magnetization of the new component satisfies the condition of the design of a yokeless magnet expressed by Eq. 3.4.2. Since side

196 HYBRID MAGNETS

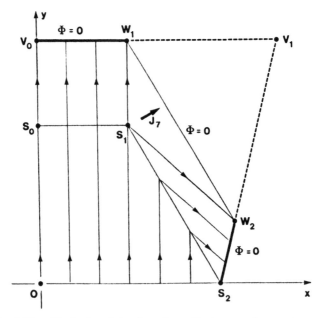

Fig. 6.1.7. Elimination of the air wedges with elements of magnetic material.

$(W_1 W_2)$ is a $\Phi = 0$ equipotential line, side $(W_1 W_2)$ can be the interface between the material of remanence $\vec{J}_7$ and an external nonmagnetic medium.

The value of remanence $\vec{J}_7$ is provided by the vector diagram of Fig. 6.1.8. The diagram shows that, in general, its magnitude is different from J_0. Only in the particular case

$$K = \frac{1}{2} \tag{6.1.6}$$

one has

$$J_7 = J_0, \tag{6.1.7}$$

and the magnetic structure has the geometry shown in Fig. 6.1.9, where the ordinate y_1 of V_0 and the ordinate y_0 of S_0 are

$$y_1 = 2y_0. \tag{6.1.8}$$

It is important to point out that the fluxes of $\vec{B}$ across segments $(V_0 W_1)$, $(S_2 W_2)$ can be closed through the fourth quadrant of the magnet cross-section independently of each other. This arrangement is shown in the schematic of Fig. 6.1.9.

In the limit of infinite magnetic permeability, one can arbitrarily choose the closed geometry of the yoke that contains segment $(V_0 W_1)$. The dotted line in Fig. 6.1.9

COMPARISON OF YOKED AND YOKELESS STRUCTURES   197

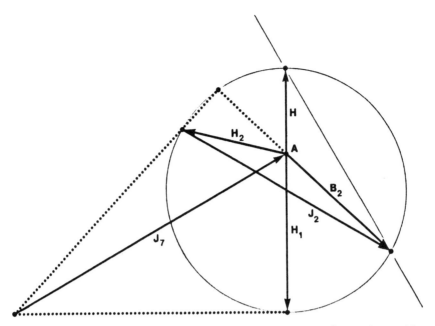

Fig. 6.1.8. Computation of the remanence of the yokeless element of magnetic material.

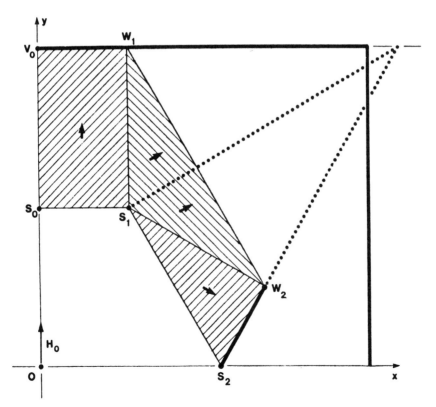

Fig. 6.1.9. Geometry of the magnet for $K = 0.5$.

corresponds to the triangular component of remanence $\vec{J}_2$ in the original yoked structure. The reduction in the area of magnetized material resulting from the insertion of the component of remanence $\vec{J}_7$ is quite apparent.

In general the region $(S_1 W_2 V_1 W_1)$ of the structure of Fig. 6.1.6 can be replaced with a material having the same remanence J_0 as the rest of the magnet. This results in the structure of Fig. 6.1.10 composed of two elements $(S_1 W_3 W_1)$ and $(S_1 W_2 W_3)$ of remanences $\vec{J}_8$ and $\vec{J}_9$ whose magnitude is equal to J_0. The calculation of $\vec{J}_8$ and $\vec{J}_9$ is provided by the vector diagram of Fig. 6.1.11. $\vec{J}_8$ and $\vec{J}_9$ are perpendicular to sides $(W_1 W_3)$ and $(W_3 W_2)$, respectively. The two remanences satisfy the conditions

$$\vec{J}_8 = -\mu_0 \vec{H}_8, \qquad \vec{J}_9 = -\mu_0 \vec{H}_9, \qquad (6.1.9)$$

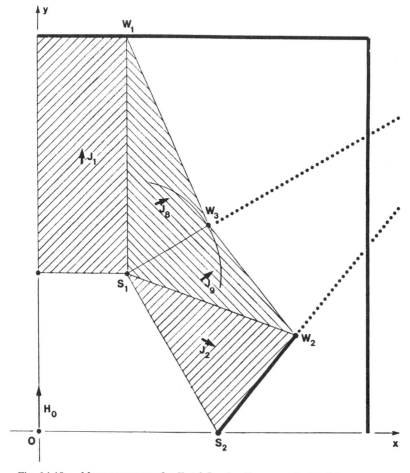

Fig. 6.1.10. Magnet geometry for $K \neq 0.5$ and uniform magnitude of the remanence.

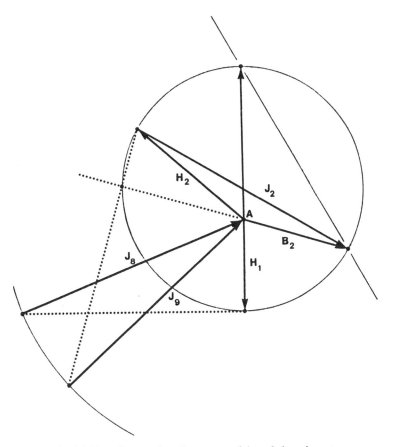

Fig. 6.1.11. Computation of geometry of the yokeless elements.

where $\vec{H}_8$, $\vec{H}_9$ are the intensities of the magnetic field in the corresponding components. Again the dotted line in Fig. 6.1.10 emphasizes the reduction of the area of magnetic material resulting from the elimination of the air wedges of the yoked magnet.

The general configuration of a hybrid structure is shown in Fig. 6.1.12. Each component of magnetized material that interfaces with the cavity is a trapezoid whose parallel sides are parallel to the magnetic induction within the material. The side $(S_{h-1} S_h)$ is the interface between the magnetized trapezoid and the cavity. Side $(W_h W_{h+1})$ is the interface between the trapezoid and the $\mu = \infty$ yoke. The transition from one trapezoid to another is established by a yokeless structure of magnetized material whose remanence satisfies the condition $\vec{B} = 0$, as indicated in Fig. 6.1.12, where the axis x is chosen to coincide with the equipotential line $\Phi = 0$ within the cavity. If this line passes through vertex S_h, vertex W_{h+1} coincides with S_h and the trapezoid reduces to a triangle, as in Fig. 6.1.7.

200 HYBRID MAGNETS

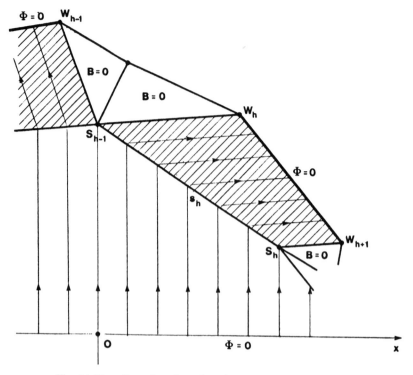

Fig. 6.1.12. General configuration of a hybrid magnetic structure.

6.2 FIGURES OF MERIT OF HYBRID STRUCTURES

Consider a yoked magnet designed around the quadrangular cavity of Fig. 6.1.1. Assume that the magnetic structure is designed with a single wedge of nonmagnetic material between the prisms of remanences $\vec{J}_1$ and $\vec{J}_4$ and another single wedge of nonmagnetic material between the prisms of remanences $\vec{J}_2$ and $\vec{J}_3$. In Fig. 6.1.1, let $2x_0$, $2y_0$ be the lengths of the two diagonals of the cavity and let y_a, y_b be the ordinates of vertices S_4, S_2, respectively. With the assumptions that point F is located at the origin of the frame of reference selected in Fig. 6.1.1 and that $\vec{H}_0$ is oriented in the direction of the axis y, the figure of merit of the yoked structure is

$$M = 2K(1-K) \frac{x_0^2 y_0}{y_a(x_0^2 + y_a^2) + y_b(x_0^2 + y_b^2)}, \tag{6.2.1}$$

where

$$2y_0 = y_a + y_b. \tag{6.2.2}$$

Equation 6.2.1 shows that M is a parabolic function of K, which attains its maximum value at

$$K = 0.5, \qquad (6.2.3)$$

independent of the dimensions of the quadrangle. One observes that the value of M given by Eq. 6.2.1 increases as x_0 increases. M attains its maximum value at $x_0 \to \infty$ when the magnet becomes the one-dimensional structure confined between two $\mu = \infty$ parallel plane surfaces located at $y = y_0$ and $y = -y_b$. In the limit $x_0 \to \infty$, the maximum value of M at $K = 0.5$ is

$$M_{max} = 0.250. \qquad (6.2.4)$$

Value 6.2.4 is the upper bound of the figure of merit that can be achieved in a yoked magnet, regardless of its geometry and field orientation. For the values of K and M given by Eqs. 6.2.3 and 6.2.4 the area ratio A_m/A_c is equal to unity. Equation 6.2.4 shows that a yoked magnet can achieve a figure of merit substantially larger than that of a yokeless magnet whose upper bound is given by Eq. 4.3.6.

At constant values of x_0, y_0 the rate of change of M as a function of y_a is

$$\frac{dM}{dy_a} = -24 K (1-K) x_0^2 y_0^2 \frac{y_a - y_0}{\left[y_a (x_0^2 + y_a^2) + y_b (x_0^2 + y_b^2) \right]^2}. \qquad (6.2.5)$$

Thus the optimum of the figure of merit is achieved at $y_a = y_0$, i.e., when the cavity is symmetric with respect to points F. The cavity then becomes a rhombus, and Eq. 6.2.1 reduces to

$$M = K(1-K) \frac{x_0^2}{x_0^2 + y_0^2}. \qquad (6.2.6)$$

As shown in Chapter 5, at the optimum value $K = 0.5$, the intensity of the magnetic field is equal everywhere to $J_0/2\mu_0$. In each region of the magnetized material, intensity and remanence are oriented in opposite direction with respect to each other. Thus the magnetic material operates at the peak of the energy product curve, as in the case of the optimum design of a traditional magnet.

It is of importance to point out that if the figure of merit were defined as the fraction of the energy used to generate the magnetic field in all of the nonmagnetic regions of a yoked magnet, the value of M given by Eq. 6.2.4 would be achieved at $K = 0.5$ regardless of the geometry of the quadrangular cavity. This situation is conceptually identical to that of the design of a traditional magnet where part of the energy is spent in the fringe field outside the region of interest, as discussed in Section 7 of this chapter.

202 HYBRID MAGNETS

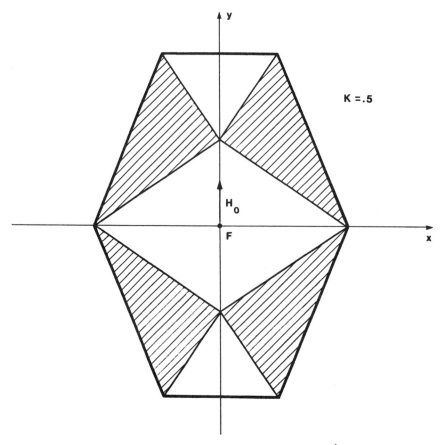

Fig. 6.2.1. Geometry of a yoked magnet with a rhombic cavity for $\vec{H}_0$ oriented along the smaller diagonal of the rhombus.

As shown in Chapter 5, the theorem of geometric invariance does not apply to a yoked magnet where the orientation of $\vec{H}_0$ is a design parameter that affects the figure of merit. Equation 6.2.6 shows that to optimize the value of M, $\vec{H}_0$ must be oriented along the shortest diagonal of the rhombus.

The geometry of the yoked structure around a rhombic cavity is shown in Fig. 6.2.1 in the particular case of a 2:3 ratio of the diagonals along the y and x axis and a value of $K = 0.5$. The configuration of the lines of flux of $\vec{B}$ in the first quadrant of the magnet cross-section is shown in Fig. 6.2.2. Line $(S_4 U_1)$ separates the flux of $\vec{B}$ within the magnet cavity from the flux of $\vec{B}$ within the triangular air wedge between magnetized components.

In Fig. 6.2.1 the ratio A_m/A_c of the cross-sectional areas of magnetic material and magnet cavity is

$$\frac{A_m}{A_c} \approx 1.44 , \tag{6.2.7}$$

which yields a value of the figure of merit

$$M \approx 0.173 . \tag{6.2.8}$$

FIGURES OF MERIT OF HYBRID STRUCTURES 203

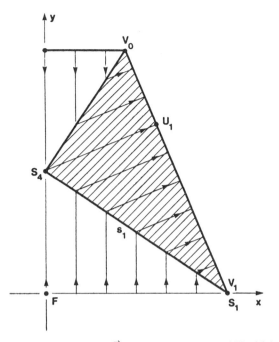

Fig. 6.2.2. Lines of flux of $\vec{B}$ in the yoked structure of Fig 6.2.1.

Consider now a yokeless, single layer magnet designed around the same cavity of Fig. 6.2.1 with point F selected at the center of the rhombus. The geometry of the yokeless magnet is shown in Fig. 6.2.3 again for $K = 0.5$. The computation of the distribution of magnetization in the first quadrant is provided by the vector diagram of

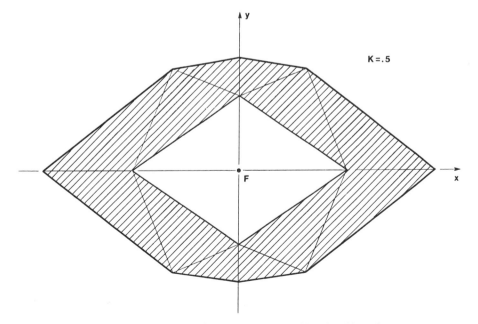

Fig. 6.2.3. Geometry of a yokeless magnet with a rhombic cavity.

204 HYBRID MAGNETS

Fig. 6.2.4. The configuration of the equipotential lines is presented in Fig. 6.2.5. As shown by the vector diagram of Fig. 6.2.4, at $K = 0.5$ one has

$$\vec{J}_1 = 2\vec{B}_1 = -2\mu_0 \vec{H}_1, \quad \vec{J}_{1,1} = -\mu_0 \vec{H}_{1,1}, \quad \vec{J}_{1,2} = \vec{B}_{1,2}. \tag{6.2.9}$$

Thus, orienting $\vec{H}_0$ along the y axis results in triangle $(S_1 U_1 T_1)$ being a $\Phi = 0$ equipotential region and in triangle $(S_4 T_4 U_1)$ being a region of zero flux of $\vec{B}$.

At $K = 0.5$ the value of the figure of merit of the yokeless magnet of Fig. 6.2.3 is

$$M \approx 0.102. \tag{6.2.10}$$

Interface $(S_1 U_1)$ in Fig. 6.2.5 is a $\Phi = 0$ equipotential line common to both the yoked structure of Fig. 6.2.2 and the yokeless structure of Fig. 6.2.5. In the yokeless structure, the area of the four $\Phi = 0$ equipotential regions is equal to approximately 46% of

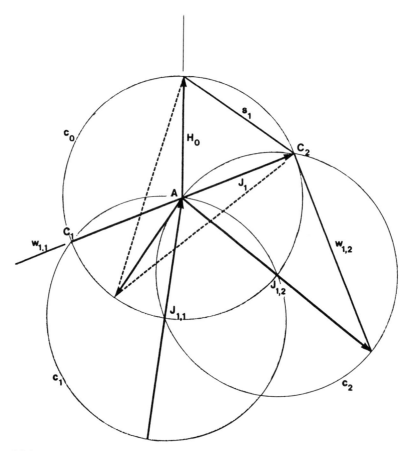

Fig. 6.2.4. Computation of the distribution of magnetization in the yokeless magnet of Fig 6.2.3.

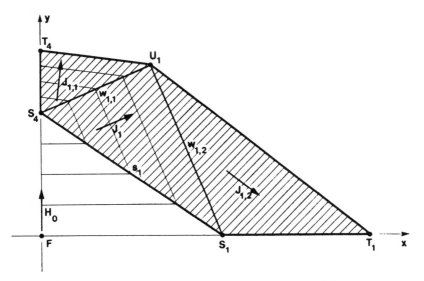

Fig. 6.2.5. Remanence of the magnetized components of the yokeless magnet.

the total area of the cross-section of the magnetized prisms at $K = 0.5$. Conversely, in the yoked structure of Fig. 6.2.1, the total area of the air gaps between magnetized prisms is equal to approximately 44% of the area of the cavity cross-section. Thus, a significant improvement of the figure of merit can be expected if these regions can be either eliminated or reduced in size. The area of the triangle $(S_1 U_1 T_1)$ in Fig. 6.2.5 can be eliminated by assuming that side $(U_1 S_1)$ is the interface between triangle $(S_4 S_1 U_1)$ and a medium of infinite magnetic permeability. Conversely in the yoked structure of Fig. 6.2.2, the wedge of nonmagnetic material can be eliminated by replacing it with the triangle of magnetized material $(S_4 U_1 T_4)$ of Fig. 6.2.5. The resulting hybrid structure is shown in Fig. 6.2.6, where the external boundary is assumed to be the interface between the magnetic structure and an ideal $\mu = \infty$ yoke. The figures of merit of the yoked, yokeless, and hybrid structures designed around the rhombic cavity are plotted in Fig. 6.2.7 versus K. For the hybrid structure the optimum value of K is

$$K \approx 0.5, \qquad (6.2.11)$$

and the maximum value of M is

$$M_{max} = 0.191, \qquad (6.2.12)$$

which corresponds to an area ratio $A_m/A_c \approx 1.31$.

Equation 6.2.12 shows that the combination of yoked and yokeless structures results in a more efficient magnet whose figure of merit is higher than the values 6.2.8 and 6.2.10 of both yoked and yokeless magnets.

Assume now that the three types of magnets are designed around the same rhombic cavity of Fig. 6.2.1 and that $\vec{H}_0$ is rotated by an angle $\pi/2$ to become aligned with the

206 HYBRID MAGNETS

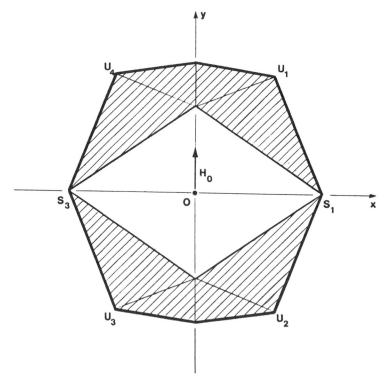

Fig. 6.2.6.   Hybrid magnet structure.

major diagonal of the rhombus. By virtue of the geometric invariance theorem, the yokeless magnet has the same geometry of Fig. 6.2.3 and its figure of merit is not affected by the change of orientation of $\vec{H}_0$. By virtue of theorem 4 of Section 3.5, the rotation of $\vec{H}_0$ by an angle $\pi/2$ results in the rotation of both remanences $\vec{J}_{1,1}$ and $\vec{J}_{1,2}$ in Fig. 6.2.5 by an angle $-\pi/2$. Thus $\vec{J}_{1,1}$ becomes parallel to the external boundary of triangle $(S_4 T_4 U_1)$ which becomes the new $\Phi = 0$ equipotential region. The remanence $\vec{J}_{1,2}$ becomes perpendicular to the external boundary of triangle $(S_1 T_1 U_1)$ which becomes the new region of zero flux of $\vec{B}$. In Fig. 6.2.5 the area of triangle $(S_4 T_4 U_1)$ is substantially smaller than the area of triangle $(S_1 T_1 U_1)$. Thus the replacement of the magnetized medium of triangle $(S_4 T_4 U_1)$ with a $\mu = \infty$ yoke yields a smaller increase of the figure of merit compared to the case of Fig. 6.2.5, where $\vec{H}_0$ is oriented along the y axis. This is reflected by the results presented in Fig. 6.2.8, which shows the figures of merit of yoked, yokeless, and hybrid structures for $\vec{H}_0$ oriented along the major diagonal of the rhombic cavity. Both yoked and hybrid structures in Fig. 6.2.8 exhibit substantially lower figures of merit compared to Fig. 6.2.7. The main difference between the results of Fig. 6.2.7 and 6.2.8 is that in Fig. 6.2.8 the yoked magnet exhibits the lowest figure of merit.

In the particular case $x_0 = y_0$, i.e., for a square cross-sectional cavity, the figures of merit of the three types of magnets are shown in Fig. 6.2.9. The difference between yoked and yokeless structures is less pronounced than in Fig. 6.2.8 and the hybrid structure again exhibits the highest value of M.

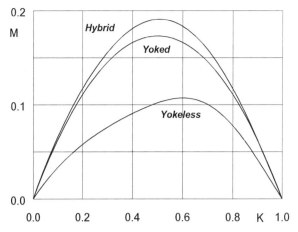

Fig. 6.2.7. Figures of merit of yoked, yokeless, and hybrid structures with a rhombic cavity ($3y_0 = 2x_0$) and $\vec{H}_0$ oriented along the minor diagonal of the rhombus.

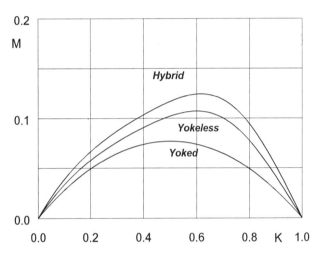

Fig. 6.2.8. Figures of merit of yoked, yokeless, and hybrid structures with a rhombic cavity ($3y_0 = 2x_1$) and $\vec{H}_0$ oriented along the major diagonal of the rhombus.

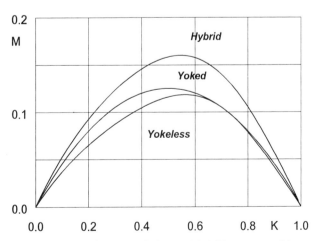

Fig. 6.2.9. Figures of merit of yoked, yokeless, and hybrid structures with a square cross-sectional cavity.

208 HYBRID MAGNETS

In the hybrid structure illustrated in Fig. 6.2.6, the yoke between points U_1, U_4 and between points U_2, U_3 can be removed without perturbing the boundary conditions at the external surface of the magnetic structure. Thus the hybrid design approach may result in a partially yoked magnet, where the yoke is confined to the regions of the boundary of the magnetic structure crossed by the flux of $\vec{B}$.

6.3 INSERTION OF THIN LAYERS OF HIGH MAGNETIC PERMEABILITY MATERIALS IN A MAGNETIC STRUCTURE

The approach of the previous section of replacing a region of magnetic material with a high permeability medium can be extended to any equipotential surface within the magnetic structure. The field configuration within the structure remains unchanged as long as the equipotential surface becomes the interface between the magnetic structure and a medium of infinite magnetic permeability.

The insertion of the $\mu = \infty$ material within the boundary of an equipotential surface has no effect on the field computed under the ideal conditions of uniform value of the remanence and perfectly transparent magnetic media. It has the effect, however, of reducing both the volume of magnetized material as well as the volume of the cavity. Thus the insertion of the $\mu = \infty$ material may affect the figure of merit.

Consider, for instance, the single layer yokeless magnet with a square cross-sectional cavity depicted in Fig. 4.2.6. The configuration of the equipotential lines in the first quadrant of the magnet cross-section has been shown in Fig. 4.2.8 for the particular value of K

$$K = 1 - \frac{1}{\sqrt{2}}, \qquad (6.3.1)$$

with point F located at the center of the square, and $\vec{H}_0$ oriented in the direction of the y axis. The geometries of all equipotential lines in Fig. 4.2.8 are similar to each other and the area of the magnetic material is equal to the area of the cavity. The first quadrant of the magnet cross-section is shown in Fig 6.3.1 which shows the value Φ_0 of the scalar potential at point S_4 and an equipotential line of potential $\overline{\Phi}$. The lined area of Fig. 6.3.1 shows the region of the magnetic structure where the scalar potential varies between $\Phi = \overline{\Phi}$ and $\Phi = \Phi_0$. Assume that this region is replaced by a $\mu = \infty$ material. This cavity of the magnet is transformed in a hexagonal cavity, and the area of its cross-section in the quadrant of Fig. 6.3.1 is

$$A_c = \frac{\overline{\Phi}}{\Phi_0} x_0^2 \left[1 - \frac{1}{2} \frac{\overline{\Phi}}{\Phi_0} \right], \qquad (6.3.2)$$

where x_0 is the abscissa of point S_1. The area of the cross-section of the magnetic material outside the lined region of Fig. 6.3.1 is

$$A_m = \frac{1}{2} \left[1 - \frac{1}{\sqrt{2}} \left(1 - \frac{\overline{\Phi}}{\Phi_0} \right)^2 \right] x_0^2 . \qquad (6.3.3)$$

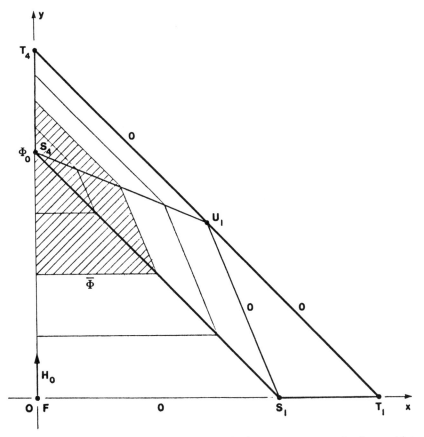

Fig. 6.3.1. Transformation of an equipotential line in a square cross-sectional magnet into the surface of a $\mu = \infty$ material.

Thus the figure of merit M is

$$\frac{M}{M_0} = 2 \frac{\overline{\Phi}}{\Phi_0} \frac{1 - \frac{1}{2}\frac{\overline{\Phi}}{\Phi_0}}{1 - \frac{1}{\sqrt{2}}\left[1 - \frac{\overline{\Phi}}{\Phi_0}\right]^2}, \qquad (6.3.4)$$

where

$$M_0 = \left[1 - \frac{1}{\sqrt{2}}\right]^2 \qquad (6.3.5)$$

is the figure of merit of the square cross-sectional magnet of Fig. 4.2.6 without the

$\mu = \infty$ material. Obviously, Eq. 6.3.4 reduces to

$$\frac{M}{M_0} = 1 \tag{6.3.6}$$

for $\overline{\Phi} = \Phi_0$. Equation 6.3.4 shows that the insertion of the $\mu = \infty$ material reduces the figure of merit. As the value of $\overline{\Phi}$ decreases, the figure of merit decreases. One has

$$M = 0 \tag{6.3.7}$$

at $\overline{\Phi} = 0$, because, as shown by Eqs. 6.3.2 and 6.3.3, A_c vanishes at $\overline{\Phi} = 0$ while A_m reduces to the area of triangle $(S_1 T_1 U_1)$.

Following the approach of Section 6.1, the area $(S_1 U_1 T_1)$ of magnetic material in Fig. 6.3.1 can be replaced by a $\mu = \infty$ material, and the $\Phi = 0$ equipotential line $(T_4 U_1 S_1)$ becomes the external boundary of the magnetic structure of a hybrid magnet. The area of its magnetic material in the first quadrant of Fig. 6.3.1 is

$$A_m = \frac{1}{\sqrt{2}} \frac{\overline{\Phi}}{\Phi_0} \left[1 - \frac{1}{2} \frac{\overline{\Phi}}{\Phi_0} \right] x_0^2 , \tag{6.3.8}$$

and by virtue of Eq. 6.3.3, the figure of merit of the hybrid magnet is

$$\frac{M}{M_0} = \sqrt{2} \tag{6.3.9}$$

independent of the selection of $\overline{\Phi}$. Thus the figure of merit of the hybrid magnet is not affected by the insertion of the $\mu = \infty$ material. The figure of merit is higher by a factor $\sqrt{2}$ than the maximum value of the figure of merit of the yokeless magnet ($\overline{\Phi} = \Phi_0$) in agreement with the results presented in Fig. 6.2.9.

Figure 6.3.2 shows the distribution of the equipotential lines in the first quadrant of the cross-section of the two-layer yokeless structure of Fig. 4.2.10. $\vec{H}_0$ in Fig. 6.3.2 is oriented along a diagonal of the square, and its magnitude corresponds to the value of K equal to twice the value given by Eq. 6.3.1. Again, point F is located at the center O of the cavity. One observes that the equipotential lines in the two-layer structure of Fig. 6.3.2 do not exhibit similar geometries. Consequently, if $\Phi = \overline{\Phi}$ is transformed into the boundary of a $\mu = \infty$ material the figure of merit is a function of $\overline{\Phi}$. This occurs even in the case of a hybrid structure where the area $(S_{1,2} T_{2,1} U_{2,1})$ layer is replaced by a partial yoke. As $\overline{\Phi}$ approaches zero, the area of the cavity vanishes, while the area of magnetized material reduces to the area of triangle $(S_{1,2} T_{2,1} U_{2,1})$. Thus the figure of merit vanishes, as in the case of a single layer magnet.

Another example is the transformation of Fig. 5.5.3 into the structure of Fig. 6.3.3 where the equipotential line $(P_0 P_1 P_2 P_3 P_4)$ encloses the $\mu = \infty$ medium. Assume that Φ_0 is the value of the scalar potential at $y = y_0$ in Fig. 5.5.3 and $\overline{\Phi}$ is the potential

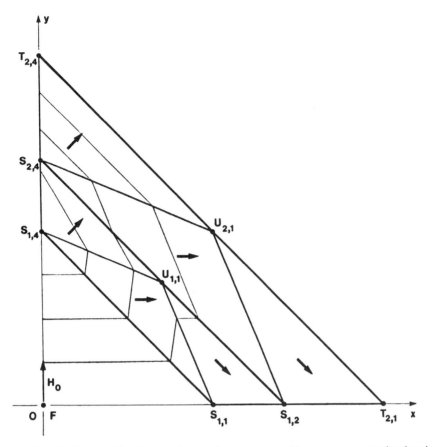

Fig. 6.3.2. Equipotential lines in a two-layer, yokeless magnet with a square cross-sectional cavity.

of the $\mu = \infty$ surface in Fig. 6.3.3. The area of the magnetic material in the region of remanence $\vec{J}_1$ between the $\Phi = 0$ and $\Phi = \overline{\Phi}$ equipotential lines in Fig. 6.3.3 is

$$A_{m,1} = \frac{K}{1-K} x_0 y_0 \frac{\overline{\Phi}}{\Phi_0}. \tag{6.3.10}$$

The area of the two regions of remanence $\vec{J}_{1,2}$ and $\vec{J}_{2,1}$ between the $\Phi = 0$ and $\Phi = \overline{\Phi}$ lines in Fig. 6.3.3 is

$$A_{m,2} = \frac{K}{1-K} \frac{y_0^2}{\sqrt{(1-K)^3(1+K)} + \sqrt{K^3(2-K)}} \frac{\overline{\Phi}}{\Phi_0} \left[1 - \frac{1}{2}\frac{\overline{\Phi}}{\Phi_0}\right], \tag{6.3.11}$$

and the area of the cavity between the $\Phi = 0$ and $\Phi = \overline{\Phi}$ lines is

$$A_c = x_0 y_0 \frac{\overline{\Phi}}{\Phi_0}. \tag{6.3.12}$$

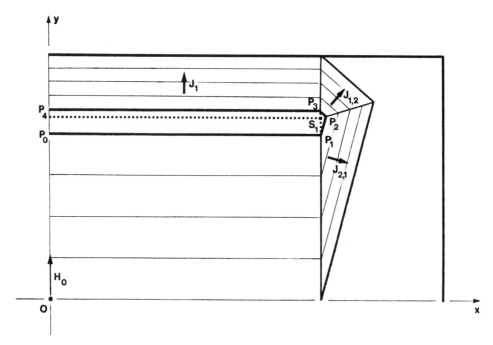

Fig. 6.3.3. Insertion of μ = ∞ material in the structure of Fig. 5.5.3.

Thus the figure of merit of the two-dimensional structure of Fig. 6.3.3 is

$$M = \frac{K(1-K)}{1 + \dfrac{y_0}{x_0} \dfrac{1}{\sqrt{(1-K)^3(1+K)} + \sqrt{K^3(2-K)}} \left[1 - \dfrac{1}{2}\dfrac{\overline{\Phi}}{\Phi_0}\right]} ; \quad (6.3.13)$$

i.e., the transformation of the $\overline{\Phi}$ line into the μ = ∞ boundary results in a decrease of the figure of merit and the minimum value of M occurs at $\overline{\Phi} = 0$. Thus, to minimize the reduction of the figure of merit, the plate of μ = ∞ material in Fig. 6.3.3 should be thin enough so that

$$\Phi_0 - \overline{\Phi} \ll \Phi_0 . \quad (6.3.14)$$

As previously stated, a μ = ∞ magnetic material can be used as a spatial filter to reduce the field distortion caused by a nonuniform distribution of the magnetic properties due to fabrication and magnetization tolerances. With finite values of the permeability of available ferromagnetic materials, in order to be an effective filter, a minimum thickness of the material is required such that the induction within the material is well below the saturation value.

To illustrate the filter effect, consider the schematic of Fig. 6.3.4 where the half plane $x < 0$ is a magnetized material with a uniform remanence $\vec{J}_0$ oriented in the posi-

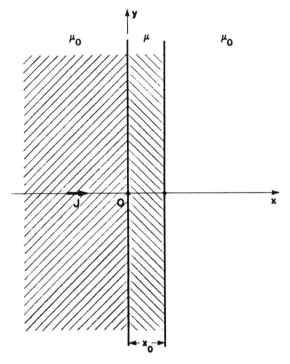

Fig. 6.3.4. Layer of high permeability material inserted between two media.

tive direction of the axis x. Assume first that the medium of region $x > 0$ is either air or a nonmagnetic material. The uniform remanence $\vec{J}_0$ generates in region $x > 0$ a uniform field $\vec{H}_0$ oriented in the positive direction of the axis x and of magnitude

$$H_0 = \frac{J_0}{2\mu_0}. \tag{6.3.15}$$

Assume now that in the region $x < 0$ a periodic distribution of remanence

$$\vec{J} = J_1 \cos ky \, \vec{x} \tag{6.3.16}$$

is superimposed to $\vec{J}_0$. In Eq. 6.3.16, $\vec{x}$ is a unit vector oriented in the positive direction of the axis x, and the period $2y_0$ of the distribution of $\vec{J}$ is

$$2y_0 = \frac{2\pi}{k}. \tag{6.3.17}$$

The scalar potential Φ generated by remanence 6.3.16 can be readily determined by solving Laplace's equation in the two regions $x \gtrless 0$ with the boundary condition at

214 HYBRID MAGNETS

$x = 0$ based on the distribution of an equivalent surface charge σ,

$$\sigma = J_1 \cos ky , \qquad (6.3.18)$$

induced by remanence 6.3.16. In the region $x > 0$, the scalar potential is

$$\Phi = \Phi_3 \cos ky \, e^{-kx} , \qquad (6.3.19)$$

where the constant of integration is

$$\Phi_3 = +\frac{J_1}{2\mu_0 k} , \qquad (6.3.20)$$

and the magnitude of the magnetic induction in the $x > 0$ region is

$$B = \frac{J_1}{2} e^{-kx} . \qquad (6.3.21)$$

Thus the magnitude of the magnetic induction generated by $\vec{J}_1$ decreases exponentially with the distance from the interface between air and magnetic material. As a consequence, in a practical magnet design, increasing the distance between the magnetic material and the region of interest is an effective way of reducing the field distortion within the region of interest.

Assume now that a layer of material of high magnetic permeability μ and thickness x_0 is inserted between the material of remanence

$$\vec{J} = \vec{J}_0 + \vec{J}_1 \qquad (6.3.22)$$

and the nonmagnetic region, as shown in the schematic of Fig. 6.3.4. The high magnetic permeability material confined between the parallel surfaces $x = 0$ and $x = x_0$ has no effect on the intensity generated by the uniform component $\vec{J}_0$ of the remanence. The potential generated by the $\vec{J}_1$ component in the $x > x_0$ region is

$$\Phi = \frac{2J_1}{k} \frac{\mu}{(\mu + \mu_0)^2 e^{+kx_0} - (\mu - \mu_0)^2 e^{-kx_0}} \cos ky \, e^{-k(x - x_0)} , \qquad (6.3.23)$$

which reduces to Eq. 6.3.19 in the limit $\mu = \mu_0$ or $x_0 = 0$.

Assume a thickness x_0 of the layer of high magnetic permeability μ small compared to the period 6.3.17 of $\vec{J}_1$. Then $k x_0 \ll 1$ and Eq. 6.3.23 yields

$$\Phi \approx \frac{J_1}{\mu_0 k} \frac{1}{2 + \frac{\mu}{\mu_0} k x_0} \cos ky \, e^{-kx} . \qquad (6.3.24)$$

Even if kx_0 is sufficiently small compared to unity, it is always possible to select a combination of x_0 and μ such that

$$\frac{\mu}{\mu_0} kx_0 \gg 1 ; \tag{6.3.25}$$

i.e., a thin layer of permeability $\mu \gg \mu_0$ whose thickness satisfies Eq. 6.3.25 results in

$$\Phi \ll \Phi_{\mu = \mu_0} , \tag{6.3.26}$$

and a magnitude of the magnetic induction in the region $x > x_0$

$$B \approx \frac{J_1}{2} \frac{\mu_0}{\mu k x_0} e^{-kx} , \tag{6.3.27}$$

which is small compared to the value 6.3.21. The filtering effect described by Eq. 6.3.27 can be achieved in a prismatic magnetic structure (such as the double layer yokeless magnet of Fig. 6.3.2) by inserting thin layers of high permeability material along the equipotential surfaces of the field generated by the main component of the remanence. This is shown in the schematic of Fig. 6.3.5 where the broken lines are

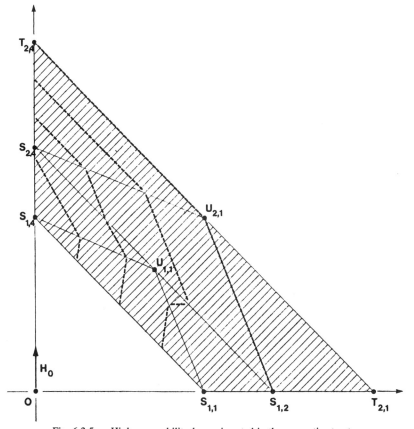

Fig. 6.3.5. High permeability layers inserted in the magnetic structure.

216 HYBRID MAGNETS

the $\mu \gg \mu_0$ layers inserted along several equipotential surfaces. Because of the exponential factor in Eq. 6.3.27, the layers inserted in the components of the magnetic structure closest to the cavity have the most beneficial filtering effect.

6.4 THREE-DIMENSIONAL MAGNETS

The design of a two-dimensional hybrid structure can be extended to a three-dimensional magnet following the same procedure developed in Chapters 4 and 5.

Consider the octahedral cavity shown in Fig. 6.4.1 and assume that the cross-sections in the planes $z = 0$ and $x = 0$ are squares whose diagonals coincide with the coordinate axes. Consider the design of a yokeless single layer magnet for the particular case

$$K = \frac{1}{2} \tag{6.4.1}$$

and $\vec{H}_0$ oriented along the axis y with the assumption that point F coincides with the center of the octahedron. The design evolves from the two-dimensional structures in

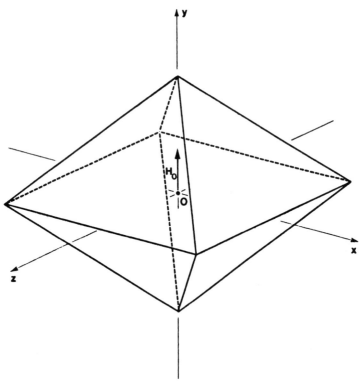

Fig. 6.4.1. Octahedral cavity.

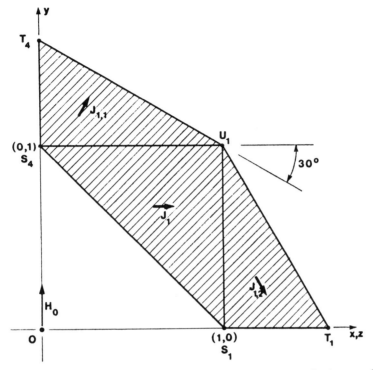

Fig. 6.4.2. Cross-section of the three-dimensional yokeless structure in the two planes $z = 0$, $x = 0$.

the planes $z = 0$ and $x = 0$ shown in Fig. 6.4.2. The external boundaries of the components with remanences $\vec{J}_{1,1}, \vec{J}_{1,2}$ are oriented at $\pi/6$ with respect to the coordinate axes x, y, respectively. The area of the component with remanence $\vec{J}_1$ in Fig. 6.4.2 is equal to the area of the cavity in the first quadrant. Thus at $K = 0.5$ the figure of merit of the two-dimensional yokeless magnet is

$$M = \frac{1}{4}\left[1 + 2\tan\frac{\pi}{6}\right]^{-1} \approx 0.116 \,. \tag{6.4.2}$$

The geometry of Fig. 6.4.2 is a particular case of the structure of Fig. 6.2.5, which was transformed into the hybrid structure of Fig. 6.2.6. The full cross-section of the magnetic structure in the planes $z = 0$ and $x = 0$ is shown in Fig. 6.4.3. The intensity of the magnetic field is zero within the two regions $(U_1 U_2 T_1)$ and $(U_3 U_4 T_3)$ that can be replaced by a material of infinite magnetic permeability. The elimination of these two regions results in a two-dimensional hybrid magnet whose figure of merit at $K = 0.5$ is

$$M = \frac{1}{4}\left[1 + \tan\frac{\pi}{6}\right]^{-1} \approx 0.158 \,, \tag{6.4.3}$$

218 HYBRID MAGNETS

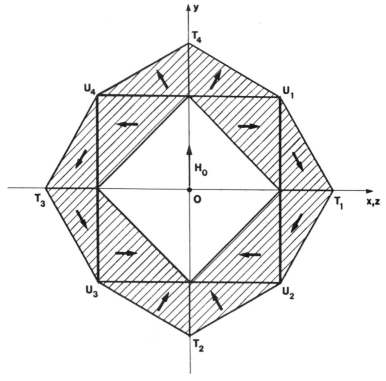

Fig. 6.4.3. Two-dimensional yokeless structure with a square cross-sectional cavity and $\vec{H}_0$ oriented along a diagonal of the square.

which is substantially higher than the value given by Eq. 6.4.2. As shown in Fig. 6.2.9, the value $K = 0.5$ is somewhat lower than the optimum value of K of both yokeless and hybrid magnets.

Following the same procedure of Section 4.5, the magnetic structure around the cavity of Fig. 6.4.1 is divided in four identical wedges separated by the planes $z = x$, $z = -x$. Within each wedge, the magnetic structure of Fig. 6.4.3 is limited to the region that contains the flux of the magnetic induction that flows within the fraction of the cavity contained within the wedge. Figure 6.4.4 shows the structure formed by the four wedges of the magnetized components with remanence parallel to the z and x axes which carry the total flux of $\vec{B}$ within the cavity. These components, which correspond to the region of remanence $\vec{J}_1$ shown in Fig. 6.4.2, are tetrahedrons with one face perpendicular to either plane $x = 0$ or plane $z = 0$. The four tetrahedrons in the $y > 0$ region have point S_4 in common, and the four tetrahedrons in the $y < 0$ region have point S_2 in common with each other.

In the eight tetrahedrons no flux of the magnetic induction crosses the faces that have point S_4 in common and the faces that have point S_2 in common with each other. Because of the orientation of $\vec{H}_0$, the intensity of the field within the component $(S_4 T_4 U_1)$ of Fig. 6.4.2 is equal and opposite to $\vec{J}_{1,1}$. This component must be replaced in the three-dimensional magnet by a magnetized material whose remanence $\vec{J}_i$ does not generate an additional flux of $\vec{B}$. By virtue of the design condition presented in Eq.

6.4.1 and because the field intensity $\vec{H}_i$ in this new component of magnetized material is

$$\mu_0 \vec{H}_i = -\vec{J}_i \,, \tag{6.4.4}$$

remanence $\vec{J}_i$ can be chosen arbitrarily as long as its magnitude satisfies the condition

$$J_i \geq \frac{J_0}{2} \,. \tag{6.4.5}$$

If $J_i = J_0/2$, the external surfaces of the new component of magnetized material become parallel to the y axis and the volume of the three-dimensional structure diverges.

The geometry of the layer of the new components whose remanences satisfy condition 6.4.5 is shown in Fig. 6.4.5 superimposed to the structure of Fig. 6.4.4. The layer

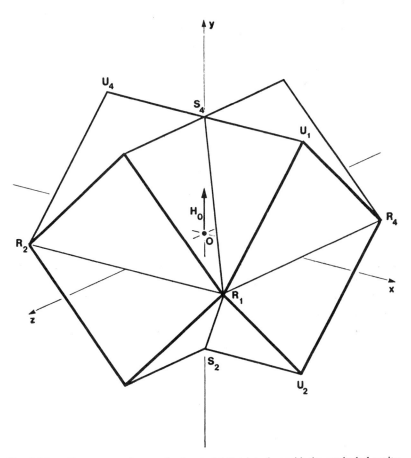

Fig. 6.4.4. Components of magnetized material that interface with the octahedral cavity.

220 HYBRID MAGNETS

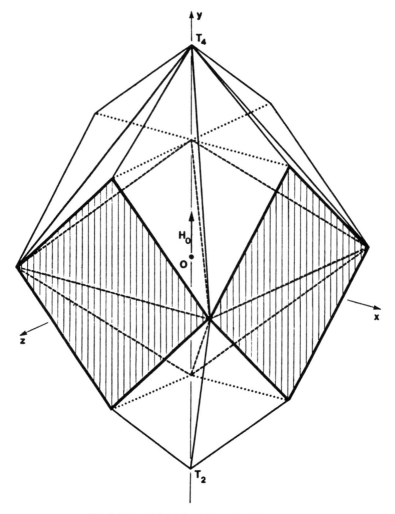

Fig. 6.4.5. Hybrid three-dimensional magnet.

consists of eight tetrahedrons in the region $y > 0$ that have point T_4 in common and eight identical tetrahedrons in the region $y < 0$ that have T_2 in common. The remanence of each tetrahedron is perpendicular to its external face. If the magnitude of $\vec{J}_i$ is chosen to be equal to J_0, points T_4 and T_2 in Fig. 6.4.5 do not coincide with points T_4 and T_2 of the cross-section of the two-dimensional magnet of Fig. 6.1.5. The ordinates of points T_4, T_2 in Fig 6.4.5 are

$$y = \pm \left[1 + \sqrt{\frac{2}{3}} \right] x_0 , \qquad (6.4.6)$$

where $2x_0$ is the diagonal of the square cross-section of the cavity. The external surfaces of the tetrahedrons in the region $y > 0$ are tangent to a sphere with radius r_i and center at S_4. The external surfaces of the tetrahedrons in the region $y < 0$ are tangent to a sphere of the same radius and centered at S_2. The radius of the two spheres is

$$r_i = \frac{x_0}{2}. \tag{6.4.7}$$

If the lined faces in Fig. 6.4.5 are assumed to be the surface of a $\mu = \infty$ material, the structure of Fig. 6.4.5 is a hybrid three-dimensional magnet.

In the two-dimensional structure of Fig. 6.4.2, the flux of $\vec{B}$ generated by the magnetic material with remanence $\vec{J}_1$ closes through the equipotential region occupied by the material of remanence $\vec{J}_{1,2}$, whose external boundary $(T_1 U_1)$ is parallel to $\vec{J}_{1,2}$ as shown in Fig. 6.4.2. If the region of remanence $\vec{J}_{1,2}$ is replaced by a $\mu = \infty$ passive ferromagnetic material, one obtains a two-dimensional hybrid structure similar to the magnet shown in Fig. 6.2.6.

In a yokeless three-dimensional magnet, the component $(S_1 T_1 U_1)$ of Fig. 6.4.2 must be replaced by a structure of equipotential magnetized components which confine the flux of $\vec{B}$ inside the magnet. Thus the remanence $J_{i,2}$ of these new components must be equal to the magnetic induction and parallel to their external surface.

The magnitude of the magnetic induction in component $(S_1 S_4 U_1)$ of the structure of Fig. 6.4.4 is equal to $J_1/2$. Thus the magnitude of remanence $\vec{J}_{i,2}$ must satisfy the condition

$$J_{i,2} \geq \frac{1}{2} J_0. \tag{6.4.8}$$

The geometry of the external layer of the components of remanence $\vec{J}_{i,2}$ is shown in Fig. 6.4.6 superimposed to the structure of Fig. 6.4.5. This external layer consists of four structures, each of them composed of four identical tetrahedrons. The four tetrahedrons that interface with face $(R_1 U_1 R_4 U_2)$ have point T_1 of the axis x in common with each other. To achieve the field confinement, the remanence of the two tetrahedrons in the region $y > 0$ must be parallel to side $(U_1 T_1)$ and the remanence of the two tetrahedrons in the region $y < 0$ must be parallel to side $(T_1 U_2)$ as in the two-dimensional structure of Fig. 6.4.3. In particular, if the magnitude of the remanences $\vec{J}_{i,2}$ is chosen to be equal to J_0, the abscissa of point T_1 in Fig. 6.4.5 is

$$x = \left[1 + \frac{1}{\sqrt{3}}\right] x_0. \tag{6.4.9}$$

It is of interest to compare the structure of Fig. 6.4.6 with the three-dimensional yokeless magnet of Fig. 4.5.6 which generates a field $\vec{H}_0$ perpendicular to a face of its cubic cavity. Because the external surface of the structure of Fig. 4.5.2 is not equipotential, the confinement of the field in the structure of Fig. 4.5.6 requires the additional transition layer depicted in Fig. 4.5.5 between the structure of Fig. 4.5.2 and the external medium. The additional transition layer is not required in the magnet of Fig. 6.4.6 because of the particular situation of Fig. 6.4.2, where the external layer $(S_1 T_1 U_1)$ is

222 HYBRID MAGNETS

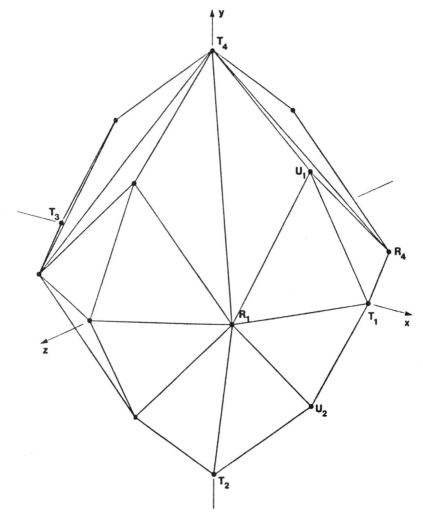

Fig. 6.4.6. Yokeless three-dimensional magnet.

an equipotential region. Thus the three-dimensional magnet of Fig. 6.4.6 is derived simply by modifying the geometry of the equipotential regions of the two-dimensional magnet of Fig. 6.4.3.

The volume of the octahedral cavity of the magnet of Fig. 6.4.6 is

$$V_c = \frac{8}{3} x_0^3 . \tag{6.4.10}$$

The volume of the structure of the eight tetrahedrons of remanence $\vec{J}_1$ shown in Fig. 6.4.4 is

$$V_{m,1} = V_c = \frac{8}{3} x_0^3 , \tag{6.4.11}$$

and the volumes of the two structures of tetrahedrons of remanences $\vec{J}_{1,1}$ and $\vec{J}_{1,2}$ shown in Figs. 6.4.5 and 6.4.6 are

$$V_{m,1,1} = \frac{8}{3}\sqrt{\frac{2}{3}}\, x_0^3 \,, \qquad V_{m,1,2} = \frac{8}{3\sqrt{3}}\, x_0^3 \,, \qquad (6.4.12)$$

respectively. Thus the figure of merit of the hybrid magnet shown in Fig. 6.4.4 is

$$M = \frac{K^2}{1+\sqrt{\frac{2}{3}}} \approx 0.138 \,. \qquad (6.4.13)$$

The figure of merit of the yokeless magnet of Fig. 6.4.6 is

$$M = \frac{K^2}{1+\frac{\sqrt{2}+1}{\sqrt{3}}} \approx 0.104 \,, \qquad (6.4.14)$$

which is lower than value in Eq. 6.4.13 because of the additional volume $V_{m,1,2}$ of magnetized material.

6.5 MAGNETS WITH RECTANGULAR PRISMATIC CAVITIES

Consider again the three-dimensional structure of Fig. 6.3.3, derived from the hybrid magnet of Fig. 5.5.3 by transforming an equipotential line of potential $\overline{\Phi}$ into the surface of a $\mu = \infty$ material. Let us extend the structure of either Fig. 5.5.3 or Fig. 6.3.3 to a three-dimensional geometry designed to generate a uniform field $\vec{H}_0$ in a rectangular prismatic cavity. The schematic of Fig. 6.5.1 shows the lines of flux of $\vec{B}$ in a cross-section parallel to a face of the cavity. Because $\vec{B}$ is identically zero in the regions of remanence $\vec{J}_2$ and $\vec{J}_3$ in Fig. 6.3.3, the design of the three-dimensional magnet is much simpler than the design of the three-dimensional yokeless and yoked structures discussed in the preceding chapters.

Assume a rectangular prismatic cavity with dimensions $2x_0$, $2y_0$, $2z_0 = 2x_0$ and assume that $\vec{H}_0$ is oriented in the direction of the axis y. The vector diagram of Fig. 5.5.2 provides the two identical geometries of the two-dimensional structures with axes z and x. If the two structures are cut in the two planes

$$x = \pm z \,, \qquad (6.5.1)$$

the $\pi/2$ wedges can be joined together. The boundary conditions on the two planes 6.5.1 are automatically satisfied because the magnetic induction is oriented everywhere along the axis y. Thus no additional magnetic structure is required to establish the tran-

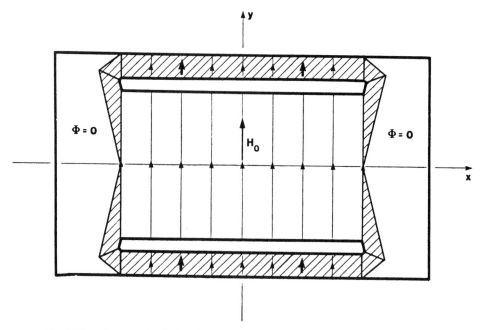

Fig. 6.5.1. Cross-section in the plane $z = 0$ of a hybrid magnet with a rectangular cavity.

sition between the structures computed in the $z = 0$ and $x = 0$ planes. This conclusion is valid also if the $z_0 \neq x_0$ as long as the magnetic components of remanences $\vec{J}_2$ and $\vec{J}_3$ are cut in planes oriented at an angle $\pi/4$ with respect to the $z = 0$, $x = 0$ planes. In the more general case of a prismatic cavity with a polygonal base, the magnetic components of remanences $\vec{J}_2$, $\vec{J}_3$ are cut in planes that bisect the angles between the lateral faces of the prism. Because of the assumption of a $\mu = \infty$ yoke, a geometrical continuity between the closed yokes selected for the individual wedges is not required to close the flux of $\vec{B}$.

The external boundary of the three-dimensional magnetic structure, which encloses the rectangular prismatic cavity, is shown in Fig. 6.5.2 in the half space $y > 0$. The schematic of an ideal yoke is shown in Fig. 6.5.3 again in the half space $y > 0$. As indicated in the figure, the yoke can be partially open because no field is present in the region between the yoke and the magnetic structure. Thus the basic configuration of the magnet can be either one of the two schematics shown in Fig. 6.5.4(a) and 6.5.4(b) [2].

In the limit $\Phi - \overline{\Phi} \ll \Phi_0$, the volume of magnetic material in the structure of Fig. 6.5.2 is

$$V_m = \frac{4K}{1 - K} y_0 x_0^2 \left\{ 1 + \frac{y_0}{x_0} \frac{1}{\sqrt{(1-K)^3(1+K)} + \sqrt{K^3(2-K)}} \right.$$

$$\left. \times \left[1 + \frac{K}{3} \frac{y_0}{x_0} \frac{1}{\sqrt{(1-K)^3(1+K)} + \sqrt{K^3(2-K)}} \right] \right\},$$

(6.5.2)

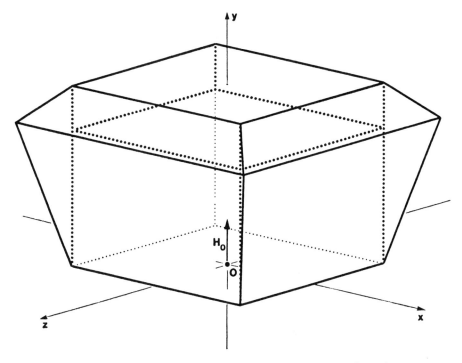

Fig. 6.5.2. Structure of the magnetic material enclosing a rectangular cavity.

and the volume in the $y > 0$ region of the cavity is

$$V_c = 4y_0 x_0^2 . \tag{6.5.3}$$

Thus the figure of merit of the three-dimensional magnet with the rectangular prismatic cavity of dimensions $2x_0$, $2y_0$, $2z_0 = 2x_0$ is

$$M = K(1 - K)\left\{1 + \frac{y_0}{x_0} \frac{1}{\sqrt{(1 - K)^3(1 + K)} + \sqrt{K^3(2 - K)}}\right.$$

$$\left. \times \left[1 + \frac{K}{3} \frac{y_0}{x_0} \frac{1}{\sqrt{(1 - K)^3(1 + K)} + \sqrt{K^3(2 - K)}}\right]\right\}^{-1} . \tag{6.5.4}$$

In practice, a yoke of finite magnetic permeability must have a minimum thickness in order to effectively confine the flux of $\vec{B}$. The finite value of μ automatically generates a magnetic field within the yoke. Thus the yoke of ferromagnetic material is not an equipotential region. In principle, a leakage of the field is generated in the surrounding medium regardless of the yoke geometry.

A passive $\mu = \infty$ yoke can always be replaced by an "active" yoke of magnetized

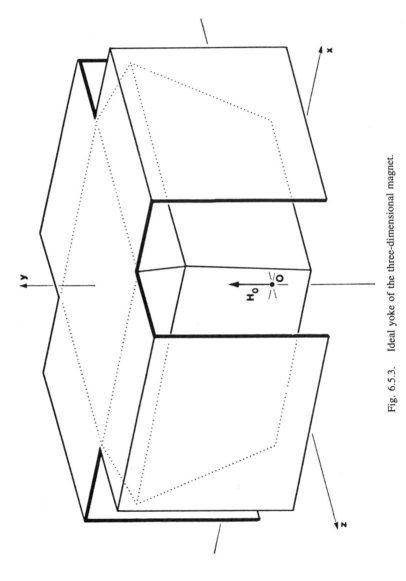

Fig. 6.5.3. Ideal yoke of the three-dimensional magnet.

226

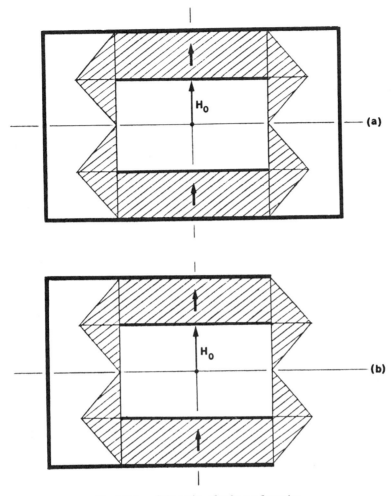

Fig. 6.5.4. Schematics of yoke configuration.

material that satisfies the condition

$$\vec{B} = \vec{J}. \tag{6.5.5}$$

A possible configuration of such a yoke is shown in Fig. 6.5.5 where the lined area is the equipotential magnetic material. The heavy arrows indicate the orientation of the remanence that is parallel to the external boundary of each component of the structure. The geometry of Figure 6.5.5 corresponds to the selection of a remanence equal to that of the rectangular region ($V_0 V_1 S_1 S_0$). ($V_0 V_1$) is the interface between this region and the triangular component of the new structure whose external boundary forms an angle θ with respect to the axis x given by

$$\sin \theta = K. \tag{6.5.6}$$

228 HYBRID MAGNETS

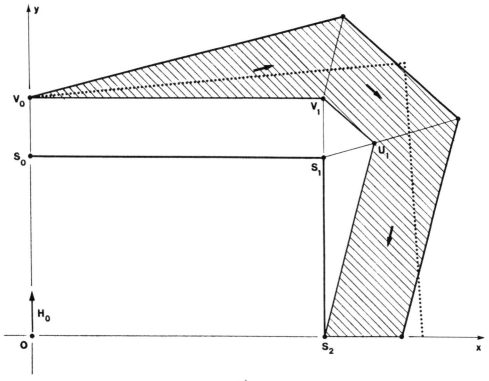

Fig. 6.5.5. Closing of the flux of $\vec{B}$ by means of magnetized material.

As in the case of an ideal $\mu = \infty$ ferromagnetic yoke, one can arbitrarily select the internal boundary of the other two components of the active yoke. Figure 6.5.5 corresponds to the particular case where the $(V_1 U_1 S_2)$ boundary is the interface between the regions of remanences $\vec{J}_{1,2}, \vec{J}_{2,1}$ and the two components of the active yoke.

It is of interest to compare the structure of Fig. 6.5.5 with a yokeless, single layer magnet designed with the method developed in Chapter 4. The dotted line in Fig. 6.5.5 shows the external boundary of such a magnet designed around the rectangular cavity $(O S_0 S_1 S_2)$ for the same value of K. The difference in total area between the two magnetic structures is quite apparent. The yokeless magnet designed with the method of Chapter 4, where each component generates and channels the flux of $\vec{B}$, exhibits a substantially smaller volume of magnetic material.

A dual solution of the problem of generating the uniform field within the rectangular cavity is derived from the same vector diagram of Fig. 5.5.2 used to design the structure of Fig. 5.5.3. This second solution is known as the clad magnet [3,4,5]. The first quadrant of its cross-section is shown in Fig. 6.5.6. The rectangular region $(S_1 T_1 U_2 S_2)$ has a remanence

$$\vec{J}_1'' = -\vec{J}_1 , \qquad (6.5.7)$$

where $\vec{J}_1$ is the remanence of the rectangular region of the structure of Fig. 5.5.3. Thus

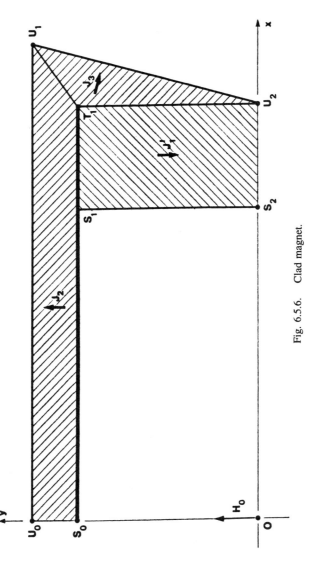

Fig. 6.5.6. Clad magnet.

$\vec{J}_1''$ is oriented in the direction opposite to $\vec{H}_0$, and its magnitude is

$$K J_1' = \mu_0 H_0. \tag{6.5.8}$$

The flux of the magnetic induction $\vec{B}_1''$ generated by $\vec{J}_1''$ closes into the cavity through the $\mu = \infty$ surface $(S_0 T_1)$ of Fig. 6.5.6. Thus the abscissas x_0, x_1 of points S_2 and U_2 are related to each other by the equation

$$x_1 = \frac{x_0}{1 - K}, \tag{6.5.9}$$

and the area of the region of remanence $\vec{J}_1''$ is

$$A_{m,1} = \frac{K}{1 - K} x_0 y_0. \tag{6.5.10}$$

The area $A_{m,1}$ is equal to the value of the area of region $(S_0 V_0 V_1 S_1)$ of the structure of Fig. 5.5.3, as given by Eq. 6.3.10 for $\overline{\Phi} = \Phi_0$. In other words, it takes the same amount of magnetic material to generate the intensity $\vec{H}_0$ in the same cavity of the two structures of Fig. 5.5.3 and Fig. 6.5.6.

In Fig. 6.5.6, interface $(T_1 U_2)$ is not an equipotential surface, and the potential of the $\mu = \infty$ surface $(S_0 T_1)$ is not zero. As a consequence, boundary $(S_0 T_1 U_2)$ cannot be the interface between the magnetic structure and the surrounding medium, and surface $(S_0 T_1)$ cannot be part of a closed surface of a $\mu = \infty$ material. Following the procedure of Section 5.5, a structure of magnetized material that satisfies condition 5.5.2 has to be inserted between boundary $(S_0 T_1 U_2)$ and the surrounding medium, as shown in Fig. 6.5.6. If the remanences $\vec{J}_2$ and $\vec{J}_3$ have the same magnitude of $\vec{J}_1''$, boundary $(U_0 U_1)$ is located at a distance y_1 from the x axis,

$$y_1 = (1 + K) y_0. \tag{6.5.11}$$

The orientation of boundary $(U_1 U_2)$ is the same as that of interface $(U_1 S_2)$ in Fig. 5.5.3 as derived from the vector diagram of Fig. 5.5.2, Thus the abscissa of point U_1 is

$$x_1' = x_1 + K \sqrt{\frac{1 + K}{1 - K}} y_0, \tag{6.5.12}$$

and the total area of the magnetic material between boundary $(S_0 T_1 U_2)$ and the external boundary $(U_0 U_1 U_2)$ is

$$A_{m,2}' = \frac{1}{2} \frac{K}{1 - K} y_0^2 \left[2 \frac{x_0}{y_0} + (1 + K) \sqrt{1 - K^2} \right]. \tag{6.5.13}$$

Thus, by virtue of Eq. 6.3.11, the ratio of area 6.5.13 to the area 5.5.10 of the two regions of remanences $\vec{J}_{1,2}$ and $\vec{J}_{2,1}$ of the structure of Fig. 5.5.3 is

$$\frac{A'_{m,2}}{A_{m,2}} = \left[\sqrt{1-K^2}(1+K) + 2\frac{x_0}{y_0}\right]\left[\sqrt{(1-K)^3(1+K)} + \sqrt{K^3(2-K)}\right]. \quad (6.5.14)$$

For values of x_0 small compared to y_0 and for large values of K, in the range

$$0.839 < K < 1, \quad (6.5.15)$$

ratio 6.5.14 becomes smaller than unity. Thus a clad magnet exhibits a better figure of merit in a situation where a strong magnetic field is required in a long prismatic cavity whose dimension along the magnetic field is large compared to the other dimensions. An example is the cavity of a magnet designed to focus an electron beam. On the other hand, for values of x_0 of the order of y_0, ratio 6.5.14 becomes larger than unity, and it becomes larger and larger as x_0/y_0 increases, and the figure of merit of the clad magnet is lower than the figure of merit of the structure of Fig. 5.5.3.

The essential differences between the structure of Fig. 6.5.6 and the structure of Section 5.5 is that in Fig. 6.5.6, the flux of the magnetic induction closes inside rather than outside. The $\vec{B} = 0$ layer derives the name of cladding layer from the fact that it totally encloses the magnetic structure [3].

6.6 CLOSURE OF THE CAVITY IN TRADITIONAL PERMANENT MAGNETS

The magnetic structures analyzed so far are characterized by cavities totally enclosed by structures of magnetic material. In principle the field in the entire volume of the cavity can be used in applications that require a uniform field. It is of interest to compare these structures with the traditional magnets that are exemplified by the schematic of Fig. 6.6.1, where the lined regions V_m represent the magnetic materials, the regions denoted by V_p are the pole pieces, and region V_y represents the yoke. The particular magnet shown in Fig. 6.6.1 is an axisymmetric structure, with axis y as its axis of symmetry, designed to generate a highly uniform field in a region of the gap V_c between the pole pieces, close to the center O of the gap [6].

The obvious difference between the structure of Fig. 6.6.1 and the yoked and yokeless structures analyzed in the previous sections is the absence of a physical boundary between the region of interest and the rest of the nonmagnetic region enclosed by the yoke.

The approach to the design of a traditional magnet like the one of Fig. 6.6.1 can be described by considering the schematic of Fig. 6.6.2, where cylinder V_m is the volume of magnetic material of uniform remanence $\vec{J}$ and V_c is the region of interest where a uniform field intensity $\vec{H}_0$ is generated by $\vec{J}$. The design approach is defined

232 HYBRID MAGNETS

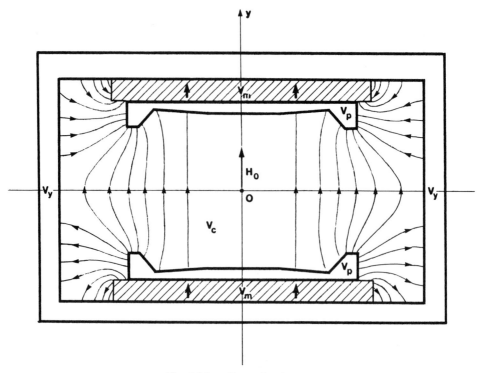

Fig. 6.6.1. Conventional magnet.

by the fundamental equations of magnetostatics 1.1.9 and 1.1.10. Equation 1.1.9 yields

$$\int_l \vec{H} \cdot \vec{dl} = 0, \tag{6.6.1}$$

where l is an arbitrary closed line selected within the magnetic structure. Furthermore from Eq. 1.1.10 one derives the equation of continuity

$$\Psi_m = \int_S \vec{B} \cdot \vec{dS}, \tag{6.6.2}$$

where Ψ_m is the total flux of $\vec{B}$ generated by J, and S is an arbitrary cross-section of the total flux of $\vec{B}$.

As shown by the dotted lines in Fig 6.6.2, the lines of flux of $\vec{B}$ must close upon themselves through V_m, V_c, and the region outside V_m and V_c which include the pole pieces and the external yoke. Equation 6.6.2 is equivalent to the equation of continuity of the electric current in a DC electric circuit and Eq. 6.6.1 is equivalent to the condi-

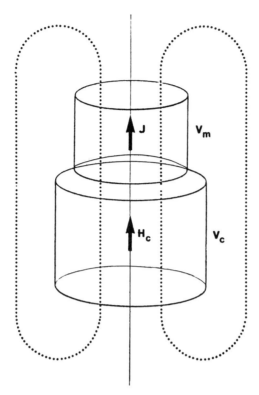

Fig. 6.6.2. Schematic of the magnetic material V_m and region of interest V_c.

tion satisfied by the electric potential across the components of a DC closed electric circuit. Because of this equivalence, in a traditional magnet design, the magnet is considered as a circuit whose components are defined by their magnetic reluctance in a way equivalent to the definition of the electric resistance of the components of a DC electric circuit [7].

In an ideal design of a permanent magnet based on Eqs. 6.6.1 and 6.6.2 volumes V_m, V_c should be the major contributors to the integral in Eq. 6.6.1. This can be achieved by minimizing the dimensions of the other components of the structure and by fabricating these components with materials of high magnetic permeability. Furthermore, the major portion of the total flux of $\vec{B}$ generated by volume V_m should be channeled through the region of interest V_c. Assume that the field is uniform in both volumes V_m and V_c, and let $\vec{H}_m$, $\vec{H}_c$ be the field intensities within V_m and V_c, respectively. In the ideal limit where Eq. 6.6.1 reduces to the contribution of V_m, V_c, vectors $\vec{H}_m$, $\vec{H}_c$ are oriented in opposite directions and their magnitudes satisfy the condition

$$H_m l_m = H_c l_c , \qquad (6.6.3)$$

where l_m, l_c are the heights of cylinders V_m, V_c. Assume that $\vec{B}_m$ is the magnetic induction within volume V_m. In the ideal limit, where the total flux of $\vec{B}$ given by Eq.

6.6.2 is channelled through V_c, the magnitudes of $\vec{B}_m$ and $\vec{H}_c$ satisfy the equation

$$B_m S_m = \mu_0 H_c S_c, \qquad (6.6.4)$$

where S_m, S_c are the areas of the cross-sections of V_m, V_c, respectively. Assume a magnetic material with linear demagnetization characteristics and zero magnetic susceptibility and define as a design parameter

$$K_m = \frac{\mu_0 H_m}{J}, \qquad (6.6.5)$$

rather than the parameter K that relates $\vec{H}_c$ to J. By virtue of Eq. 6.6.5 the energy stored in the magnetic material is

$$W_m = \frac{B_m H_m V_m}{2K_m(1 - K_m)}, \qquad (6.6.6)$$

and the energy of the magnetic field within volume V_c is

$$W_c = \frac{1}{2} \mu_0 H_c^2 V_c. \qquad (6.6.7)$$

Hence, by virtue of Eqs. 6.6.3 and 6.6.4, the figure of merit of the ideal magnet is

$$M = \frac{W_c}{W_m} = K_m(1 - K_m), \qquad (6.6.8)$$

which is independent of the value of the field intensity within volume V_c [8]. The maximum value of M is achieved at

$$K_m = \frac{1}{2}, \qquad (6.6.9)$$

which corresponds to the maximum value of the energy product given by Eq. 1.7.6 for ideal magnetic materials.

In practice, the components of a magnetic circuit outside of V_m, V_c also contribute to the integral in Eq. 6.6.1 and the flux of $\vec{B}$ is not confined to V_c, i.e.,

$$\frac{H_c l_c}{H_m l_m} = F_h < 1, \qquad \frac{\mu_0 H_c S_c}{B_m S_m} = F_f < 1, \qquad (6.6.10)$$

where the two parameters F_h, F_g depend upon the geometry and the selection of the materials. Thus the figure of merit of a practical magnet can be written in the form

$$M \approx K_m (1 - K_m) F_h F_f . \tag{6.6.11}$$

The elementary considerations leading to Eq. 6.6.11 define the approach to the traditional design of a permanent magnet. If parameters F_h, F_f can be assumed to be independent of K_m, the design starts by selecting the operating point on the demagnetization curve that optimizes the energy product. Thus the figure of merit of the magnet is

$$M \approx \frac{1}{4} F_h F_f \tag{6.6.12}$$

in the ideal situation defined by Eq. 6.6.9. The design evolves to determine the optimum geometry that yields a maximum value of parameters F_h and F_f. An essential task is the computation of the pole pieces which must be designed to provide the desired value of $\vec{H}_c$ and the desired distribution of $\vec{H}_c$ within the region of interest. In other words the pole pieces make it possible to uncouple the selection of $\vec{H}_c$ from the value of $\vec{H}_m$ that optimizes the energy product of the magnetic material. This degree of freedom defines the function of the pole pieces and is the conceptual difference between a structure like the magnet of Fig. 6.6.1 and a structure like the magnet of Fig. 6.5.1, where K_m and parameter K are related to each other by the equation

$$K + K_m = 1 . \tag{6.6.13}$$

The lack of a physical boundary of region V_c in a traditional magnet makes it impossible to confine the field within the region of interest and may result in a value of parameter F_f small compared to unity. This is apparent in Fig. 6.6.1, which shows a large fraction of the flux of $\vec{B}$ outside the region between the pole pieces.

The closing of the region of interest in a traditional magnet can be achieved by following the approach defined in Section 6.1 with components of magnetic material which satisfy the condition

$$\vec{B} = 0 \tag{6.6.14}$$

like in the structure of Fig. 6.5.1. For instance, consider the schematics of the two-dimensional magnet of Fig. 6.6.3 whose pole pieces are designed to generate a uniform field intensity $\vec{H}_0$ oriented along the y axis,

$$K = \frac{\mu_0 H_0}{J_1} = 0.4 . \tag{6.6.15}$$

The field intensity H_m within the rectangular region ($V_0 V_1 W_1 W_0$) of the magnetic

236 HYBRID MAGNETS

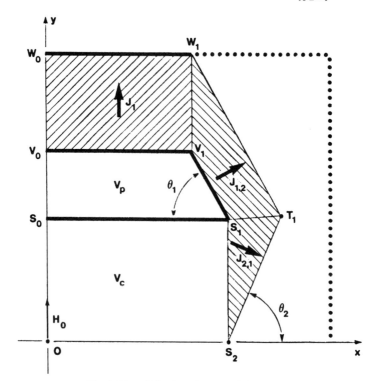

Fig. 6.6.3. Pole piece for $K_m = 0.5$, $K = 0.4$.

material of remanence J_1 is chosen to satisfy the optimum condition 6.6.9. The pole pieces as well as the external yoke in Fig. 6.6.3 are assumed to be composed of infinite permeability material.

The flux of $\vec{B}_m$ within the region $(V_0 V_1 W_1 W_0)$ is equal to the flux of $\mu_0 \vec{H}_0$ within the region $(O\, S_0 S_1 S_2)$ of the cavity. Consequently the abscissa x_1 of point V_1 is related to the abscissa x_2 of point S_1 by the equation

$$x_1 = 0.8\, x_2 . \tag{6.6.16}$$

The thickness of the pole pieces in the schematic of Fig. 6.6.3 corresponds to an angle θ_1,

$$\theta_1 = 60^o . \tag{6.6.17}$$

The magnitude of the remanences $\vec{J}_{1,2}$, $\vec{J}_{2,1}$ is assumed to be equal to J_1. The triangular region $(S_1 S_2 T_1)$ closes the rectangular cavity of the magnet. The remanence $\vec{J}_{2,1}$ in this region satisfies condition 6.6.14 and is perpendicular to side $(S_2 T_1)$. Thus, because of Eq. 6.6.15, side $(S_2 T_1)$ forms an angle θ_2 with respect to the axis x, given by

$$\cos\theta_2 = \frac{\mu_0 H_0}{J_1} = 0.4, \qquad \theta_2 \approx 66.4°. \qquad (6.6.18)$$

The trapezoidal region $(V_1 S_1 T_1 W_1)$ also satisfies condition 6.6.14, and its remanence is perpendicular to side $(T_1 W_1)$ which is parallel to side $(S_1 V_1)$. The heavy line $(W_0 W_1)$ is part of the external yoke of $\mu = \infty$ material. Because no flux of $\vec{B}$ is present in the regions of remanence $\vec{J}_{1,2}$, $\vec{J}_{2,1}$, sides $(T_1 W_1)$, $(S_2 T_1)$ do not have to be part of the interface with the external yoke whose geometry may be arbitrarily selected, as indicated by the dotted line in the schematic of Fig. 6.6.3.

The figure of merit of the magnet of Fig. 6.6.3 is

$$M \approx 0.16. \qquad (6.6.19)$$

The presence of the components of magnetic material of remanences $\vec{J}_{1,2}$, $\vec{J}_{2,1}$ is essential to generate a uniform field in the rectangular cavity of the magnet. Figure 6.6.4 shows the configuration of the equipotential lines that results from the removal of the two components of remanences $\vec{J}_{1,2}$, $\vec{J}_{2,1}$.

The dotted lines show the equipotential lines in the structure of Fig. 6.6.3. In par-

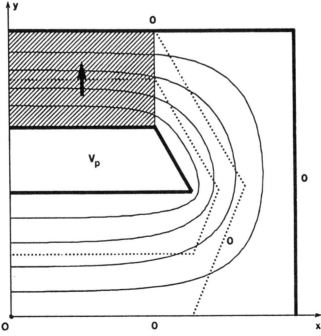

Fig. 6.6.4. Equipotential lines in the permanent magnet without the components of remanences $\vec{J}_{1,2}, \vec{J}_{2,1}$.

238 HYBRID MAGNETS

ticular the dotted line $\Phi = 0$ establishes the boundary of the field generated by the structure. With the removal of the components of remanences $\vec{J}_{1,2}, \vec{J}_{2,1}$, the $\Phi = 0$ boundary line coincides with the yoke boundary.

In Fig. 6.6.3 one has

$$0 < \theta_2 < \frac{\pi}{2}. \tag{6.6.20}$$

Thus the type of structure depicted in Fig. 6.6.3 is limited to the generation of values of K

$$K \leq 1, \tag{6.6.21}$$

i.e., $\mu_0 H_0$ cannot exceed the remanence of the magnetic material of the triangular component $(S_1 S_2 T_1)$.

The structure of Fig. 6.6.3 is designed for $K < 0.5$. If the value of K is selected in the range

$$\frac{1}{2} < K < 1, \tag{6.6.22}$$

the field intensity within the cavity becomes larger than the field intensity in the region of remanence $\vec{J}_1$. Thus the pole pieces must be shaped in a way to focus the field generated by the magnetic material into the cavity. This focusing effect is obtained, for instance, with the shape of the pole pieces shown in the schematics of Fig. 6.6.5 that is designed to generate a value

$$K = 0.8. \tag{6.6.23}$$

Again, the rectangular component $(V_0 S_1 W_2 W_1)$ of remanence J_1 is designed to satisfy the optimum condition 6.6.9. The angle θ_1 of the triangular pole piece of Fig. 6.6.5 is given by

$$\cos \theta_1 = \frac{K_m}{K}. \tag{6.6.24}$$

Thus by virtue of Eqs. 6.6.9 and 6.6.23 one has

$$\cos \theta_1 = \frac{1}{1.6}, \quad \theta_1 \approx 51.3^o. \tag{6.6.25}$$

The other components of remanence $\vec{J}_{1,1}, \vec{J}_{1,2}, \vec{J}_{2,1}$ are all designed to satisfy condition 6.6.14, and the geometry shown in Fig. 6.6.5 corresponds to remanences with the

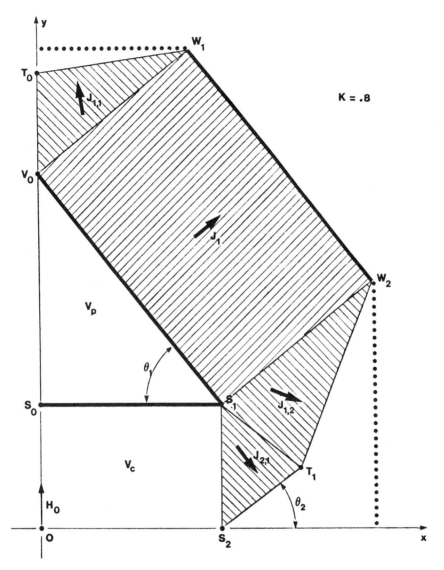

Fig. 6.6.5. Pole piece for $K_m = 0.5$, $K = 0.8$.

same magnitude throughout the entire structure. Thus angle θ_2 is given by

$$\cos\theta_2 = K = 0.8, \qquad \theta_2 \approx 36.9°. \tag{6.6.26}$$

As in the case of Fig. 6.6.3, the material of remanence $\vec{J}_1$ must interface with the external yoke through side $(W_1 \ W_2)$, and again, because of condition 6.6.14 in the other components of magnetic material, the geometry of the yoke can be arbitrarily chosen as indicated by the dotted line in Fig. 6.6.5.

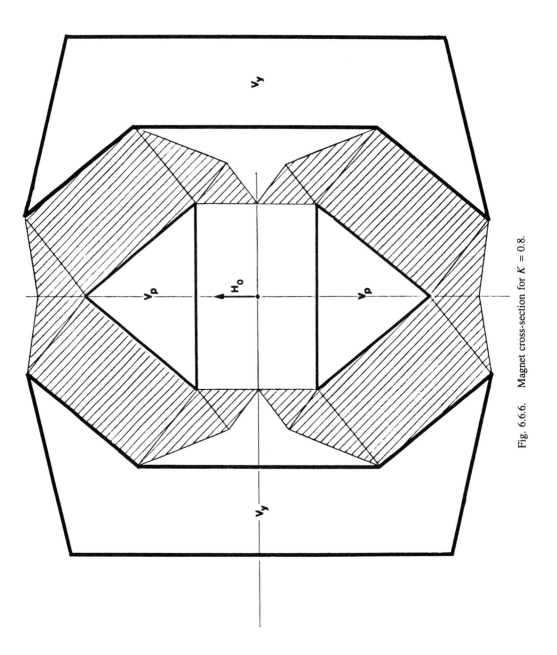

Fig. 6.6.6. Magnet cross-section for $K = 0.8$.

The figure of merit of the magnet of Fig. 6.6.5 is

$$M \approx 0.18 , \qquad (6.6.27)$$

which is larger than the value 6.6.19 of the $K = 0.4$ magnet because, as K increases, the area of the material of remanence $\vec{J}_1$ becomes increasingly larger than the area of the other components of the structure. A schematic of the full cross-section of the $K = 0.8$ magnet is shown in Fig. 6.6.6. One observes that the yoke does not have to totally enclose the magnetic structure.

The closure of the cavity in the structures of Figs. 6.6.3 and 6.6.4 is achieved with magnetized components that cannot operate at the peak of the energy product curve, because they must satisfy condition 6.6.14. Therefore the optimum design condition 6.6.9 does not extend to the entire structure of the magnet. The confinement of the flux within the closed cavity is achieved at the cost of reducing the figure of merit below the ideal limit $M = 0.25$.

It is of importance to point out that the geometries and the field configuration of the magnet with pole pieces like the magnet of Figs. 6.6.3 and 6.6.5 are always exact solutions of the problem of generating a uniform field within a closed cavity. Thus the magnets with pole pieces enjoy the same property of the yokeless, yoked, and hybrid structures developed according to the methodology of Chapter 3. However, the analytical solution of a uniform field is valid only outside of the $\mu = \infty$ boundary of the pole pieces, in contrast to the solution of the other categories of magnetic structures, where the solution is valid everywhere regardless of the presence of the $\mu = \infty$ components.

REFERENCES

[1] M. G. Abele, Optimum design of two-dimensional permanent magnets, Technical Report No. 21, New York University, Oct 15, 1989.
[2] M. G. Abele, A high efficiency yoked permanent magnet, Technical Report No. 23, New York University, Oct 1, 1990.
[3] W. Neugebauer, E. M. Branch, Applications of Cobalt-Samarium magnets to microwave tubes, Technical Report, Microwave Tube Operations, General Electric, March 1972, Schenectady, New York.
[4] J. P. Clarke, H. A. Leupold, *IEEE Transactions Magnetism*, MAG-22, No.5 (1986), p.1063.
[5] H. A. Leupold, E. Potenziani, *IEEE Transactions Magnetism*, MAG-22, No.5 (1986), p.1078.
[6] T. Miyamoto, H. Sakurai, M. Aoki, A permanent magnet assembly for MRI devices, *Proceedings of the Tenth International Workshop on Rare Earth Magnets and their Applications*, Tokyo, Japan, pp. 113-120, May 1989.
[7] R. J. Parker and R. J. Studders, Permanent Magnets and their Applications, John Wiley, New York 1961.
[8] H. A. Leupold, E. Potenziani, An overview of modern permanent magnet design, US Army SLCET-TR-90-6, August 1990, Forth Monmouth, New Jersey, USA.

CHAPTER 7

Materials with Linear Magnetic Characteristics

INTRODUCTION

The structures analyzed so far are based on an ideal model of zero magnetic susceptibility χ_m of the magnetic materials and infinite magnetic permeability μ of the ferromagnetic materials.

As indicated in Chapter 1, magnetic materials of practical interest for the development of high field permanent magnets exhibit quasi-linear demagnetization characteristics whose slopes correspond to values of the magnetic susceptibility χ_m of the order of 5×10^{-2} or smaller. Even with a straight line approximation of the demagnetization curve, each element of the magnetic structure loses the property of being perfectly transparent to the magnetic field, and in principle, only approximate solutions can be sought for the calculation of field and geometry of complex magnetic structures. However, in the limit of $\chi_m \ll 1$, one can expect χ_m to result in a small perturbation of the field obtained in the ideal situation of a perfectly transparent material. Thus the magnetic structures can be analyzed by the perturbation of the exact solution obtained for $\chi_m = 0$.

Similarly, if the magnetic permeability μ is finite, the boundaries of the ferromagnetic material cease to be equipotential surfaces and again one has to resort to approximate solutions. However, with ferromagnetic materials characterized by large values of μ ($\mu \gg \mu_0$), as long as the materials do not work close to saturation, in general, the finite value of μ results also in a small perturbation of the solution obtained for $\mu = \infty$. As a consequence the effects of $\chi_m \neq 0$ and $\mu_0/\mu \neq 0$ can be analyzed independently of each other, and the combined effect of μ and χ_m can be obtained by adding the two independent solutions [1,2].

7.1 FIELD COMPUTATION IN MAGNETIC STRUCTURES WITH SMALL, NONZERO SUSCEPTIBILITY

Consider first the case of magnetic materials with $\chi_m \neq 0$ and $\mu = \infty$ ferromagnetic materials. Assume that the magnetic materials exhibit a linear demagnetization curve

$$\vec{B} = \vec{J} + \mu_0(1 + \chi_m)\vec{H}, \qquad (7.1.1)$$

where the positive magnetic susceptibility χ_m satisfies the condition

$$\chi_m \ll 1 . \qquad (7.1.2)$$

The solution of the field equation in each region of a magnetic structure can be written in the form

$$\vec{B} = \vec{B}(0) + \delta\vec{B} , \quad \vec{H} = \vec{H}(0) + \delta\vec{H} , \qquad (7.1.3)$$

where $\vec{B}(0)$, $\vec{H}(0)$ are the magnetic induction and the field intensity which satisfy Eq. 7.1.1 in the limit $\chi_m = 0$. By virtue of Eq. 7.1.2 one can assume

$$|\delta\vec{B}| \ll |\vec{B}(0)| , \quad |\delta\vec{H}| \ll |\vec{H}(0)| . \qquad (7.1.4)$$

By neglecting higher order terms, Eq. 7.1.1 yields

$$\delta\vec{B} = \mu_0 \chi_m \vec{H}(0) + \mu_0 \delta\vec{H} . \qquad (7.1.5)$$

Thus, once the magnet geometry and the field configuration have been computed for the ideal condition $\chi_m = 0$, the field distortion caused by a small value of χ_m can be computed by means of Eq. 7.1.5, with the assumption of a structure of perfectly transparent materials magnetized with a remanence

$$\delta\vec{J} = \mu_0 \chi_m \vec{H}(0) . \qquad (7.1.6)$$

Intensity $\vec{H}(0)$ varies from component to component of the structure. Although $\vec{H}(0)$ is uniform within each component, in general vectors $\delta\vec{B}$ and $\delta\vec{H}$ generated by $\delta\vec{J}$ are not.

Consider the region of a two-dimensional yokeless magnet shown in Fig. 7.1.1, where the distribution of $\vec{H}$ and the geometry of the magnet components are computed in the ideal case $\chi_m = 0$ for the indicated position of point F and an arbitrary value of parameter K. The vector diagram of Fig. 7.1.2 provides the remanences $\vec{J}$ and the intensities $\vec{H}$ in the magnetic structure with the assumption that intensity $\vec{H}_0$ is oriented along the line that contains point F and vertex S_h as shown in Fig. 7.1.1.

Equation 7.1.6 yields the polarization induced by intensity $\vec{H}(0)$ in each region of the magnet. By means of Eq. 7.1.5 one can compute the equivalent surface charge densities $\delta\sigma_k$ induced by $\delta\vec{J}$ on the interfaces between magnetized prisms. According to definition 3.1.8,

$$\delta\sigma_k = \mu_0 \chi_m (\vec{H}_{k-1} - \vec{H}_k) \cdot \vec{n}_k . \qquad (7.1.7)$$

In general, the values of $\delta\sigma_k$ given by Eq. 7.1.7 and the orientations $\vec{\tau}_k$ of the inter-

244 MATERIALS WITH LINEAR CHARACTERISTICS

faces computed for the ideal case $\chi_m = 0$ do not satisfy condition 3.1.10 of existence of a uniform field solution. A distortion of the field results in each region of the magnet if at some of the vertices U_h, S_h, T_h, U_{h+1}, ... one has

$$\sum_k \delta\sigma_k \vec{\tau}_k \neq 0. \tag{7.1.8}$$

Inequality 7.1.8 means that a singularity of the field intensity is generated by the nonzero value of χ_m at some of the vertices of the polygonal contours. Because of Eq. 7.1.8 and the logarithmic singularity of the field, condition 7.1.4 can be assumed to be valid everywhere with the exception of points in close proximity of these vertices.

Consider first the vertices of the external boundary of the magnetic structure of Fig. 7.1.1. Vertex T_h is common to sides $u_{h,2}$, $u_{h+1,1}$ and the interface t_h between the regions where the intensities computed for $\chi_m = 0$ are equal to $\vec{H}_{h,2}$ and $\vec{H}_{h+1,1}$. As shown in Section 1 of Chapter 4,

$$|\vec{H}_{h,2}| = |\vec{H}_{h+1,1}|. \tag{7.1.9}$$

Vectors $\vec{H}_{h,2}$ and $\vec{H}_{h+1,1}$ are perpendicular to sides $u_{h,2}$ and $u_{h+1,1}$, respectively, and they have the same tangential components along interface t_h that bisects the angle between $u_{h,2}$ and $u_{h+1,1}$. The surface charge density $\delta\sigma_3$ induced by $\vec{H}_{h,2}$ and $\vec{H}_{h+1,1}$ on interface t_h is

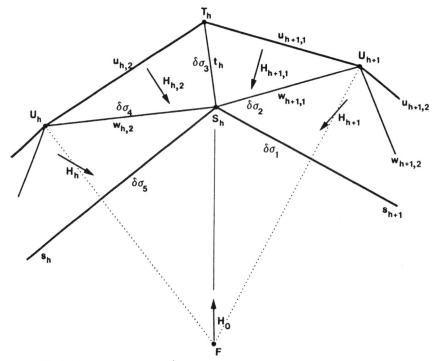

Fig. 7.1.1. Distribution of $\vec{H}$ in a section of a two-dimensional yokeless magnet.

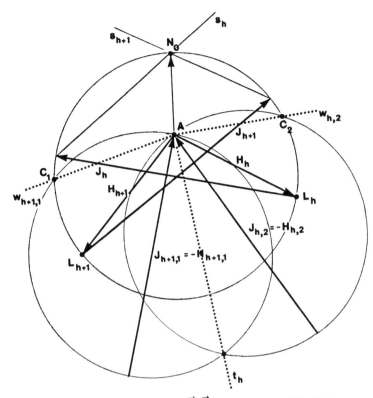

Fig. 7.1.2. Computation of $\vec{J}$, $\vec{H}$ in the structure of Fig 7.1.1.

$$\delta\sigma_3 = 2\mu_0\chi_m \vec{H}_{h+1,1} \cdot \vec{n}_3 = -2\mu_0\chi_m \vec{H}_{h,2} \cdot \vec{n}_3, \qquad (7.1.10)$$

where unit vector $\vec{n}_3$ perpendicular to t_h is oriented from the region of intensity $\vec{H}_{h+1,1}$ to the region of intensity $\vec{H}_{h,2}$. At vertex T_h one has

$$\sum_k \delta\sigma_k \vec{\tau}_k = 0. \qquad (7.1.11)$$

The cancellation of the singularity at T_h is shown in Fig. 7.1.3, where all unit vectors $\vec{\tau}_k$ are rotated by $-\pi/2$ with respect to the corresponding interfaces.

Consider vertex U_{h+1} and the two interfaces $u_{h+1,1}$ and $w_{h+1,1}$ in Fig. 7.1.1. The sum of vectors $\delta\sigma_k \vec{\tau}_k$ over the two lines $u_{h+1,1}$ and $w_{h+1,1}$ is equal and opposite to intensity $\vec{H}_{h+1}$, as shown in Fig. 7.1.4. The diagram of the dotted vectors in Fig. 7.1.4 gives the sum of vectors $\delta\sigma_k \vec{\tau}_k$ over the interfaces $u_{h,2}$, $w_{h,2}$ which have vertex U_h in common. One observes that the sum of these two vectors is equal in magnitude and orientation to intensity $\vec{H}_h$. As a consequence, the sum of the two vectors $\delta\sigma_k \vec{\tau}_k$ over the two sides $u_{h+1,2}$ and $w_{h+1,2}$, which have vertex U_{h+1} in common, must be equal in magnitude and orientation to $\vec{H}_{h+1}$. Therefore, condition 3.1.10 is always satisfied at

246 MATERIALS WITH LINEAR CHARACTERISTICS

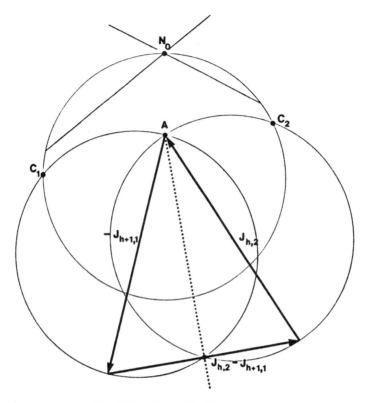

Fig. 7.1.3. Sum of $\delta\sigma_k \vec{\tau}_k$ at vertex T_h.

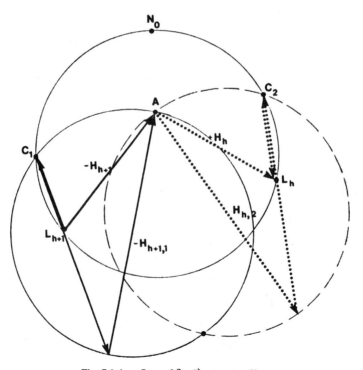

Fig. 7.1.4. Sum of $\delta\sigma_k \vec{\tau}_k$ at vertex U_{h+1}.

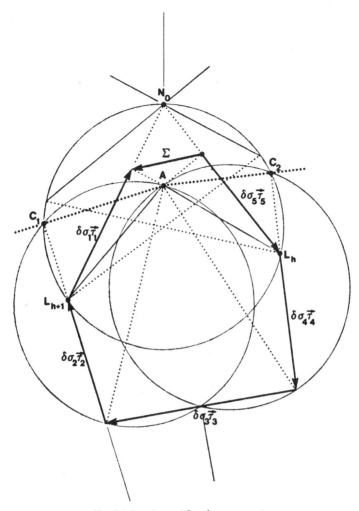

Fig. 7.1.5. Sum of $\delta\sigma_k\vec{\tau}_k$ at vertex S_h.

vertex U_{h+1} and no singularity of the field intensity generated by the induced polarization is found at the vertices of the external boundary of the magnet.

Consider now vertex S_h of the internal boundary. The sum of the three vectors $\delta\sigma_k\vec{\tau}_k$ over interfaces $w_{h,2}$, t_h and $w_{h+1,1}$ is a vector whose origin is located at point L_h of Fig. 7.1.5 and whose tip is located at point L_{h+1}. As indicated in Fig. 7.1.2, points L_h, L_{h+1} are the origins of vectors $\vec{J}_h$, $\vec{J}_{h+1}$, respectively. The two vectors $\delta\sigma_k\vec{\tau}_k$ over interfaces s_h, s_{h+1} are the projections of vectors $\vec{H}_h$, $-\vec{H}_{h+1}$ over lines (N_0L_h), $(L_{h+1}N_0)$, respectively, as shown in Fig. 7.1.5. Thus the sum of vectors $\delta\sigma_k\vec{\tau}_k$ at point S_h is independent of the geometry and magnetization of the magnet components that interface with the outside medium.

Vector $\Sigma\sigma_k\vec{\tau}_k$ at point S_h is determined by the value of K and remanences $\vec{J}_h$, $\vec{J}_{h+1}$. As indicated in Fig. 7.1.6, let α_h be the angle between the two unit vectors $\vec{n}_h$, $\vec{n}_{h+1}$ perpendicular to s_h, s_{h+1}, respectively, and let α be the angle of $\vec{n}_h$ relative to

248 MATERIALS WITH LINEAR CHARACTERISTICS

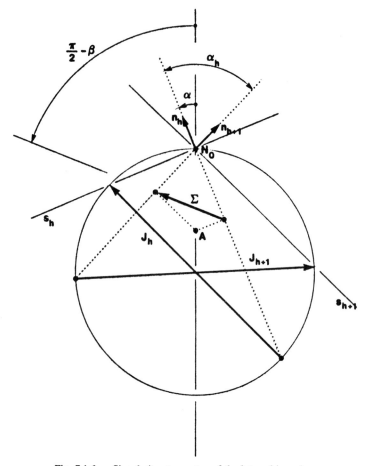

Fig. 7.1.6. Singularity at a vertex of the internal boundary.

the vector $\vec{H}_0$. The distance of point A from point N_0 is equal to KJ_0. Thus the vector defined by Eq. 7.1.8 has a magnitude

$$\left|\sum \delta\sigma_k \vec{\tau}_k\right| = \chi_m KJ_0 \sin\alpha_h \,, \tag{7.1.12}$$

and its components parallel and perpendicular to $\vec{H}_0$ are

$$\begin{aligned}\Sigma_{\parallel} &= \chi_m \frac{KJ_0}{2}[\cos 2(\alpha_h-\alpha) - \cos 2\alpha]\\ \Sigma_{\perp} &= \chi_m \frac{KJ_0}{2}[\sin 2(\alpha_h-\alpha) + \sin 2\alpha]\,,\end{aligned} \tag{7.1.13}$$

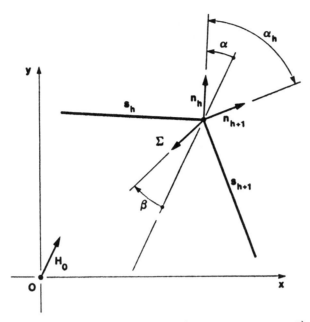

Fig. 7.1.7. Orientation of sum of $\delta\sigma_k \vec{\tau}_k$ at point S_h relative to $\vec{H}_0$.

and the angle γ of vector 7.1.8 relative to $\vec{H}_0$ is

$$\gamma = \tan^{-1} \frac{\Sigma_\perp}{\Sigma_{||}} . \tag{7.1.14}$$

Thus in the yokeless structure of Fig. 7.1.1 all singularities of the field intensity are limited to the vertices of the internal boundary.

As previously stated, in the vector diagram of Fig. 7.1.5 all unit vectors $\vec{\tau}_k$ are rotated by $-\pi/2$ with respect to the corresponding interfaces. Thus in the frame of reference of the magnet cross-section, at each vertex S_h of the internal boundary, vector 7.1.8 is oriented at the angle

$$\beta = \frac{\pi}{2} - \gamma \tag{7.1.15}$$

with respect to $\vec{H}_0$ as indicated in the schematic of Fig. 7.1.7.

7.2 SINGULARITIES IN MULTILAYER MAGNETS

As shown in Chapter 4, multilayered structures of concentric yokeless magnets can be used to generate increasingly large values of the field within the cavity. The magnetic

250 MATERIALS WITH LINEAR CHARACTERISTICS

material of each individual magnet is subjected not only to its own field, but also to the field generated by all the other layers of the structure that surround it. Thus, in a general case, with the exception of the outside surface of the multilayer structure, the nonzero magnetic susceptibility of the material will yield a singularity of the intensity at all the vertices of each individual layer.

A particular situation arises when internal and external boundaries of all magnets of the multilayer structure have similar geometry and all layers are designed for the same value of K and the same orientation of $\vec{H}_0$. Consider the schematic of Fig. 7.2.1, where S_e is a vertex of the internal boundary of a magnet which generates intensity $\vec{H}_0$. Vector $\vec{\Sigma}$ is the sum of vectors $\delta\sigma_k\vec{\tau}_k$ at S_e. Assume that a second magnet of similar geometry is located inside the first one, as indicated in Fig. 7.2.1. If both magnets have identical distribution of the remanence, the total intensity within the common cavity is equal to $2\vec{H}_0$.

As shown in the previous section, the sum of vectors $\delta\sigma_k\vec{\tau}_k$ generated by the field of the second magnet at vertex T_i of its external boundary is always zero. On the other hand, the field of the external magnet at point T_i generates a vector

$$\delta\sigma_6\vec{\tau}_6 + \delta\sigma_7\vec{\tau}_7 = -\vec{\Sigma} ; \qquad (7.2.1)$$

i.e., vector 7.2.1 is equal and opposite to the vector $\vec{\Sigma}$ generated at vertex S_e by the field of the external magnet. As a consequence, if the air gap between the two magnets

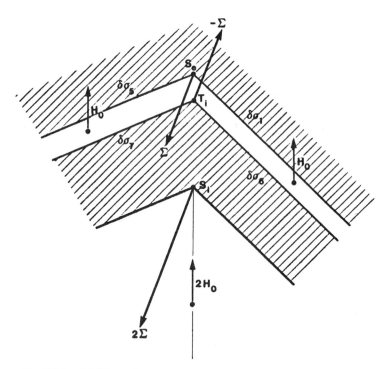

Fig. 7.2.1. Multilayered structure of concentric magnets of similar geometries.

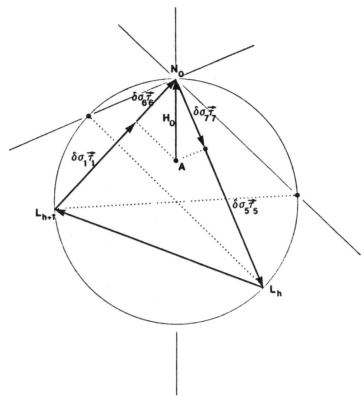

Fig. 7.2.2. Cancellation of the field singularities at the vertices common to two layers of a multilayered structure.

is eliminated, points T_i and S_e coincide and the total sum of vectors $\delta\sigma_k\vec{\tau}_k$ cancels, as shown by the vector diagram of Fig. 7.2.2.

Conversely, at vertex S_i of the internal boundary, the fields of both magnets generate vectors of equal magnitude and orientation. Consequently at point S_i one has

$$\sum \delta\sigma_k\vec{\tau}_k = 2\vec{\Sigma}. \tag{7.2.2}$$

Thus in a multilayer structure of n magnets of similar geometries, with no gaps in between, singularities are confined to the vertices of the internal cavity where the magnitude of vector $\vec{\Sigma}$ is

$$\left|\sum \delta\sigma_k\vec{\tau}_k\right| = n\chi_m K J_0 \sin\alpha_0, \tag{7.2.3}$$

and its orientation is given by Eq. 7.1.14.

As an example consider the square cross-sectional magnet analyzed in Chapter 4,

252 MATERIALS WITH LINEAR CHARACTERISTICS

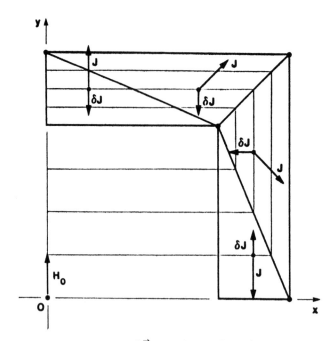

Fig. 7.2.3. Induced polarizations $\delta \vec{J}$ in the first quadrant of a square cross-sectional magnet for $\vec{H}_0$ parallel to a side of the square.

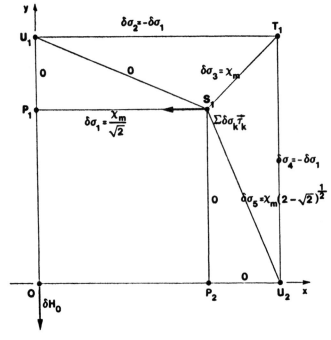

Fig. 7.2.4. Values of charges $\delta \sigma_k$ in a single layer square cross-sectional magnet.

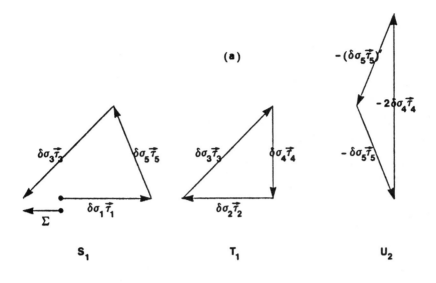

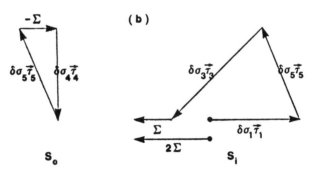

Fig. 7.2.5. Computation of $\sum \sigma_k \vec{\tau}_k$ in a single layer (a) and a double layer (b) square cross-sectional magnet.

designed for

$$K = 1 - \frac{1}{\sqrt{2}}, \qquad (7.2.4)$$

and an orientation of vector $\vec{H}_0$ parallel to a side of the square. Figure 7.2.3 shows the distributions of the induced polarizations $\delta \vec{J}$ as well as the remanences $\vec{J}$ in the first quadrant of the magnet cross-section. The values of the equivalent surface charge densities $\delta \sigma_k$ induced by $\delta \vec{J}$ are shown in Fig. 7.2.4. The computation of the sum of vectors $\delta \sigma_k \vec{\tau}_k$ at points S_1, T_1, U_2 is shown in Fig. 7.2.5(a). One has

$$\left[\sum \delta \sigma_k \vec{\tau}_k \right]_{T_1} = \left[\sum \delta \sigma_k \vec{\tau}_k \right]_{U_2} = 0 \qquad (7.2.5)$$

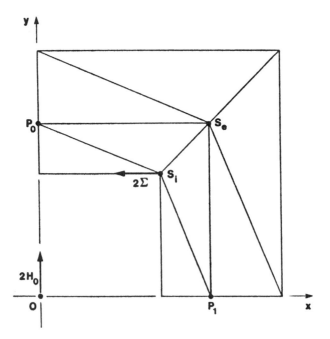

Fig. 7.2.6. Double layer square cross-sectional magnet.

and

$$\left[\sum \delta \sigma_k \vec{\tau}_k\right]_{S_1} = -\left[1 - \frac{1}{\sqrt{2}}\right]\chi_m J_0 \vec{x} \tag{7.2.6}$$

at point S_1. In Eq. 7.2.6, $\vec{x}$ is a unit vector oriented in the direction parallel to the x axis. Vector 7.2.6 is shown in Fig. 7.2.4.

Assume now that a second square cross-sectional magnet designed for the same value of K and the same orientation of $\vec{H}_0$ is inserted in the cavity of Fig. 7.2.3, as indicated in the schematic of Fig. 7.2.6. As shown by the vector diagrams of Fig. 7.2.5(b), the singularity at point S_e vanishes, and at point S_i one has

$$\left[\sum \delta \sigma_k \vec{\tau}_k\right]_{S_i} = -2\left[1 - \frac{1}{\sqrt{2}}\right]\chi_m J_0 \vec{x}. \tag{7.2.7}$$

7.3 YOKED MAGNETS

Consider the region of a two-dimensional yoked magnet shown in Fig. 7.3.1 where S_h is a vertex common to sides s_h, s_{h+1} of the polygonal contour of the cavity. Both

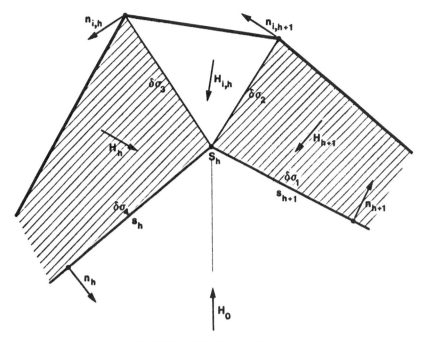

Fig. 7.3.1. Yoked magnetic structure.

configurations of the schematics of Figs. 7.1.1 and 7.3.1 correspond to the same value of parameter K and the same orientation of $\vec{H}_0$. Intensity $\vec{H}_{i,h}$ within the air wedge between the magnetized components of Fig. 7.3.1 results from the selection of point D shown in the vector diagram of Fig. 7.3.2. At point S_h one has

$$\sum_k \delta\sigma_k \vec{\tau}_k = \mu_0 \chi_m \left[-(\vec{H}_{h+1} \cdot \vec{n}_{h+1})\vec{\tau}_1 + (\vec{H}_{h+1} \cdot \vec{n}_{i,h+1})\vec{\tau}_2 \right. \\ \left. - (\vec{H}_h \cdot \vec{n}_{i,h})\vec{\tau}_3 + (\vec{H}_h \cdot \vec{n}_h)\vec{\tau}_4 \right], \quad (7.3.1)$$

where $\vec{n}_h$, $\vec{n}_{i,h}$, $\vec{n}_{i,h+1}$, $\vec{n}_{h+1}$ are the unit vectors perpendicular to the interfaces having point S_h in common, with the orientations defined in Fig. 7.3.1. The values $k = 1, 2, 3, 4$ of subindices k of unit vectors $\vec{\tau}_k$ correspond to the interfaces where the charge densities are $\delta\sigma_1$, $\delta\sigma_2$, $\delta\sigma_3$, $\delta\sigma_4$ shown in Fig. 7.3.1.

In general, as shown in the vector diagram of Fig. 7.3.2, one has

$$\sum_k \delta\sigma_k \vec{\tau}_k \neq 0 \quad (7.3.2)$$

at each vertex of the internal boundary of the yoked magnet, regardless of the position

256 MATERIALS WITH LINEAR CHARACTERISTICS

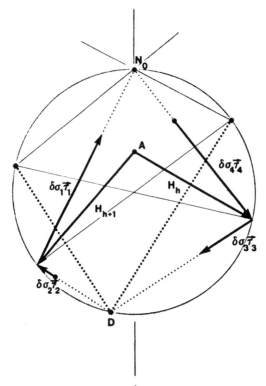

Fig. 7.3.2. Computation of vectors $\vec{H}$ and $\delta\sigma_k \vec{\tau}_k$ in the structure of Fig. 7.3.1.

of point D and the value of parameter K, with the exception of the particular value

$$K = \frac{1}{2}, \tag{7.3.3}$$

which corresponds to the vector diagram if Fig. 7.3.3, where point A coincides with the center of the circle. In Fig. 7.3.3 one has

$$\delta\sigma_1\vec{\tau}_1 + \delta\sigma_2\vec{\tau}_2 = -\delta\sigma_3\vec{\tau}_3 - \delta\sigma_4\vec{\tau}_4, \tag{7.3.4}$$

i.e., vector $\vec{\Sigma}$ cancels at point S_h.

Thus, in a yoked magnet, for the particular value $K = 1/2$, no singularity is generated at the corners of the internal boundary of the magnetized material. No distortion and no change of orientation are suffered by the field within the cavity as a result of the nonzero value of χ_m. In each region of the magnet the magnitude of the field intensity decreases by the same amount,

$$\delta H = -\frac{1}{2}\chi_m H_0. \tag{7.3.5}$$

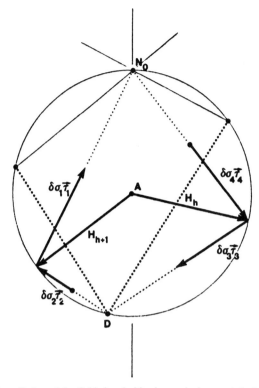

Fig. 7.3.3. Cancellation of the field singularities in a yoked magnet designed for $K = 0.5$.

7.4 FIELD DISTORTION IN A TWO-DIMENSIONAL SQUARE CROSS-SECTIONAL MAGNET

Chapter 3 has shown that the geometry of a two-dimensional yokeless magnet is independent of the orientation of intensity $\vec{H}_0$ of the field inside the cavity. However, the value of $\vec{H}$ in each region of the magnetic structure changes with the orientation of $\vec{H}_0$, and as a consequence, a nonzero value of χ_m results in a change $\delta \vec{J}$ of the remanence of each region which depends on the orientation of $\vec{H}_0$. Thus the distortion of the field within each region of the magnet, including its cavity, can be expected to be a function of the orientation of $\vec{H}_0$.

As an example consider again the square cross-sectional yokeless magnet with $\vec{H}_0$ oriented parallel to a side of the square, as shown in Fig. 7.2.3. In Fig. 7.2.4 the sum of vectors $\delta \sigma_k \vec{\tau}_k$ at point S_1 cancels if on side $(P_1 S_1)$ one adds a surface charge density

$$\delta \sigma = \chi_m \left[1 - \frac{1}{\sqrt{2}} \right] J_0 . \qquad (7.4.1)$$

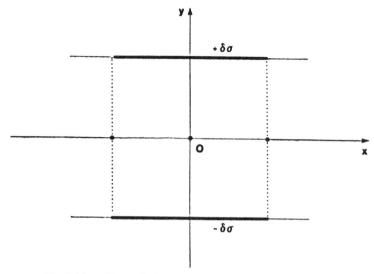

Fig. 7.4.1. Charge distribution on two strips in the plane $y = \pm x_0$.

Thus the structure of Fig. 7.2.4 may be considered as the superposition of two surface charge distributions: a structure identical to Fig. 7.2.4 where the total charge density on side $(P_1 S_1)$ is

$$\delta \sigma'_1 = \chi_m J_0 \tag{7.4.2}$$

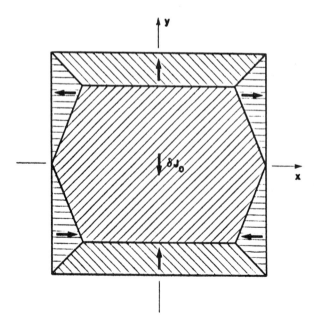

Fig. 7.4.2. Magnetic structure that generates a zero value of $\delta \vec{B}$.

and the geometry of Fig. 7.4.1 with a distribution of charges $\pm\delta\sigma$ on two strips on the planes $y = \pm x_0$ bound by planes $x = \pm x_0$, where x_0 is half the side of the square cross-section of the magnet cavity, as indicated in Fig. 7.4.1. The total charge density 7.4.2 on side $(P_1 S_1)$ is achieved by assuming that the region of nonmagnetic material $(O\, P_1 S_1 P_2)$ in Fig. 7.2.4 is replaced by a magnetic material of remanence

$$\delta \vec{J}_0 = \chi_m \left[1 - \frac{1}{\sqrt{2}} \right] J_0 \vec{y}, \qquad (7.4.3)$$

which coincides with the induced remanence $\delta \vec{J}$ of the triangular region $(P_2 U_2 S_1)$ in Fig. 7.2.4. With the addition of the surface charge density 7.4.1, the values of $\delta \sigma_k$ on the interfaces of Fig. 7.2.4 are

$$\delta \sigma_k = -\chi_m \sigma_k, \qquad (7.4.4)$$

where σ_k are the surface charge densities on the same interfaces induced by the magnetic material in the ideal limit $\chi_m = 0$. Thus because the field generated by σ_k is uniform everywhere, the field generated by $\delta \sigma_k$ is also uniform.

In the limit $\chi_m = 0$, the value of $\vec{H}_0$ generated by the magnetic material within the cavity is

$$\vec{H}_0 = \frac{1}{\mu_0} \left[1 - \frac{1}{\sqrt{2}} \right] J_0 \vec{y}. \qquad (7.4.5)$$

Thus the uniform value of $\delta \vec{H}_0$ generated within the cavity by the distribution of charge $\delta \sigma_k$ given by Eq. 7.4.4 is

$$\delta \vec{H}_0 = -\frac{\chi_m}{\mu_0} \left[1 - \frac{1}{\sqrt{2}} \right] J_0 \vec{y}, \qquad (7.4.6)$$

i.e., by virtue of Eq. 7.4.3

$$\delta \vec{H}_0 = -\frac{\delta \vec{J}_0}{\mu_0}. \qquad (7.4.7)$$

Thus the magnetic induction $\delta \vec{B}$ generated by charges 7.4.4 is zero within the region $(O\, P_1 S_1 P_2)$, and as a consequence, $\delta \vec{B}$ must be zero everywhere.

The substitution of the cavity with the magnetic material of remanence 7.4.3 results in the structure of Fig. 7.4.2 which shows the distribution of polarizations $\delta \vec{J}$ induced by χ_m throughout the structure of the magnet. Vectors $\delta \vec{J}$ have the same magnitude

$$|\delta \vec{J}| = \frac{\chi_m}{\sqrt{2}} J_0. \qquad (7.4.8)$$

260 MATERIALS WITH LINEAR CHARACTERISTICS

Because the magnetic induction generated by the structure of Fig. 7.4.2 is identically zero, the perturbation generated by χ_m in the square cross-sectional magnet is equal and opposite to the magnetic induction generated everywhere by charges $\pm\delta\sigma$ in Fig. 7.4.1.

In the region

$$0 < x < x_0, \qquad 0 < y < x_0 \qquad (7.4.9)$$

the two-dimensional distribution of charges $\pm\delta\sigma$ of Fig. 7.4.1 generates a scalar potential

$$\Phi(x,y) = -\frac{\delta\sigma}{2\pi\mu_0}\left[(x_0 - x)\ln\frac{r_{1+}}{r_{1-}} + (x_0 + x)\ln\frac{r_{2+}}{r_{2-}}\right.$$

$$+ (x_0 - y)(\phi_{2+} - \phi_{1+}) \qquad (7.4.10)$$

$$\left. - (x_0 + y)(\phi_{2-} - \phi_{1-})\right],$$

where

$$r_{1+} = \left[(x_0 - x)^2 + (x_0 - y)^2\right]^{1/2}$$

$$r_{2+} = \left[(x_0 + x)^2 + (x_0 - y)^2\right]^{1/2}$$

$$r_{1-} = \left[(x_0 - x)^2 + (x_0 + y)^2\right]^{1/2}$$

$$r_{2-} = \left[(x_0 + x)^2 + (x_0 + y)^2\right]^{1/2}$$

$$\tan\phi_{1+} = \frac{x_0 - y}{x_0 - x} \qquad (7.4.11)$$

$$\tan(\pi - \phi_{2+}) = \frac{x_0 + x}{x_0 - y}$$

$$\tan\phi_{1-} = \frac{x_0 + y}{x_0 - x}$$

$$\tan(\pi - \phi_{2-}) = \frac{x_0 + x}{x_0 + y}.$$

Figure 7.4.3 shows the equipotential lines

$$\frac{\Phi}{\Phi_0} = \text{constant} \qquad (7.4.12)$$

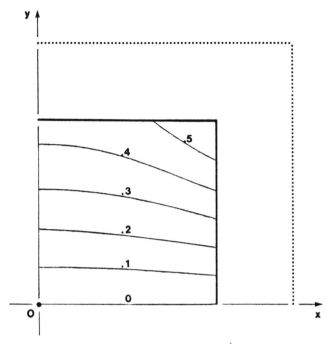

Fig. 7.4.3. Equipotential lines of the perturbation field for $\vec{H}_0$ oriented parallel to a side of the square.

in the first quadrant of the geometry of Fig. 7.4.1. Potential Φ_0 is defined as

$$\Phi_0 = \frac{\delta\sigma}{\mu_0} x_0 ; \qquad (7.4.13)$$

i.e., Φ_0 is the scalar potential generated at $y = x_0$ by a uniform distribution of surface charges $\pm\delta\sigma$ on two infinite parallel planes at $y = \pm x_0$.

From Eq. 7.4.10 one can compute the components of the field distortion generated by the equivalent charge distribution of Fig 7.4.1. In particular the y component δH_y of the field perturbation at $x = y = 0$ is given by

$$\frac{1}{\chi_m} \frac{\delta H_y}{H_0} = -\frac{1}{2} , \qquad (7.4.14)$$

where H_0 is the magnitude of $\vec{H}_0$ given by Eq. 7.4.5 in the ideal case $\chi_m = 0$. Thus the field at the center of the cavity decreases proportionally to χ_m. By comparing Eq. 7.4.14 with Eq. 7.4.6 one observes that the value of δH_y at $x = y = 0$ is half the value of the intensity δH_0 generated by the structure of Fig. 7.4.2.

262 MATERIALS WITH LINEAR CHARACTERISTICS

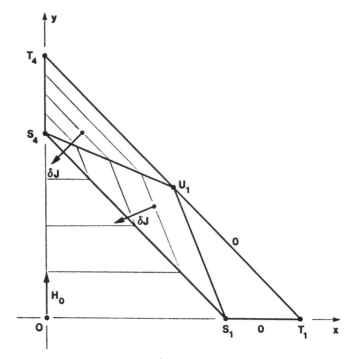

Fig. 7.4.4. Induced polarizations $\delta \vec{J}$ for H_0 parallel to a diagonal of the square.

Assume now an orientation of $\vec{H}_0$ along a diagonal of the square cross-sectional cavity. The distribution of the induced polarization $\delta \vec{J}$ is shown in Fig. 7.4.4, and the values of the equivalent surface charge densities $\delta\sigma_k$ induced by $\delta \vec{J}$ are shown in Fig. 7.4.5. At points S_1 and S_4 the sums of vectors $\delta\sigma_k \vec{\tau}_k$ are oriented along the y axis and are given by

$$\vec{\Sigma}_1 = -\vec{\Sigma}_2 = \left[\sum \delta\sigma_k \vec{\tau}_k\right]_1 = \left[1 - \frac{1}{\sqrt{2}}\right] \chi_m J_0 \vec{y}, \qquad (7.4.15)$$

where $\vec{y}$ is a unit vector oriented in the positive direction of the y axis. The magnitude of vectors $\vec{\Sigma}_1, \vec{\Sigma}_2$ coincides with the magnitudes of vectors 7.2.6 in Fig 7.2.4.

Vectors 7.4.15 cancel if surface charge densities

$$\pm \delta\sigma = \pm \frac{\chi_m}{\sqrt{2}} \left[1 - \frac{1}{\sqrt{2}}\right] J_0 \qquad (7.4.16)$$

are added on the sides of the cavity, as indicated in Fig. 7.4.6. Thus as in the case of Fig. 7.2.4, the charge distribution of Fig. 7.4.5 may be considered as the superposition of two distributions of charges: a structure identical to Fig. 7.4.5 where the total charge

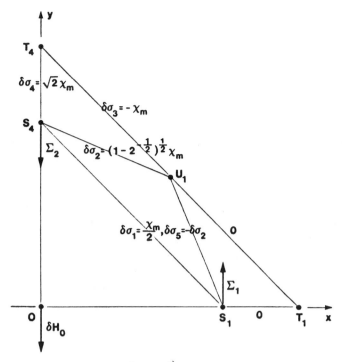

Fig. 7.4.5. Charge densities $\delta\sigma_k$ for $\vec{H}_0$ parallel to a diagonal of the square.

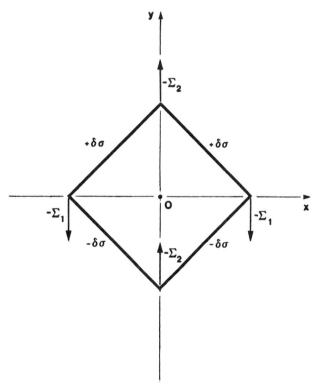

Fig. 7.4.6. Charge distribution $\delta\sigma$ to cancel the field singularities.

264 MATERIALS WITH LINEAR CHARACTERISTICS

density on side (S_1, S_4) is

$$\delta\sigma_1' = \frac{\chi_m}{\sqrt{2}} J_0 , \qquad (7.4.17)$$

and the geometry of Fig 7.4.6 with the distribution of charges $\pm\delta\sigma$ given by Eq 7.4.16. The equipotential lines generated by the equivalent charge distribution of Fig. 7.4.6 is shown in Fig. 7.4.7, where the value of Φ is normalized to the same constant Φ_0 defined by Eq. 7.4.13. The computation of the y component of the field generated by the equivalent charge distribution 7.4.16 yields a value of δH_y at $x = y = 0$ given again by Eq. 7.4.14. Thus the decrease of the field at the center of the cavity is not affected by the orientation of vector $\vec{H}_0$. However, as apparent from Figs. 7.4.3 and 7.4.7, the field configuration depends on the orientation of $\vec{H}_0$. Figure 7.4.7 shows the y component δH_y increasing on the y axis and decreasing on the x axis, while in Fig. 7.4.3, δH_y decreases on the y axis and increases on the x axis.

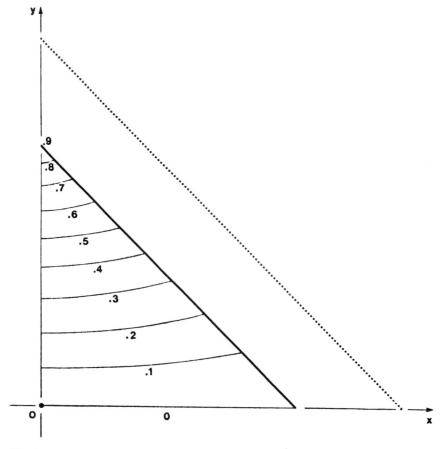

Fig. 7.4.7. Equipotential lines of the perturbation field for $\vec{H}_0$ parallel to a diagonal of the square.

As previously stated the nonuniform field distortion within the cavity caused by χ_m implies the presence of a magnetic field in the nonmagnetic medium that surrounds the yokeless magnet. From the induced polarization $\delta\vec{J}$ in Figs. 7.2.3 and 7.4.4 one can readily compute the dipole moment per unit length $\vec{p}_y$ of the two-dimensional structures. In both cases of the two orientations of $\vec{H}_0$ one has

$$\vec{p}_y = -4\chi_m x_0^2 \left[1 - \frac{1}{\sqrt{2}}\right] J_0 \vec{y}. \tag{7.4.18}$$

Thus at a distance $r \gg x_0$, the asymptotic value of the potential Φ_e generated by $\vec{p}_y$ is

$$\Phi_e = \frac{\vec{p}_y \cdot \vec{r}}{2\pi\mu_0 r}, \tag{7.4.19}$$

where $\vec{r}$ is the unit vector oriented in the direction of the radial distances from the axis $x = y = 0$. By virtue of Eq. 7.4.18,

$$\Phi_e = -\frac{2\chi_m}{\pi\mu_0} x_0^2 \left[1 - \frac{1}{\sqrt{2}}\right] J_0 \frac{\cos\psi}{r}. \tag{7.4.20}$$

7.5 FIELD COMPUTATION IN THE PRESENCE OF FERROMAGNETIC MEDIA WITH LINEAR MAGNETIC CHARACTERISTICS

Let us analyze the effect of a finite value of the magnetic permeability of ferromagnetic media present in a yoked or hybrid magnetic structure. Assume that the magnetic materials are perfectly transparent ($\chi_m = 0$) and that the ferromagnetic media exhibit a linear characteristic

$$\vec{B} = \mu\vec{H}, \tag{7.5.1}$$

where the magnetic permeability μ satisfies the condition

$$\frac{\mu}{\mu_0} \gg 1. \tag{7.5.2}$$

Consider the schematic of Fig. 7.5.1 where S_h is the interface between a medium of magnetic permeability μ and an external nonmagnetic medium ($\mu = \mu_0$). Assume that a magnetic field is generated by an arbitrary distribution of magnetic charges m_i located in the nonmagnetic medium. At each point P of S_h the field generated by charges m_i in the presence of the medium of permeability μ must satisfy the boundary conditions expressed by Eqs. 1.1.35 and 1.1.39 in the absence of electric currents.

266 MATERIALS WITH LINEAR CHARACTERISTICS

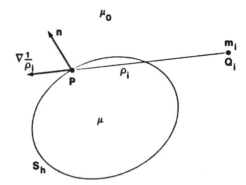

Fig. 7.5.1. Medium of permeability μ in a nonmagnetic medium.

Thus at each point P of the interface the tangential component of the field intensity $\vec{H}$ is continuous and the normal component of $\vec{H}$ satisfies the equation

$$\frac{\mu_0}{\mu}\vec{H}_e \cdot \vec{n} - \vec{H}_i \cdot \vec{n} = 0, \qquad (7.5.3)$$

where $\vec{H}_e, \vec{H}_i$ are the values of $\vec{H}$ at two points P_e, P_i at an infinitesimal distance from P outside and inside S_h, respectively. Unit vector $\vec{n}$ is perpendicular to S_h at point P and oriented outwards with respect to the medium of permeability μ, as indicated in Fig. 7.5.1.

The boundary conditions on surface S_h can be satisfied by replacing the medium of permeability μ with a surface charge density σ distributed on the surface S_h and by assuming that

$$\mu = \mu_0 \qquad (7.5.4)$$

everywhere. The intensity of the field generated by the element of charge $\sigma\, dS$ located at P is perpendicular to S_h and it is given by

$$\vec{H} = \pm \frac{\sigma(P)}{2\mu_0}\vec{n} \qquad (7.5.5)$$

at points P_e and P_i, respectively. Thus, the normal components of $\vec{H}_e$ and $\vec{H}_i$ at P_e and P_i suffer a discontinuity given by

$$(\vec{H}_e - \vec{H}_i)\cdot\vec{n} = 2H = \frac{\sigma(P)}{\mu_0}, \qquad (7.5.6)$$

and because of Eq. 7.5.3, Eq. 7.5.6 becomes

$$\left[1 - \frac{\mu_0}{\mu}\right]\vec{H}_e \cdot \vec{n} = \frac{\sigma(P)}{\mu_0}. \qquad (7.5.7)$$

At point P the normal component of $\vec{H}_e$ satisfies the boundary condition

$$\frac{\sigma(P)}{2\mu_0} - \frac{1}{4\pi\mu_0} \int_{S_h} \sigma \nabla_P \left[\frac{1}{\rho}\right] \cdot \vec{n}\, dS - \frac{1}{4\pi\mu_0} \sum_i m_i \nabla_P \left[\frac{1}{\rho_i}\right] \cdot \vec{n} = \vec{H}_e \cdot \vec{n}\,. \quad (7.5.8)$$

The second term on the left hand side of Eq. 7.5.8 is the normal component of the field intensity generated at P by the distribution of σ on S_h. Point P is at the distance ρ from charge $\sigma\, dS$ located on surface S_h, and at the distance ρ_i from charge m_i. As indicated in Fig. 7.5.1, the gradients are computed at point P. By virtue of 7.5.7, Eq 7.5.8 transforms to

$$\left[1 - \frac{2}{1 - \dfrac{\mu_0}{\mu}}\right] \sigma(P) - \frac{1}{2\pi} \int_{S_h} \sigma \nabla_P \left[\frac{1}{\rho}\right] \cdot \vec{n}\, dS = \frac{1}{2\pi} \sum_i m_i \nabla_P \left[\frac{1}{\rho_i}\right] \cdot \vec{n}\,. \quad (7.5.9)$$

The integration of the terms of Eq. 7.5.9 over the closed surface S_h yields

$$\frac{1}{2\pi} \int_{S_h} \sigma \left[\int_{S_h} \nabla_P \left[\frac{1}{\rho}\right] \cdot \vec{n}\, dS\right] dS = -\int_{S_h} \sigma\, dS\,, \quad (7.5.10)$$

and

$$\frac{1}{2\pi} \sum_i m_i \int_{S_h} \nabla_P \left[\frac{1}{\rho_i}\right] \cdot \vec{n}\, dS = -\frac{1}{2\pi} \sum_i m_i \Omega_i(Q_i)\,, \quad (7.5.11)$$

where $\Omega(Q_i)$ is the solid angle of view of the closed surface S_h from point Q_i, where charge m_i is located. If Q_i is located outside S_h, one has

$$\Omega_i(Q_i) = 0\,. \quad (7.5.12)$$

Hence, by virtue of Eqs. 7.5.10, 7.5.11, and 7.5.12 the integration of Eq. 7.5.9 over S_h yields

$$\int_{S_h} \sigma\, dS = 0\,, \quad (7.5.13)$$

which reflects the fact that the material of permeability μ in the presence of the field generated by external sources is polarized by the field, but it cannot acquire a nonzero charge.

In the limit of a large value of μ such that condition 7.5.2 is satisfied, the normal component of $\vec{H}_e$ on surface S_h may be written in the form

$$H_{e,n} \approx H_{e,0}\left[1 - G\frac{\mu_0}{\mu}\right], \qquad (7.5.14)$$

where $H_{e,0}$ is the field intensity in the limit $\mu = \infty$ and G is a nondimensional factor that depends on the geometry of S_h. In general G is a function of position of point P. By virtue of Eqs. 7.5.7 and 7.5.14, the surface charge density $\sigma(P)$ may be written in the form

$$\sigma(P) \approx \sigma_0(P) + \delta\sigma, \qquad (7.5.15)$$

where σ_0 is the solution of Eq. 7.5.9 in the limit $\mu = \infty$. One has

$$\sigma_0(P) = \mu_0 H_{e,0}, \qquad (7.5.16)$$

and Eq. 7.5.15 yields

$$\delta\sigma \approx -(1 + G)\frac{\mu_0}{\mu}\sigma_0. \qquad (7.5.17)$$

By substituting the value of σ given by Eq. 7.5.15 in Eq. 7.5.9, within the limits of assumption 7.5.2, $\delta\sigma$ must satisfy the equation

$$\delta\sigma + \frac{1}{2\pi}\int_{S_h}(\delta\sigma)\nabla_P\left[\frac{1}{\rho}\right]\cdot \vec{n}\,dS = -2\frac{\mu_0}{\mu}\sigma_0; \qquad (7.5.18)$$

i.e., by virtue of Eq. 7.5.17 function G is a solution of the integral equation

$$G - 1 + \frac{1}{2\pi\sigma_0}\int_{S_h}(1 + G)\sigma_0\nabla_P\left[\frac{1}{\rho}\right]\cdot \vec{n}\,dS = 0. \qquad (7.5.19)$$

The surface charge $\delta\sigma$ obtained by solving Eq. 7.5.18 generates at each point a scalar potential given by

$$\delta\Phi = \frac{1}{4\pi\mu_0}\int_{S_h}\frac{\delta\sigma}{\rho}\,dS. \qquad (7.5.20)$$

Equation 7.5.20 provides the perturbation of the field computed outside S_h under the assumption $\mu = \infty$. Inside S_h, Eq. 7.5.20 provides the field generated within the high permeability material, because of the finite value of μ. In the limit $\mu = \infty$, one has $\delta\sigma = 0$, and the potential $\delta\Phi$ given by Eq. 7.5.20 is identically zero inside surface S_h.

7.6 FIELD INSIDE A FERROMAGNETIC MEDIUM

By virtue of Eq. 7.5.17, Eq. 7.5.20 transforms to

$$\delta\Phi = -\frac{1}{4\pi\mu}\int_{S_h}\frac{1+G}{\rho}\sigma_0 dS \qquad (\frac{\mu}{\mu_0} \gg 1). \qquad (7.6.1)$$

Thus the magnetic induction $\vec{B}$ inside the high permeability material is

$$\vec{B}(P) \approx \frac{1}{4\pi}\int_{S_h}(1+G)\sigma_0 \nabla_P\left[\frac{1}{\rho}\right]dS, \qquad (7.6.2)$$

where the gradient is computed at point P. Equation 7.6.2 shows that, in the limit 7.5.2, the magnetic induction inside S_h is independent of μ and is determined only by the geometry of S_h and the distribution of σ_0. In some particular cases, G is a constant independent of the position on surface S_h, and as a consequence, the field generated by $\delta\sigma$ is proportional to the external field in the absence of the medium of permeability μ. This is the case, for instance, of a spherical body of permeability μ located in an external uniform field. The field inside the sphere is uniform and oriented in the same direction of the external field in the absence of the sphere.

In general, Eqs. 7.5.18 and 7.5.19 cannot be solved in a closed form, and one must resort to a numerical solution. This is accomplished by dividing the closed surface S_h in a number N_h of surface elements δS_h where the charge density σ is replaced by its average value $\overline{\sigma}_h$. Thus the integral in Eq. 7.5.18 is replaced by a sum over the N_h elements δS_h. By writing Eq. 7.5.18 for each element of surface, one derives a system of N_h equations in N_h unknown variables $\overline{\sigma}_h$.

An example of numerical solution of Eq. 7.5.18 is the field computation in the two-dimensional problem of a high permeability material whose cross-section is the hexadecagon shown in Fig. 7.6.1 with sides tangent to an ellipse with a 2:1 ratio between principal axes. The hexadecagon is assumed to be located in a uniform external field of intensity $\vec{H}_0$ oriented at an angle $\pi/4$ with respect to the axes of the ellipse. The center of the ellipse is assumed to be positioned on the $\Phi = 0$ equipotential plane of the external field.

Figure 7.6.2 shows the equipotential lines outside the hexadecagon in the limit $\mu = \infty$. Because of symmetry, the hexadecagonal boundary is a $\Phi = 0$ equipotential line in the limit $\mu = \infty$ only. A finite value of μ results in a distribution of nonzero

270 MATERIALS WITH LINEAR CHARACTERISTICS

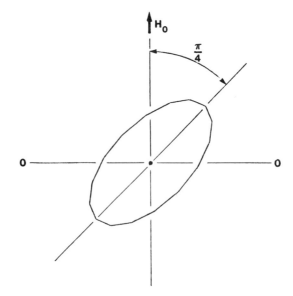

Fig. 7.6.1. Two-dimensional problem of a material of permeability μ and a hexadecagonal cross-section.

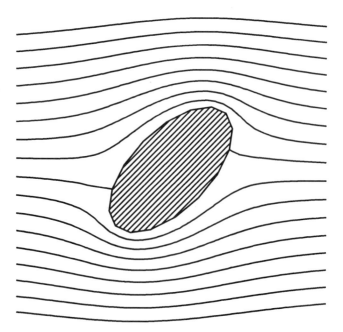

Fig. 7.6.2. Equipotential lines for $\mu = \infty$.

potential on the boundary. A $\Phi = 0$ line within the hexadecagon separates the regions of positive and negative values of Φ, as shown in Fig. 7.6.3, which is computed for $\mu/\mu_0 = 10$. As μ/μ_0 approaches unity, the field distortion caused by the presence of the hexadecagon decreases, as shown by Fig. 7.6.4, which is computed for $\mu/\mu_0 = 2$.

Figure 7.6.5 shows the distribution of

$$\frac{\delta\sigma}{\sigma_0} = -(1 + G)\frac{\mu_0}{\mu} \tag{7.6.3}$$

versus the angular position of a point of the hexadecagonal boundary relative to the orientation of its major axis. The numerical solutions shown in Fig. 7.6.5 have been obtained by dividing each side of the hexadecagon in eight equal intervals and by assuming that σ_0 and $\delta\sigma$ are constant within each interval.

From the values of $\delta\sigma/\sigma_0$ one obtains

$$G = -\lim_{\mu \to \infty}\left[\frac{\mu}{\mu_0}\frac{\delta\sigma}{\sigma_0}\right] - 1. \tag{7.6.4}$$

The values of 7.6.4 are plotted in Fig. 7.6.6, which shows that G is a function of position on the boundary of the hexadecagon.

It is of interest to compare the numerical results plotted in Fig. 7.6.5 with the values of $\delta\sigma/\sigma_0$ on a cylinder of radius r_0 and permeability μ immersed in a uniform field of intensity H_0 perpendicular to its axis. Assume a system of polar coordinates

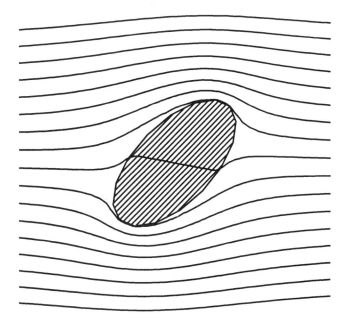

Fig. 7.6.3. Equipotential lines for $\mu/\mu_0 = 10$.

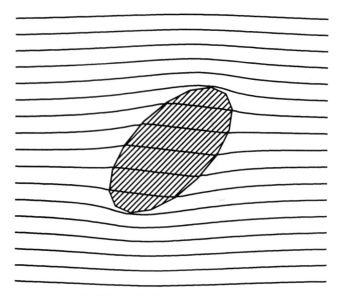

Fig. 7.6.4. Equipotential lines for $\mu/\mu_0 = 2$.

r, ψ where r is the distance from the axis of the cylinder and ψ is the angle between r and the direction of $\vec{H}_0$. The scalar potential inside the cylinder is

$$\Phi = -2\mu_0 \frac{H_0}{\mu + \mu_0} r \cos \psi, \qquad (7.6.5)$$

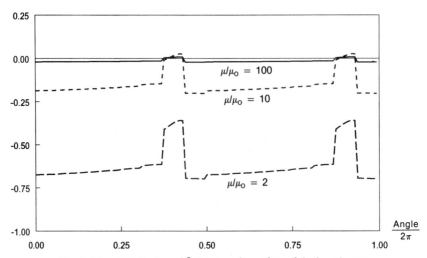

Fig. 7.6.5. Distribution of $\delta\sigma/\sigma_0$ over the surface of the hexadecagon.

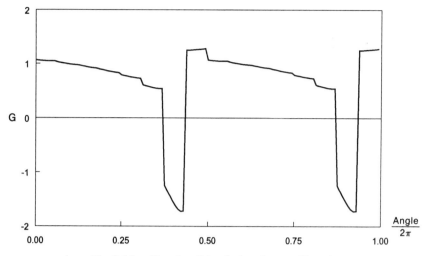

Fig. 7.6.6. Function G for the hexadecagonal boundary.

and the surface charge density is

$$\sigma = 2\mu_0 H_0 \frac{\mu - \mu_0}{\mu + \mu_0} \cos\psi . \tag{7.6.6}$$

In the limit $\mu \to \infty$, one has

$$\sigma = \sigma_0 = 2\mu_0 H_0 \cos\psi , \tag{7.6.7}$$

and for $\mu \gg \mu_0$, one has

$$\delta\sigma \approx -2\frac{\mu_0}{\mu}\sigma_0 . \tag{7.6.8}$$

Thus, by virtue of Eq. 7.6.4 function G for the cylinder is a constant equal to

$$G = 1 , \tag{7.6.9}$$

and the intensity $\delta\vec{H}$ of the field inside the cylinder is

$$\delta\vec{H} \approx 2\frac{\mu_0}{\mu}\vec{H}_0 , \tag{7.6.10}$$

which yields the value of the magnetic induction

$$\vec{B} = \mu\,\delta\vec{H} = 2\mu_0\vec{H}_0 \qquad (\frac{\mu}{\mu_0} \gg 1) \qquad (7.6.11)$$

independent of the value of μ.

REFERENCES

[1] M.G. Abele, H. Rusinek, Effects of demagnetization characteristics in two-dimensional permanent magnets, Technical Report No. 22, New York University, June 1, 1990.

[2] M.G. Abele, H. Rusinek, Field configuration in permanent magnets with linear characteristics of magnetic media and ferromagnetic materials, Technical Report No. 24, New York University, August 15, 1991.

CHAPTER 8

Open Magnetic Structures

INTRODUCTION

The exact mathematical solutions of direct and inverse boundary value problems developed in the preceding chapters have been obtained in structures where the magnetized material fully encloses the internal cavity. In practice, a magnet is an open structure with no physical separation between the surrounding medium and the region of interest. In such a structure one cannot assign arbitrarily a field configuration within any region of the magnet and expect such a configuration to result from the geometries and the distributions of remanences discussed in the preceding chapters. For instance, one cannot expect an open geometry of uniformly magnetized polyhedrons to generate a perfectly uniform field within the region of interest.

It is of great importance to determine how the opening of a magnetic structure alters the properties of the ideal closed geometries. The approach followed in this chapter is to solve the direct boundary value problems formulated with assigned geometries of the magnet opening.

The field properties of an open yokeless permanent magnet can be analyzed with exact mathematical solutions obtained by direct integration over the volume of magnetized material. However, in the case of yoked and hybrid magnets, in general, closed form solutions cannot be obtained and one must resort to approximate numerical methods, as discussed in Sections 5 and 6 of this chapter.

8.1 CYLINDRICAL MAGNET OF FINITE LENGTH

Consider a section of finite length $2z_0$ of the hollow cylindrical structure analyzed in Section 4 of Chapter 2 in the particular case of the solenoidal remanence defined by Eq. 2.4.11. The corresponding distribution of surface charge density on the two cylindrical surfaces of radii r_1 and r_2 is

$$\sigma(r_1,\psi) = -J_0 \cos\psi, \qquad \sigma(r_2,\psi) = J_0 \left[\frac{r_1}{r_2}\right]^2 \cos\psi. \qquad (8.1.1)$$

If the section of the hollow cylindrical structure is confined between the planes $z = \pm z_0$, as shown in Fig. 8.1.1, no surface charge is generated on the two end surfaces. Furthermore, because of the condition $\nabla \cdot \vec{J} = 0$, no volume charge density is

276 OPEN MAGNETIC STRUCTURES

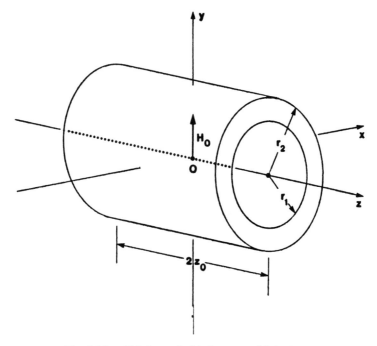

Fig. 8.1.1. Yokeless cylindrical magnet of finite length.

induced in the magnetized material. The medium surrounding the open structure of Fig. 8.1.1 is assumed to be non-magnetic.

The scalar potential generated by the distribution of charges 8.1.1 can be computed by means of Eq. 1.4.15 of Chapter 1. In cylindrical polar coordinates, by virtue of Eq. 8.1.1., the scalar potential Φ at a point P of coordinates r, ψ, z is

$$\Phi(r,\psi,z) = -\frac{J_0 r_1}{4\pi\mu_0} \int_{-z_0}^{+z_0} d\zeta \int_0^{2\pi} \left[\frac{1}{\rho_1} - \frac{r_1}{r_2}\frac{1}{\rho_2}\right] \cos\gamma \, d\gamma, \qquad (8.1.2)$$

where ρ_1, ρ_2 are the distances of point P from the points of coordinates r_1, γ, ζ and r_2, γ, ζ, respectively,

$$\rho_i = \left[(z-\zeta)^2 + r_i^2 + r^2 - 2r_i r \cos(\gamma - \psi)\right]^{1/2}, \qquad i = 1, 2. \qquad (8.1.3)$$

In Eq. 8.1.2 one has

$$\int_0^{2\pi} \frac{\cos\gamma \, d\gamma}{\left[(z-\zeta)^2 + r_i^2 + r^2 - 2r_i r \cos(\gamma-\psi)\right]^{1/2}}$$
$$= -\frac{4\cos\psi}{\left[(z-\zeta)^2 + (r_i+r)^2\right]^{1/2}} \left[E(k_i) - (2-k_i^2)D(k_i)\right], \qquad (8.1.4)$$

where

$$k_i^2 = \frac{4r_i r}{(z-\zeta)^2 + (r_i + r)^2}, \quad (i = 1, 2), \tag{8.1.5}$$

and

$$E(k_i) = \int_0^{\pi/2} \left[1 - k_i^2 \sin^2\gamma\right]^{1/2} d\gamma. \tag{8.1.6}$$

$E(k_i)$ is the complete elliptic integral of the second kind. Function $D(k_i)$ is defined as

$$D(k_i) = \frac{1}{k_i^2}\left[F(k_i) - E(k_i)\right], \tag{8.1.7}$$

where $F(k_i)$ is the complete elliptic integral of the first kind (App. III),

$$F(k_i) = \int_0^{\pi/2} \frac{d\gamma}{\left[1 - k_i^2 \sin^2\gamma\right]^{1/2}}. \tag{8.1.8}$$

At a small distance r from the z axis, i.e., for $k_i^2 \ll 1$, functions E and D can be written as series of ascending powers of k_i^2,

$$\begin{aligned}E &\approx \frac{\pi}{2}\left[1 - 2\left[\frac{k_i^2}{8}\right] - 3\left[\frac{k_i^2}{8}\right]^2 - \cdots\right] \\ D &\approx \frac{\pi}{4}\left[1 + 3\left[\frac{k_i^2}{8}\right] + 15\left[\frac{k_i^2}{8}\right]^2 + \cdots\right],\end{aligned} \tag{8.1.9}$$

and Eq. 8.1.4 reduces to

$$\int_0^{2\pi} \cdots d\gamma \approx \pi \frac{r_i}{\left[(z-\zeta)^2 + r_i^2\right]^{3/2}} r\cos\psi. \tag{8.1.10}$$

Thus, in the proximity of the z axis, Eq. 8.1.2 yields

$$\Phi \approx -\frac{J_0 r_1^2}{4\mu_0} r\cos\psi \int_{-z_0}^{+z_0}\left[\frac{1}{[(z-\zeta)^2 + r_1^2]^{3/2}} - \frac{1}{[(z-\zeta)^2 + r_2^2]^{3/2}}\right]d\zeta. \tag{8.1.11}$$

The intensity $\vec{H}$ of the magnetic field generated by Eq. 8.1.11 is oriented in the direction of the y axis ($\psi = 0$), as indicated in Fig. 8.1.1. One has

$$\lim_{r \to 0} \vec{H} = \frac{J_0}{4\mu_0} \left\{ \frac{z_0 - z}{[(z_0 - z)^2 + r_1^2]^{1/2}} + \frac{z_0 + z}{[(z_0 + z)^2 + r_1^2]^{1/2}} \right.$$

$$\left. - \frac{r_1^2}{r_2^2} \left[\frac{z_0 - z}{[(z_0 - z)^2 + r_2^2]^{1/2}} + \frac{z_0 + z}{[(z_0 + z)^2 + r_2^2]^{1/2}} \right] \right\} \vec{y}, \qquad (8.1.12)$$

where $\vec{y}$ is a unit vector oriented in the positive direction of the coordinate line r in the $\psi = 0$ plane. The amplitude of Eq. 8.1.12 attains a maximum at $z = 0$, which is given by

$$H_{\substack{r=0 \\ z=0}} = \frac{J_0 z_0}{2\mu_0} \left[\frac{1}{(r_1^2 + z_0^2)^{1/2}} - \frac{r_1^2}{r_2^2} \frac{1}{(r_2^2 + z_0^2)^{1/2}} \right]. \qquad (8.1.13)$$

One observes that in the limit $z_0 \to \infty$, Eq. 8.1.13 reduces to the value of the intensity inside the hollow cylindrical magnet of infinite length as given by Eq. 2.4.14.

At large distances from the magnet of finite length $2z_0$, the asymptotic value of the intensity on the z axis is

$$\vec{H}_{r=0} \approx \frac{3J_0}{4\mu_0} \left(1 - \frac{r_1^2}{r_2^2} \right) \frac{r_1^2 r_2^2 z_0}{|z|^5} \vec{y}, \qquad |z| \gg z_0. \qquad (8.1.14)$$

Equation 8.1.14 shows that asymptotically $\vec{H}$ is the intensity of the field generated by an octupole, as defined by Eq. 1.5.24, whose potential decreases with ρ as ρ^{-4}.

In order to analyze the nature of the multipoles associated with the distribution of magnetic charges 8.1.1, consider an element of the cylindrical magnet of infinitesimal length $d\zeta$, located at $z = \zeta$. Assume a frame of spherical polar coordinates ρ, θ, ψ where ρ is the distance of a point P from the center O' of the element of the cylindrical magnet and θ is the angle between ρ and the z axis as indicated in Fig. 8.1.2. One has

$$\rho = [(z - \zeta)^2 + r^2]^{1/2}. \qquad (8.1.15)$$

The expansion in spherical harmonics of the scalar potential $d\Phi$ generated by the element of the cylindrical magnet at a distance $\rho > r_2$ can be obtained from Eq. 1.5.9 of Chapter 1 by identifying charge m_i in Eqs. 1.5.11, 1.5.12, and 1.5.13 with an element of charge

$$dm_i = \sigma r \, d\psi \, d\zeta \qquad (8.1.16)$$

on the two cylindrical surfaces of radii r_1 and r_2.

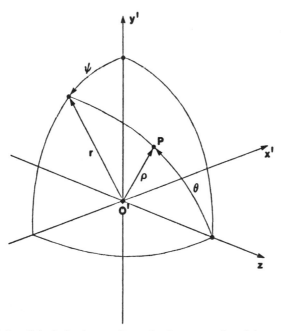

Fig. 8.1.2. Spherical polar coordinates for the computation of the multipoles.

Let us compute the scalar potential $d\Phi$ generated by the element of infinitesimal length $d\zeta$ as the sum of the potentials of multipoles located at center O'. The coefficients $dp_0^{(l)}$, $dp_{j,1}^{(l)}$, $dp_{j,2}^{(l)}$ of the multipoles are provided by Eqs. 1.5.11, 1.5.12, 1.5.13, by replacing the sum over the point charges with the integral over charge distribution 8.1.1. By virtue of the symmetry of charge distribution 8.1.1 one obtains

$$dp_0^{(l)} = 0, \qquad dp_{j,2}^{(l)} = 0, \qquad (8.1.17)$$

independent of the values of l and j. Coefficient $dp_{j,1}^{(l)}$ is given by

$$dp_{j,1}^{(l)} = 2\frac{(l-j)!}{(l+j)!} J_0 r_1^2 \left[r_2^{l-1} - r_1^{l-1}\right] P_l^j(0) \int_0^{2\pi} \cos\psi \cos j\psi \, d\psi \, d\zeta, \qquad (8.1.18)$$

where $P_l^j(0)$ is the value of $P_l^j(\cos\theta)$ at $\theta = \pi/2$. For $l = 1$, Eq. 8.1.18 yields

$$dp_{j,1}^{(1)} = 0. \qquad (8.1.19)$$

For $l > 1$, $dp_{j,1}^{(l)}$ vanishes for $j \neq 1$, and for $j = 1$ one has

$$dp_{1,1}^{(l)} = 2\pi \frac{(l-1)!}{(l+1)!} J_0 r_1^2 \left[r_2^{l-1} - r_1^{l-1}\right] P_l^1(0) \, d\zeta. \qquad (8.1.20)$$

280 OPEN MAGNETIC STRUCTURES

The Legendre associated functions $P_l^j(\cos\theta)$ are equal to zero at $\theta = \pi/2$ for even numbers l. As a consequence, the element of the cylindrical magnet has no multipole moment for multipoles of even order. Furthermore, because of Eq. 8.1.19, the element of the magnet has no dipole moment. Thus the dominant term of the expansion of the scalar potential $d\Phi$ is the potential of an octupole ($l = 3$). The value of $P_3^1(0)$ is

$$P_3^1(0) = \frac{3}{2}\left[\sin\theta(5\cos^2\theta - 1)\right]_{\theta=\frac{\pi}{2}} = -\frac{3}{2}, \qquad (8.1.21)$$

and the coefficients of the octupole moment are

$$dp_{1,1}^{(3)} = -\frac{\pi}{4}J_0\left[r_2^2 - r_1^2\right]r_1^2\,d\zeta, \qquad (8.1.22)$$

and

$$dp_0^{(3)} = dp_{1,2}^{(3)} = dp_{2,1}^{(3)} = dp_{2,2}^{(3)} = dp_{3,1}^{(3)} = dp_{3,2}^{(3)} = 0. \qquad (8.1.23)$$

Thus, by virtue of Eq. 1.5.9, for $\rho \to \infty$, the scalar potential generated by the element of the cylindrical magnet is

$$d\Phi = -\frac{J_0}{16\mu_0}\left[1 - \frac{r_1^2}{r_2^2}\right]\frac{r_1^2 r_2^2}{\rho^4}P_3^1(\cos\theta)\cos\psi\,d\zeta + \cdots. \qquad (8.1.24)$$

The higher order terms in Eq. 8.1.24 depend upon ρ, θ, ζ as

$$-\frac{1}{\rho^{2(n+1)}}P_{2n+1}^1(\cos\theta)\,d\zeta = \frac{1}{r^l}(1-\xi^2)^{\frac{l}{2}}\frac{P_l^1(\xi)}{1-\xi^2}\,d\xi$$

$$(l = 2n+1, \quad n = 1, 2, 3, \ldots), \qquad (8.1.25)$$

where

$$\xi = \cos\theta. \qquad (8.1.26)$$

In Eq. 8.1.25 one can write

$$(1-\xi^2)^{\frac{l}{2}} = \frac{l!\,2^l}{(2l)!}P_l^l(\xi). \qquad (8.1.27)$$

Since the Legendre associated functions satisfy the condition

$$\int_{-1}^{+1} \frac{P_l^m P_l^n}{1-\xi^2} d\xi = 0 \qquad (m \neq n), \qquad (8.1.28)$$

the integral of 8.1.25 over the variable ζ from $-\infty$ to $+\infty$ is

$$\int_{-\infty}^{+\infty} \frac{P_l^1(\cos\theta)}{\rho^{l+1}} d\zeta = 0. \qquad (8.1.29)$$

This property is of importance to understand the field confinement within the cylindrical magnet of infinite length. The contribution of each multipole associated with each element $d\zeta$ of the magnet cancels out in the region $r > r_2$, and as a consequence, the intensity is identically equal to zero outside the magnet, as expressed by Eqs. 2.4.13.

Assume a section of the cylindrical magnet of length $2z_0$ and center at O. At a distance ρ from O

$$\rho > (r_2^2 + z_0^2)^{1/2}, \qquad (8.1.30)$$

the scalar potential Φ_3 generated by the distribution of octupole moments over the length $2z_0$ is

$$\Phi_3 = -\frac{J_0 r_1^2}{240\mu_0} \left(r_2^2 - r_1^2\right) \frac{\cos\psi}{r^3} \int_{\xi_1}^{\xi_2} \frac{P_3^1(\xi) P_3^3(\xi)}{1-\xi^2} d\xi, \qquad (8.1.31)$$

where the limits of integration ξ_1, ξ_2 are

$$\xi_1 = \frac{z - z_0}{[(z - z_0)^2 + r^2]^{1/2}}, \qquad \xi_2 = \frac{z + z_0}{[(z + z_0)^2 + r^2]^{1/2}}. \qquad (8.1.32)$$

Thus

$$\Phi_3 = -\frac{3}{32} \frac{J_0 r_1^2}{\mu_0} (r_2^2 - r_1^2) \left[\xi_1(1-\xi_1^2)^2 - \xi_2(1-\xi_2^2)^2\right] \frac{\cos\psi}{r^3}. \qquad (8.1.33)$$

The effect of the finite length $2z_0$ on the magnetic field generated inside the cylinder of radius r_1 close to the center O of the magnet can be analyzed by means of

282 OPEN MAGNETIC STRUCTURES

the expansion in spherical harmonics given by Eq. 1.5.8 within a sphere of radius r_1 and center at O. Again, coefficients $dq_0^{(l)}$, $dq_{j,1}^{(l)}$, $dq_{j,2}^{(l)}$ are obtained by integration over the surface charge distribution 8.1.1. By virtue of Eqs. 1.5.14 and 8.1.17 one has

$$dq_0^{(l)} = 0, \qquad dq_{j,2}^{(l)} = 0. \tag{8.1.34}$$

The integration of coefficients $dq_{j,1}^{(l)}$ for $j = 1$ yields

$$q_{1,1}^{(l)} = \frac{2}{l(l+1)} \int_{-z_0}^{+z_0} d\zeta \int_0^{2\pi} \left[\frac{r_1 \sigma(r_1,\psi)}{\rho_1^{l+1}} \frac{P_l^1 \zeta}{\rho_1} + \frac{r_2 \sigma(r_2,\psi)}{\rho_2^{l+1}} \frac{P_l^1 \zeta}{\rho_2} \right] \cos\psi \, d\psi, \tag{8.1.35}$$

where

$$\rho_1 = \left[r_1^2 + \zeta^2 \right]^{1/2}, \qquad \rho_2 = \left[r_2^2 + \zeta^2 \right]^{1/2}. \tag{8.1.36}$$

For $l = 1$, Eq. 8.1.35 yields

$$q_{1,1}^{(1)} = -2\pi J_0 z_0 \left[\frac{1}{(r_1^2 + z_0^2)^{1/2}} - \frac{r_1^2}{r_2^2} \frac{1}{(r_2^2 + z_0^2)^{1/2}} \right], \tag{8.1.37}$$

and the term $l = 1$ of the expansion of the scalar potential in Eq. 1.5.8 of Chapter 1 is

$$\Phi_1 = \frac{q_{1,1}^{(1)}}{4\pi\mu_0} \rho \, P_1^1(\cos\theta) \cos\psi$$

$$= -\frac{J_0 z_0}{2\mu_0} \left[\frac{1}{(r_1^2 + z_0^2)^{1/2}} - \frac{r_1^2}{r_2^2} \frac{1}{(r_2^2 + z_0^2)^{1/2}} \right] y. \tag{8.1.38}$$

Equation 8.1.38 yields a uniform intensity whose amplitude is equal to the value of H at $r = 0$, $z = 0$ given by Eq. 8.1.13. A plotting of Φ_1 versus z_0 for the particular value $r_2 = 2r_1$ is shown in Fig. 8.1.3.

For $l = 2$, one has

$$q_{1,1}^{(2)} = 0. \tag{8.1.39}$$

For $l = 3$, the integration of $dq_{j,1}^{(3)}$ for $j = 1$ yields

$$q_{1,1}^{(3)} = \frac{\pi}{2} J_0 r_1^2 z_0 \left[(r_1^2 + z_0^2)^{-\frac{5}{2}} - (r_2^2 + z_0^2)^{-\frac{5}{2}} \right], \tag{8.1.40}$$

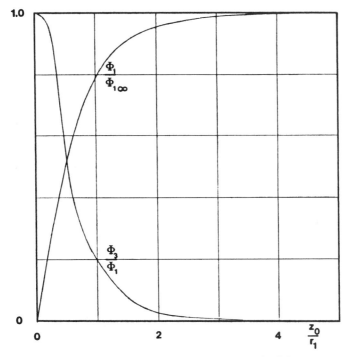

Fig. 8.1.3. Term Φ_1 and ratio Φ_3/Φ_1 versus the length of the magnet.

and the term $l = 3$ of expansion 1.5.8 is

$$\Phi_3 = \frac{q_{1,1}^{(3)}}{4\pi\mu_0} \rho^3 P_3^1 (\cos\theta) \cos\psi$$

$$= \frac{3}{16\mu_0} J_0 r_1^2 z_0 \left[(r_1^2 + z_0^2)^{-\frac{5}{2}} - (r_2^2 + z_0^2)^{-\frac{5}{2}} \right] y (4z^2 - r^2) .$$

(8.1.41)

Thus, center O is an example of the category of points discussed in Section 3 of Chapter 1 where the orientation of the equipotential surfaces is multivalued. Function Φ_3 is zero on the plane

$$y = 0 \qquad (8.1.42)$$

as well as on the conical surface

$$r = (x^2 + y^2)^{1/2} = 2z . \qquad (8.1.43)$$

Figure 8.1.4 is a plot of coefficient

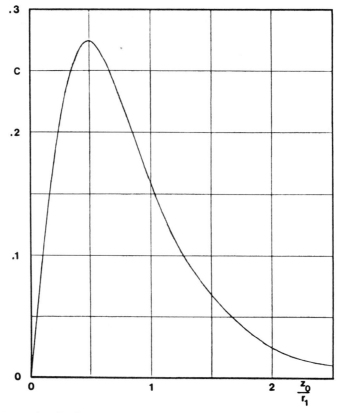

Fig. 8.1.4. Coefficient C defined by Eq. 8.1.44 versus the length of the magnet.

$$C(r_1, r_2, z_0) = z_0 \left[(r_1^2 + z_0^2)^{-\frac{5}{2}} - (r_2^2 + z_0^2)^{-\frac{5}{2}} \right] \quad (8.1.44)$$

versus z_0 for the particular value $r_2 = 2r_1$. The ratio $|\Phi_3/\Phi_1|$ normalized to unity is also shown in Fig. 8.1.3 for the points on the z axis, $r = 0$, $z = \pm r_1$. The maximum value of $|\Phi_3/\Phi_1|$ occurs at $z_0 = 0$ and it is equal to

$$\left. \frac{\Phi_3}{\Phi_1} \right|_{max} = \frac{3}{2} \frac{1 - \frac{r_1^5}{r_2^5}}{1 - \frac{r_1^3}{r_2^3}}. \quad (8.1.45)$$

The equipotential lines in the y, z plane are shown in Fig. 8.1.5 for $r_2 = 2r_1$ and $z_0 = 2r_1$. The shaded area is the region of the negative value of Φ. The $\Phi = 0$ line, which separates the two regions of positive and negative values of Φ, approaches asymptotically the line

$$y = 2z, \quad (8.1.46)$$

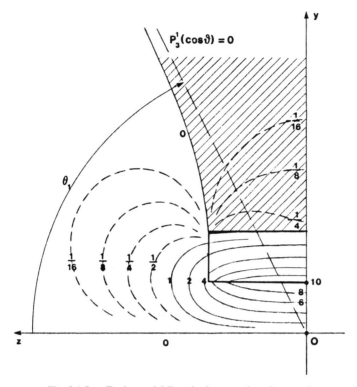

Fig. 8.1.5. Equipotential lines in the y, z plane for $z_0 = 2r_1$

which corresponds to the root of

$$P_3^1(\xi) = 0 , \tag{8.1.47}$$

i.e.,

$$\cos\theta_1 = 5^{-1/2} . \tag{8.1.48}$$

Figure 8.1.5 clearly shows the configuration of the octupole field in the space surrounding the magnet. The external cylindrical surface $r = r_2$ is within the region of negative values of Φ and Φ attains its maximum negative value at

$$r = r_2 , \quad z \approx 1.5 r_1 . \tag{8.1.49}$$

One observes that the $\Phi = 0$ line inside the magnetized material is very close to the $r = r_2$ boundary. A plotting of Φ on the line $y = r_2$ is shown in Fig. 8.1.6.

Figure 8.1.7 shows the equipotential lines for a short magnet ($z_0 = 0.2 r_1$). Again the $\Phi = 0$ line approaches asymptotically line 8.1.46. It is worthwhile pointing out that the field configuration at a distance large compared to the dimensions of the cylindrical magnet is independent of the geometrical parameters of the magnetic structure [1].

286 OPEN MAGNETIC STRUCTURES

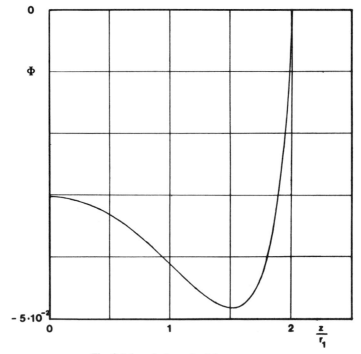

Fig. 8.1.6. Scalar potential at $y = r_2$ versus z.

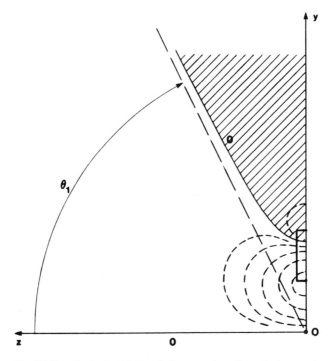

Fig. 8.1.7. Equipotential lines in the y, z plane for a short magnet.

8.2 OPEN YOKELESS SPHERICAL MAGNET

The preceding chapters have shown that closed three-dimensional geometries of magnetized material also can be designed to confine the field without the need of external magnetic shields of infinite magnetic permeability. The absence of a magnetic field in the surrounding space means that the multipole moments of these structures are all identically equal to zero.

If the internal cavity of such a three-dimensional magnet is open to the surrounding space, the property of field confinement is lost, and the field generated outside, as well as the perturbation of the field inside the cavity, depend upon the geometry of the opening and its position in the magnetic structure.

To analyze the effect of the opening of a three-dimensional magnet, consider the particular case of the spherical geometry discussed in Section 5 of Chapter 2, where the magnetized material is confined between two concentric spheres of radii ρ_1, ρ_2, with the distribution of remanence $\vec{J}$ given by Eq. 2.5.30. In a spherical frame of reference ρ, θ, ψ, whose origin coincides with the center of the spheres, the components of $\vec{J}$ are given by Eq. 2.5.30 and can be written in the form

$$J_\rho = 2J\cos\theta, \qquad J_\theta = J\sin\theta, \qquad J_\psi = 0, \qquad (8.2.1)$$

where J is an arbitrary constant. Remanence 8.2.1 generates an equivalent volume charge density υ in the region $\rho_1 < \rho < \rho_2$,

$$\upsilon = -\nabla \cdot \vec{J} = -\frac{6}{\rho} J\cos\theta, \qquad (8.2.2)$$

and a surface charge density σ on the two spheres,

$$\sigma = \begin{cases} -2J\cos\theta, & \rho = \rho_1 \\ +2J\cos\theta, & \rho = \rho_2. \end{cases} \qquad (8.2.3)$$

$\vec{J}$ is independent of the angular coordinate ψ. Consequently the terms of the expansion of the scalar potential Φ vanish for $j \neq 0$ and outside the sphere of radius ρ_2, Φ reduces to

$$\Phi(\rho, \theta) = \frac{1}{4\pi\mu_0} \sum_{l=1}^{\infty} p_0^{(l)} \frac{P_l(\cos\theta)}{\rho^{l+1}}. \qquad (8.2.4)$$

By integrating Eq. 1.5.11 over charge distributions 8.2.2 and 8.2.3, the multipole moment coefficients $p_0^{(l)}$ in Eq. 8.2.4 are

$$p_0^{(l)} = 4\pi J \frac{l-1}{l+2} (\rho_2^{l+2} - \rho_1^{l+2}) \int_{-1}^{+1} \xi\, P_l(\xi)\, d\xi. \qquad (8.2.5)$$

288 OPEN MAGNETIC STRUCTURES

$p_0^{(l)}$ is zero for $l = 1$. Furthermore in Eq. 8.2.5 one has

$$\int_{-1}^{+1} \xi \, P_l(\xi) \, d\xi = \int_{-1}^{+1} P_1(\xi) \, P_l(\xi) \, d\xi = 0 \quad (l > 1). \tag{8.2.6}$$

Thus

$$p_0^{(l)} = 0 \tag{8.2.7}$$

for all values of l. The spherical magnet does not exhibit magnetic moments.

Figure 8.2.1 shows the schematic of the spherical magnet in cartesian coordinates where the y axis coincides with the $\theta = 0$ axis of the spherical frame of reference ρ, θ, ψ. Assume now that the magnetized material contained within a half cone of semiangle ω_0 is removed from the spherical magnet and assume that the axis of the cone is perpendicular to y. In Fig. 8.2.1 the axis of the cone is selected to coincide with the axis z of the cartesian frame of reference. The surface charge density on the

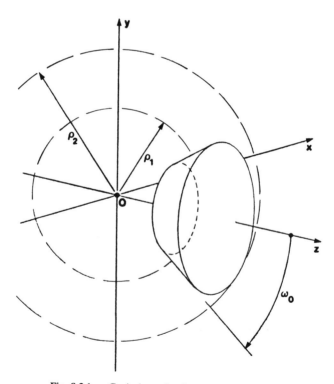

Fig. 8.2.1. Conical opening in a spherical magnet.

conical surface of Fig. 8.2.1 is

$$\sigma = -J_\theta \frac{\cos\omega_0 \cos\chi}{(1 - \sin^2\omega_0 \cos^2\chi)^{1/2}} = -J\cos\omega_0\cos\chi, \qquad (8.2.8)$$

where χ is the angle between plane $x = 0$ and the plane that contains the z axis and a point of the conical surface of Fig. 8.2.1 of coordinates ρ, θ, ψ as indicated in Fig. 8.2.2, which shows the frame of spherical polar coordinates ρ, ω, χ whose axis $\omega = 0$ coincides with the axis z. In the spherical triangle (PYZ) shown in Fig. 8.2.2, the angular coordinate θ is related to ω and χ by the equation

$$\cos\theta = \sin\omega\cos\chi. \qquad (8.2.9)$$

Surface charge density σ on the two spherical surfaces of Fig. 8.2.1 and volume charge density υ can then be written in the form

$$\begin{aligned}
\sigma &= -2J\sin\omega\cos\chi & (\rho = \rho_1) \\
\sigma &= +2J\sin\omega\cos\chi & (\rho = \rho_2) \\
\upsilon &= -\frac{6}{\rho} J\sin\omega\cos\chi & (\rho_1 < \rho < \rho_2, \; \omega < \omega_0).
\end{aligned} \qquad (8.2.10)$$

In the frame of coordinates ρ, ω, χ, expansion 1.5.9 of the scalar potential gen-

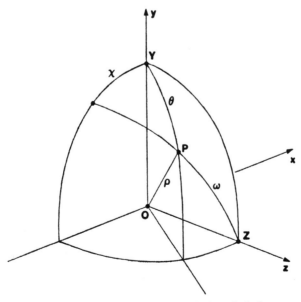

Fig. 8.2.2. Spherical frames of reference in the spherical magnet.

erated by the distribution of charges 8.2.8 and 8.2.10 contains only terms $j = 1$. Also, because of symmetry, coefficients $p_{i,j,2}^{(l)}$ are equal to zero, and by integrating Eq. 1.5.12 over charge distributions 8.2.8 and 8.2.10 for $j = 1$, the components of the multipole moments of the structure of Fig. 8.2.1 are

$$p_{1,1}^{(l)} = \frac{4\pi J (l-1)!}{(l+2)!} (\rho_2^{l+2} - \rho_1^{l+2})$$

(8.2.11)

$$\times \left\{ (l-1) \int_0^{\omega_0} \sin^2 \omega \, P_l^1(\cos \omega) \, d\omega - \frac{1}{4} \sin 2\omega_0 \, P_l^1(\cos \omega_0) \right\}.$$

In the limit $\omega_0 = \pi$ the geometry of Fig. 8.2.1 becomes again the full spherical magnet, and in Eq. 8.2.11 one has

$$\int_0^\pi \sin^2 \omega \, P_l^1(\cos \omega) \, d\omega = \int_{-1}^{+1} P_1^1(\xi) P_l^1(\xi) \, d\xi = 0 \qquad (l > 1),$$

(8.2.12)

i.e.,

$$\left[p_{1,1}^{(l)} \right]_{\omega_0 = \pi} = 0.$$

(8.2.13)

Equation 8.2.11 shows that the conical opening of the magnet generates multipoles for all values of l. In particular, for $l = 1$, only the second term within parentheses in Eq. 8.2.11 contributes to the dipole moment, due to the surface charge density 8.2.8 generated by the conical opening.

For $\omega_0 = \pi/2$, the volume of Fig. 8.2.1 becomes half the sphere. The dipole moment vanishes because no surface charge is induced on the plane which contains the $\theta = 0$ axis, and one has

$$\int_0^{\pi/2} \sin^2 \omega \, P_{2n+1}^1(\cos \omega) \, d\omega = 0 \qquad (n = 1, 2, 3, \ldots).$$

(8.2.14)

Thus the field generated in the $\rho > \rho_2$ region by the half sphere results from the even order multipoles.

Assume now that the axis of the conical opening is made to coincide with the $\theta = 0$ axis of the spherical magnet. The surface charge density on the conical surface of the opening is

$$\sigma = \left[J_\theta \right]_{\theta = \omega_0} = J \sin \omega_0.$$

(8.2.15)

Because of symmetry, it is convenient to expand the scalar potential in the same frame of spherical polar coordinates ρ, θ, ψ. For $\rho > \rho_2$, the potential generated by the volume of magnetized material removed from the spherical magnet is given by Eq. 8.2.4, where

$$p_0^{(l)} = \frac{4\pi J}{l+2} (\rho_2^{l+2} - \rho_1^{l+2}) \left\{ (l-1) \int_0^{\omega_0} \sin\theta \cos\theta \, P_l(\cos\theta) \, d\theta + \frac{1}{2} \sin^2\omega_0 \, P_l(\cos\omega_0) \right\}, \quad (8.2.16)$$

which reduces to Eq. 8.2.5 for $\omega_0 = \pi$. Again multipoles of all values of l are generated by the material removed from the spherical magnet.

It is of interest to compare the dipole moment ($l = 1$) given by Eq. 8.2.16 with the dipole moment given by Eq. 8.2.11. One has

$$\left| \frac{p_{1,1}^{(1)}}{p_0^{(1)}} \right| = \frac{1}{2}. \quad (8.2.17)$$

If the conical sector becomes equal to half the sphere ($\omega_0 = \pi/2$), Eq. 8.2.16 yields

$$p_0^{(2n+1)} = 0 \quad (n = 0, 1, 2, \ldots). \quad (8.2.18)$$

Thus, again, only multipoles whose order is an even number contribute to the field in the $\rho > \rho_2$ region.

In a way similar to the situation found in the cylindrical magnet of finite length, the opening of the spherical magnet affects the field inside the cavity of radius ρ_1. Consider first the case of the full spherical magnet. Because of symmetry and by virtue of remanence 8.2.1, Eq. 1.5.8 of the scalar potential inside the sphere of radius ρ_1 reduces to

$$\Phi(\rho, \theta) = \frac{1}{4\pi\mu_0} \sum_{l=1}^{\infty} q_0^{(l)} \rho^l P_l(\cos\theta), \quad (8.2.19)$$

where

$$q_0^{(l)} = 2\pi \int_{-1}^{+1} \left[\frac{\sigma_1}{\rho_1^{l-1}} + \frac{\sigma_2}{\rho_2^{l-1}} + \int_{\rho_1}^{\rho_2} \frac{\upsilon}{\rho^{l-1}} d\rho \right] P_l(\xi) \, d\xi, \quad (8.2.20)$$

and by virtue of Eqs. 8.2.2 and 8.2.3,

$$q_0^{(l)} = 4\pi J \left[\frac{1}{\rho_2^{l-1}} - \frac{1}{\rho_1^{l-1}} - 3\int_{\rho_1}^{\rho_2} \frac{d\rho}{\rho^l} \right] \int_{-1}^{+1} \xi P_l(\xi) \, d\xi. \quad (8.2.21)$$

For $l = 1$

$$\int_{-1}^{+1} \xi P_1(\xi) d\xi = \frac{2}{3}, \qquad (8.2.22)$$

and Eq. 8.2.21 yields

$$q_0^{(1)} = -8\pi J \ln\frac{\rho_2}{\rho_1}. \qquad (8.2.23)$$

For $l > 1$ one has

$$q_0^{(l)} = 4\pi \frac{l+2}{l-1} J \left[\frac{1}{\rho_2^{l-1}} - \frac{1}{\rho_1^{l-1}}\right] \int_{-1}^{+1} \xi P_l(\xi) d\xi = 0. \qquad (8.2.24)$$

Thus, the scalar potential 8.2.19 within the $\rho < \rho_1$ region is

$$\Phi(\rho, \theta) = -\frac{2}{\mu_0} J \rho \ln\frac{\rho_2}{\rho_1} \cos\theta, \qquad (8.2.25)$$

which coincides with the scalar potential given by the first equation of 2.5.23 if J is assumed to be equal to $2J_0/3$.

Let us confine the analysis of the effect of the opening on the field in the $\rho < \rho_1$ region to the case of the conical opening with $\theta = 0$ as its axis, which introduces the surface charge density 8.2.15 on the surface of the cone $\theta = \theta_0$. The value of the scalar potential is given by Eq. 8.2.19 where

$$q_0^{(l)} = 4\pi J \left\{ \left[\frac{1}{\rho_2^{l-1}} - \frac{1}{\rho_1^{l-1}} - 3\int_{\rho_1}^{\rho_2}\frac{d\rho}{\rho^l}\right] \int_{\xi_0}^{+1} \xi P_l(\xi) d\xi + \frac{1}{2}\sin^2\omega_0\, P_l(\xi_0) \int_{\rho_1}^{\rho_2}\frac{d\rho}{\rho^l} \right\}, \qquad (8.2.26)$$

where

$$\xi_0 = \cos\omega_0. \qquad (8.2.27)$$

For $l = 1$, Eq. 8.2.26 reduces to

$$q_0^{(1)} = -4\pi J \ln\frac{\rho_2}{\rho_1}\left[1 - \frac{1}{2}\xi_0(1+\xi_0^2)\right]. \qquad (8.2.28)$$

Thus within the $\rho < \rho_1$ region, the term $l = 1$ of the expansion of Φ generated by the magnetized material removed from the sphere is

$$\Phi_1(\rho, \theta) = -\frac{J}{\mu_0} \ln\frac{\rho_2}{\rho_1} \left[1 - \frac{1}{2}\xi_0(1 + \xi_0^2) \right] \rho \cos\theta, \quad (8.2.29)$$

and obviously, $\Phi_1 = 0$ for $\xi_0 = 1$. Equation 8.2.29 shows that the intensity of the magnetic field at center O ($\rho = 0$) is oriented along the $\theta = 0$ axis and its magnitude still increases with the radial dimension ρ_2 as $\ln(\rho_2/\rho_1)$. Thus the opening of the spherical magnet maintains the property of Eq. 2.5.24 which says that the intensity within region $\rho < \rho_1$ is not bound by an upper limit as long as the linear approximation of the magnetic characteristic is valid. A plotting of the potential given by Eq. 8.2.29 versus ω_0 is shown in Fig. 8.2.3 normalized to the value of Φ_1 for $\xi_0 = -1$.

For $l > 1$, Eq. 8.2.26 is

$$q_0^{(l)} = \frac{2\pi}{l-1} J \left[\frac{1}{\rho_2^{l-1}} - \frac{1}{\rho_1^{l-1}} \right] \left[2(l+2) \int_{\xi_0}^{1} \xi P_l(\xi) \, d\xi - (1 - \xi_0^2) P_l(\xi_0) \right]. \quad (8.2.30)$$

In particular, for $l = 2$, one has

$$q_0^{(2)} = +3\pi J \left[\frac{1}{\rho_2} - \frac{1}{\rho_1} \right] (1 - \xi_0^4), \quad (8.2.31)$$

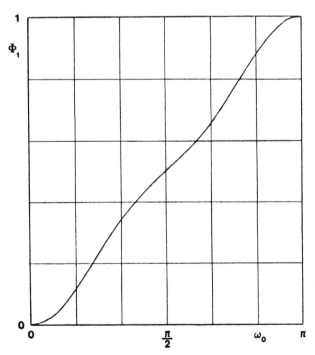

Fig. 8.2.3. Potential generated by the conical opening versus ω_0.

and, within the $\rho < \rho_1$ region, the $l = 2$ term of the expansion of Φ generated by the removed magnetized material is

$$\Phi_2 = \frac{3J}{4\mu_0} \left[\frac{1}{\rho_2} - \frac{1}{\rho_1} \right] (1 - \xi_0^4) \, \rho^2 P_2(\xi)$$

$$= \frac{3J}{8\mu_0} \left[\frac{1}{\rho_2} - \frac{1}{\rho_1} \right] (1 - \xi_0^4) \, (2y^2 - r^2) \, ,$$

(8.2.32)

where

$$y = \rho \cos \theta \, , \qquad r = \rho \sin \theta \, . \qquad (8.2.33)$$

Φ_2 is a particular case of the configurations of scalar potential discussed in Section 3 of Chapter 1. The equipotential surfaces of Φ_2 are the hyperboloids

$$r^2 - 2y^2 = constant \, , \qquad (8.2.34)$$

and the equipotential surface $\Phi_2 = 0$ is the right circular cone

$$r = 2^{1/2} y \, . \qquad (8.2.35)$$

Along the $\theta = 0$ axis, $\Phi_2(\rho, 0)$ corresponds to a magnetic field whose intensity changes linearly with the coordinate y with a rate of change

$$\left[\frac{\partial^2 \Phi_2}{\partial \rho^2} \right]_{\theta = 0} = \frac{3J}{2\mu_0} \left[\frac{1}{\rho_2} - \frac{1}{\rho_1} \right] (1 - \xi_0^4) \, . \qquad (8.2.36)$$

8.3 PRISMATIC YOKELESS MAGNETS OF FINITE LENGTH

The analysis of the finite length effects can be readily extended to the category of yokeless magnetic structures composed of uniformly magnetized polyhedrons developed in Chapter 4. As in the case of the cylindrical yokeless magnet, Eq. 1.4.14 provides the exact solution of the scalar potential. In a structure of uniformly magnetized polyhedrons, the remanence is solenoidal within each polyhedron, and the integral in Eq. 1.4.13 reduces to the contribution of the surface charges on the interfaces between polyhedrons and between the nonmagnetic medium and the magnetic structure.

In general Eq. 1.4.14 can be written in the form

$$\Phi = \frac{1}{4\pi\mu_0} \sum_{h,k} \int_{S_{h,k}} \frac{1}{\rho} (\vec{J}_h - \vec{J}_k) \cdot \vec{n}_{h,k} \, dS , \qquad (8.3.1)$$

where $S_{h,k}$ is the interface between polyhedron number h and polyhedron number k. Unit vector $\vec{n}_{h,k}$ is perpendicular to $S_{h,k}$ and oriented from polyhedron number h to polyhedron number k. The equivalent surface charge density on interface $S_{h,k}$ is

$$\sigma_{h,k} = (\vec{J}_h - \vec{J}_k) \cdot \vec{n}_{h,k} . \qquad (8.3.2)$$

Assume a section of finite length $2z_0$ of a yokeless two-dimensional structure introduced in Chapter 4. If the structure is confined between two planes $z = \pm z_0$, the two end surfaces are parallel to the remanences of the magnetized components, and as a consequence, no surface charge is induced on the end surfaces. All interfaces $S_{h,k}$ in Eq. 8.3.1 are plane surfaces parallel to the axis z.

Consider a section of $S_{h,k}$ between planes $z = \zeta$ and $z = \zeta + d\zeta$ in a system of coordinates x, y, z. The intersection of $S_{h,k}$ with plane $z = \zeta$ is a segment of line of length $2\tau_{0,h,k}$ whose center $O_{h,k}$ has coordinates $x_{h,k}, y_{h,k}, \zeta$. Let $\alpha_{h,k}$ be the angle between the segment and the $y = 0$ plane, as indicated in Fig. 8.3.1.

Surface charge $\sigma_{h,k}$ on an element of surface $S_{h,k}$ contained between planes $z = \zeta$ and $z = \zeta + d\zeta$ generates at a point P of coordinates x, y, z the scalar potential

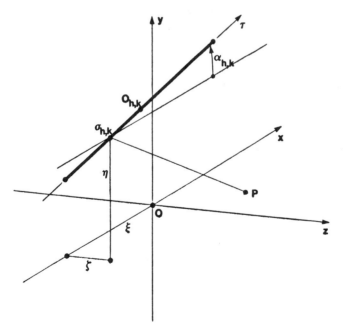

Fig. 8.3.1. Reference system of the interfaces in a prismatic magnet.

$$d\Phi_{h,k} = \frac{\sigma_{h,k}}{4\pi\mu_0} \left\{ \ln\left[\sqrt{1 + \left[\frac{\tau_{0,h,k} - \tau_{h,k}}{\rho_{h,k}}\right]^2} + \frac{\tau_{0,h,k} - \tau_{h,k}}{\rho_{h,k}}\right] \right.$$

$$\left. + \ln\left[\sqrt{1 + \left[\frac{\tau_{0,h,k} + \tau_{h,k}}{\rho_{h,k}}\right]^2} + \frac{\tau_{0,h,k} + \tau_{h,k}}{\rho_{h,k}}\right] \right\} d\zeta, \qquad (8.3.3)$$

where

$$\rho_{h,k} = \left\{ \left[(x - x_{h,k})\sin\alpha_{h,k} + (y - y_{h,k})\cos\alpha_{h,k}\right]^2 + (z - \zeta)^2 \right\}^{1/2}$$

$$\tau_{h,k} = (x - x_{h,k})\cos\alpha_{h,k} + (y - y_{h,k})\sin\alpha_{h,k}. \qquad (8.3.4)$$

The symbol $\rho_{h,k}$ denotes the distance of point P from the projection $Q_{h,k}$ of P on the line which contains a segment of length $2\tau_{0,h,k}$, and $\tau_{h,k}$ is the distance of $Q_{h,k}$ from $O_{h,k}$. Equation 8.3.1 reduces to the integral of

$$d\Phi = \sum_{h,k} d\Phi_{h,k} \qquad (8.3.5)$$

over the length $2z_0$ of the magnetic structure.

Following the procedure of Section 2 of this chapter, the properties of function 8.3.5 can be analyzed by computing the multipole moments of a section of magnet of length $d\zeta$. Again the calculation is conducted in a frame of spherical polar coordinates ρ, θ, ψ with the origin at the center O' of the section of magnet of length $d\zeta$ and $\theta = 0$ coinciding with the axis z. Outside a spherical surface which encloses the slice of the magnetic structure, the expansion of Eq. 8.3.5 in spherical harmonics is

$$d\Phi = \frac{1}{4\pi\mu_0} \sum_{l,j} \frac{1}{\rho^{l+1}} \left[dp_{j,1}^{(l)} \cos j\psi + dp_{j,2}^{(l)} \sin j\psi \right] P_l^j(\xi), \qquad (8.3.6)$$

where by virtue of Eqs. 1.5.12 and 1.5.13, coefficients $dp_{j,1}^{(l)}$, $dp_{j,2}^{(l)}$ are

$$dp_{j,1}^{(l)} = \sum_{h,k} dp_{i,j,1}^{(l)} = 2\sum_{h,k} \frac{(l-j)!}{(l+j)!} P_l^j(0) \sigma_i \int_{\tau_i} r_i^l \cos j\psi_i \, d\tau \, d\zeta$$

$$dp_{j,2}^{(l)} = \sum_{h,k} dp_{i,j,2}^{(l)} = 2\sum_{h,k} \frac{(l-j)!}{(l+j)!} P_l^j(0) \sigma_i \int_{\tau_i} r_i^l \sin j\psi_i \, d\tau \, d\zeta. \qquad (8.3.7)$$

In Eqs. 8.3.7, the index i denotes interface $S_{h,k}$ and r_i is the distance of a point of $S_{h,k}$ from the z axis. In particular for $l = j = 1$, Eqs. 8.3.7 yield

$$dp_{1,1}^{(1)} = \sum_{h,k} \sigma_i \int y_i d\tau d\zeta$$

$$dp_{1,2}^{(1)} = \sum_{h,k} \sigma_i \int x_i d\tau d\zeta ,$$
(8.3.8)

where

$$x_i = r_i \sin \psi_i , \qquad y_i = r_i \cos \psi_i$$
(8.3.9)

are the x, y coordinates of a point of the ith interface of the magnetic structure. Coefficients 8.3.8 are the two components of the dipole moment of the slice of the magnetic structure of thickness $d\zeta$. By definition dipole moment $\vec{dp}_1^{(1)}$ is the sum of the dipole moments of each element of volume of the slice of the magnetic structure, i.e.,

$$\vec{dp}_1^{(1)} = dp_{1,2}^{(1)} \vec{x} + dp_{1,1}^{(1)} \vec{y} = \int_{S_c} \vec{J} dS\, d\zeta ,$$
(8.3.10)

where S_c is the area of the cross-section of the magnetic structure perpendicular to the z axis. Thus term $l = j = 1$ of Eq. 8.3.6 is the potential of a dipole of moment $\vec{dp}_1^{(1)}$ located at point O', and, by virtue of Eq. 1.5.20, one has

$$(d\Phi)_{l=j=1} = -\frac{1}{4\pi\mu_0} \nabla \left[\frac{1}{\rho}\right] \cdot \vec{dp}_1^{(1)} .$$
(8.3.11)

The potential of the dipole has a symmetry of revolution about the axis of the dipole, and it changes sign across the plane perpendicular to $\vec{dp}_1^{(1)}$ that passes through the origin O'. The potential in the medium surrounding a prismatic yokeless magnet of infinite length is identically equal to zero. Thus, as in the case of the cylindrical magnet analyzed in Section 8.1, the contribution of each multipole associated with each element $d\zeta$ of the prismatic magnet must cancel out in the external medium. In the particular case $l = j = 1$, because of symmetry, the cancellation of the dipole moment outside the prismatic magnet of infinite length is possible only if the dipole moment of each slice of thickness dS is identically equal to zero. As a consequence, because $\vec{J}$ is uniform in each prism of the slice, by virtue of Eq. 8.3.10, the geometry and the distribution of remanence in a prismatic yokeless structure must satisfy the condition

$$\int_{S_c} \vec{J} dS = 0 .$$
(8.3.12)

298 OPEN MAGNETIC STRUCTURES

Hence, the dominant term of the scalar potential generated by a section of arbitrary length of a prismatic yokeless structure of arbitrary geometry is the field of either a quadrupole ($l=2$), or a higher order multipole.

If the magnetic structure is symmetric with respect to the plane $x = 0$ ($\psi = 0$) such that Eq. 8.3.6 satisfies the condition

$$d\Phi(\psi) = d\Phi(-\psi) ,\qquad(8.3.13)$$

one has

$$dp_{j,2}^{(l)} = 0 \qquad(8.3.14)$$

for all values of l, j. Furthermore, if the magnetic structure is symmetric with respect to the plane $y = 0$ ($\psi = \pm \pi/2$) such that

$$d\Phi(\psi) = -d\Phi(\pi - \psi) ,\qquad(8.3.15)$$

Eq. 8.3.6 must satisfy the condition

$$\cos j\pi = -1 ; \qquad(8.3.16)$$

i.e., Eq. 8.3.6 reduces to the terms with odd values of j. On the other hand, one has

$$P_l^j(0) = 0 \qquad(8.3.17)$$

for even numbers l and odd numbers j. Hence, a symmetric structure that satisfies conditions 8.3.13 and 8.3.15 has no quadrupole moment. Expansion 8.3.6 is then limited to terms with odd values of l and j, and the dominant term of the scalar potential is the field of an octupole ($l = 3$) as in the case of the cylindrical structure analyzed in Section 8.1.

8.4 SQUARE CROSS-SECTIONAL YOKELESS MAGNET

As an example, consider a section of the yokeless single layer magnet with a square cross-sectional cavity analyzed in Section 4.2 for the particular case

$$K = 1 - 2^{-\frac{1}{2}} . \qquad(8.4.1)$$

The magnet section is shown in Fig. 8.4.1, and the first quadrant of the magnet cross-section in Fig. 8.4.2 shows the orientation of the remanences of the components that correspond to an intensity $\vec{H}_0$ oriented along the y axis. By virtue of theorem 4 of Section 3.5, the orientation of the remanences is obtained from Fig. 4.2.8 by rotating vectors $\vec{J}_1, \vec{J}_{1,1}, \vec{J}_{1,2}$ by a $\pi/4$ angle.

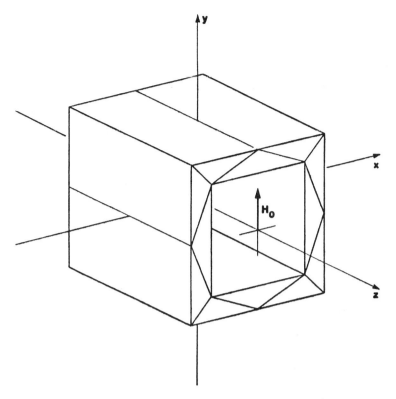

Fig. 8.4.1. Square cross-sectional magnet of finite length.

The values of the surface charge densities σ_i on the interfaces of the first quadrant in Fig. 8.4.2 are

$$\sigma_1 = \sigma_3 = -J_0$$
$$\sigma_2 = \sigma_4 = 2^{-\frac{1}{2}}J_0 \qquad (8.4.2)$$
$$\sigma_5 = -(2 - 2^{\frac{1}{2}})^{\frac{1}{2}}J_0.$$

The structure of Fig. 8.4.1 satisfies the symmetry conditions 8.3.13, 8.3.15, and in the $l = j = 1$ terms of the first equation of system 8.3.7, one has

$$P_1^1(0) = 1. \qquad (8.4.3)$$

In the first quadrant, surface charges 8.4.2 yield the values of $dp_{i,1,1}^{(1)}$

$$\begin{aligned}
dp_{1,1,1}^{(1)} &= -r_0^2 J_0 d\zeta \\
dp_{2,1,1}^{(1)} &= 2^{\frac{1}{2}} r_0^2 J_0 d\zeta \\
dp_{3,1,1}^{(1)} &= -dp_{4,1,1}^{(1)} = -2^{-\frac{1}{2}} r_0^2 J_0 d\zeta \\
dp_{5,1,1}^{(1)} &= -(2^{\frac{1}{2}} - 1) r_0^2 J_0 d\zeta,
\end{aligned} \qquad (8.4.4)$$

300 OPEN MAGNETIC STRUCTURES

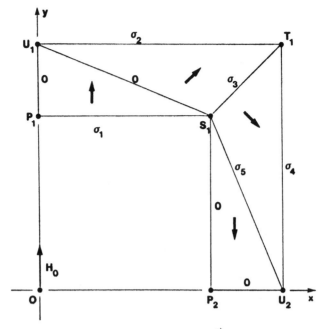

Fig. 8.4.2. Remanences and surface charge densities for $\vec{H}_0$ parallel to a side of the cavity.

where $2r_0$ is the side of the internal square of the magnet cross-section. The sum of the five terms in Eq. 8.4.4 is

$$\sum_{i=1}^{5} dp_{i,1,1}^{(1)} = 0 . \tag{8.4.5}$$

As expected, the element of magnet exhibits no dipole moment.

The quadrupole moment is zero, and the octupole moment ($l = 3$) has both terms ($j = 1$) and ($j = 3$). One has

$$P_3^1(0) = -\frac{3}{2} , \qquad P_3^3(0) = 15 . \tag{8.4.6}$$

In the first quadrant, the values of $dp_{i,1,1}^{(3)}$ are

$$\begin{aligned}
dp_{1,1,1}^{(3)} &= \frac{1}{3} r_0^4 J_0 d\zeta , & dp_{2,1,1}^{(3)} &= -\frac{2^{3/2}}{3} r_0^4 J_0 d\zeta , \\
dp_{3,1,1}^{(3)} &= 2^{-5/2} 3 \, r_0^4 J_0 d\zeta , & dp_{4,1,1}^{(3)} &= -2^{-5/2} 3 \, r_0^4 J_0 d\zeta , \\
dp_{5,1,1}^{(3)} &= 2^{-3/2} 3^{-1} (2^{1/2} - 1)(2^{3/2} + 1) r_0^4 J_0 d\zeta .
\end{aligned} \tag{8.4.7}$$

The sum of the five terms in Eq. 8.4.7 is

$$\sum_{i=1}^{5} dp_{i,1,1}^{(3)} = -\frac{1}{2^{3/2}\,3}(5 - 2^{1/2})\,r_0^4\,J_0\,d\zeta. \qquad (8.4.8)$$

The values of the $j = 3$ terms of the octupole moment in the first quadrant are

$$dp_{1,3,1}^{(3)} = dp_{2,3,1}^{(3)} = 0, \qquad dp_{3,3,1}^{(3)} = \frac{r_0^4}{2^{9/2}\,3}\,J_0\,d\zeta,$$

$$dp_{4,3,1}^{(3)} = -\frac{11}{2^{9/2}\,3}\,r_0^4\,J_0\,d\zeta, \qquad dp_{5,3,1}^{(3)} = \frac{r_0^4}{2^{5/2}\,3}(2^{1/2} - 1)(7 + 2^{1/2}\,3)\,J_0\,d\zeta, \qquad (8.4.9)$$

and the sum of the five terms in Eq. 8.4.9 is

$$\sum_{i=1}^{5} dp_{i,3,1}^{(3)} = \frac{r_0^4}{2^{7/2}\,3}(2^{7/2} - 7)\,J_0\,d\zeta. \qquad (8.4.10)$$

Thus the components $j = 1$ and $j = 3$ of the octupole moment of the full slice of the magnet section of length $d\zeta$ are

$$dp_{1,1}^{(3)} = -\frac{2^{1/2}}{3}\,r_0^4\,(5 - 2^{1/2})\,J_0\,d\zeta \qquad (j = 1)$$

$$dp_{3,1}^{(3)} = \frac{r_0^4}{2^{3/2}\,3}(2^{7/2} - 7)\,J_0\,d\zeta \qquad (j = 3). \qquad (8.4.11)$$

The ratio of the $j = 3$ to the $j = 1$ components of the octupole moment is

$$\left|\frac{dp_{3,1}^{(3)}}{dp_{1,1}^{(3)}}\right| \approx 0.3. \qquad (8.4.12)$$

Thus the square cross-sectional prismatic magnet exhibits a substantial $j = 3$ component of the octupole moment, in contrast with the cylindrical structures of Section 8.1 that have only the $j = 1$ component of their multipoles.

It is of interest to compare the value $dp_{1,1}^{(3)}$ given by Eqs. 8.4.11 with the octupole moment 8.1.22 of a cylindrical structure for comparable values of K and area of the internal cross-section. For $z_0 \to \infty$, by virtue of Eq. 8.1.13, the cylindrical magnet of Section 8.1 generates the same value of K given by Eq. 8.4.1 if its radial dimensions r_1, r_2 are

$$\frac{r_1}{r_2} = (2^{1/2} - 1)^{1/2}, \qquad (8.4.13)$$

and Eq. 8.1.22 yields the value of the octupole moment

$$\left[dp_{1,1}^{(3)}\right]_{circle} = -\frac{\pi}{2^{3/2}} r_1^4 J_0 d\zeta . \qquad (8.4.14)$$

Assuming the same area of the internal cross-sections in Figs. 8.1.1 and 8.4.1 ($\pi r_1^2 = 4r_0^2$), one has

$$\frac{\left[dp_{1,1}^{(3)}\right]_{square}}{\left[dp_{1,1}^{(3)}\right]_{circle}} \approx 0.94 , \qquad (8.4.15)$$

i.e., both structures have comparable octupole moments.

Let us analyze now the field close to the center O of the magnet by means of the series of spherical harmonics 1.5.8. The spherical polar coordinates ρ, θ, ψ are chosen with the origin at O, the axis z as $\theta = 0$, and the plane $x = 0$ as $\psi = 0$.

Because of symmetry, series 1.5.8 is limited to terms with values of l and j equal to odd numbers, and the scalar potential Φ within a region $\rho < r_0$ is

$$\Phi = \frac{1}{4\pi\mu_0} \sum_{l} \sum_{j=1}^{l} q_{j,1}^{(l)} \rho^l P_l^j(\xi) \cos j\psi . \qquad (8.4.16)$$

Coefficients $q_{j,1}^{(l)}$ are

$$q_{j,1}^{(l)} = \sum_i q_{i,j,1}^{(l)} , \qquad (8.4.17)$$

where subindex i denotes the interfaces S_i of the structure of Fig. 8.4.1. By virtue of Eqs. 1.5.12 and 1.5.14 one has

$$q_{i,j,1}^{(l)} = 2\frac{(l-j)!}{(l+j)!} \int_{S_i} \frac{\sigma_i}{\rho_i^{l+1}} P_l^j(\xi_i) \cos j\psi_i \, dS_i . \qquad (8.4.18)$$

Consider the term $l = 1$ in Eq. 8.4.16. For $j = l = 1$, in the first quadrant of the magnet cross-section, the contribution of surface charge σ_1 defined in Fig. 8.4.2 is

$$q_{1,1,1}^{(1)} = \sigma_1 \int_{-z_0}^{+z_0} d\zeta \int_0^{r_0} \frac{\cos\psi}{\rho^2} P_1^1(\cos\theta) \, dx , \qquad (8.4.19)$$

where

$$\rho = (\zeta^2 + r_0^2 + x^2)^{1/2} , \qquad P_1^1(\cos\theta) = \sin\theta = \left[\frac{x^2 + r_0^2}{\zeta^2 + r_0^2 + x^2}\right]^{1/2} . \qquad (8.4.20)$$

One has

$$q_{1,1,1}^{(1)} = \sigma_1 r_0 \int_{-z_0}^{+z_0} d\zeta \int_0^{r_0} \frac{dx}{(\zeta^2 + r_0^2 + x^2)^{3/2}} = \sigma_1 \Omega_1, \qquad (8.4.21)$$

where Ω_1 is the solid angle with vertex at O subtended by the interface in the first quadrant which supports charge density σ_1. Similarly the contribution $q_{2,1,1}^{(1)}$ of charge density σ_2 is

$$q_{2,1,1}^{(1)} = \sigma_2 \Omega_2, \qquad (8.4.22)$$

where solid angle Ω_2 is

$$\Omega_2 = 2^{1/2} r_0 \int_{-z_0}^{+z_0} d\zeta \int_0^{2^{1/2} r_0} \frac{dx}{(\zeta^2 + x^2 + 2r_0^2)^{3/2}}. \qquad (8.4.23)$$

Contribution $q_{3,1,1}^{(1)}$ of charge density σ_3 is

$$q_{3,1,1}^{(1)} = \frac{\sigma_3}{2^{1/2}} \int_{-z_0}^{+z_0} d\zeta \int_{2^{1/2} r_0}^{2 r_0} \frac{r \, dr}{(\zeta^2 + r^2)^{3/2}}$$

$$= \frac{\sigma_3}{2^{1/2}} \ln \frac{\left[(z_0^2 + 2r_0^2)^{1/2} + z_0\right] \left[(z_0^2 + 4r_0^2)^{1/2} - z_0\right]}{\left[(z_0^2 + 2r_0^2)^{1/2} - z_0\right] \left[(z_0^2 + 4r_0^2)^{1/2} + z_0\right]}. \qquad (8.4.24)$$

Contribution of surface charge σ_4 is

$$q_{4,1,1}^{(1)} = 2^{1/2} \frac{\sigma_4}{\sigma_3} q_{3,1,1}^{(1)}. \qquad (8.4.25)$$

Contribution $q_{5,1,1}^{(1)}$ of charge density σ_5 is

$$q_{5,1,1}^{(1)} = \sigma_5 \int_{-z_0}^{+z_0} d\zeta \int_{-\tau_0}^{+\tau_0} \frac{\cos\psi}{\rho^2} P_1^1(\cos\theta) \, d\tau, \qquad (8.4.26)$$

where

$$\tau_0 = 2^{1/2} r_0 \sin \frac{\pi}{8}, \quad \rho = \left[\zeta^2 + 2r_0^2 \cos^2 \frac{\pi}{8} + \tau^2 \right]^{1/2},$$

$$P_1^1(\cos\theta) = \left[\frac{2r_0^2 \cos^2 \frac{\pi}{8} + \tau^2}{\zeta^2 + 2r_0^2 \cos^2 \frac{\pi}{8} + \tau^2} \right]^{1/2}. \quad (8.4.27)$$

One has

$$q_{5,1,1}^{(1)} = \frac{\sigma_5}{2^{1/2}} (1 - 2^{-1/2})^{1/2} \, \Omega_5, \quad (8.4.28)$$

where

$$\Omega_5 = 2^{1/2} r_0 \cos \frac{\pi}{8} \int_{-z_0}^{+z_0} d\zeta \int_{-\tau_0}^{+\tau_0} \frac{d\tau}{\rho^3}. \quad (8.4.29)$$

Ω_5 is the solid angle with vertex at O subtended by the interface which supports charge density σ_5. By virtue of Eqs. 8.4.2 and 8.4.25, $q_{4,1,1}^{(1)}$ and $q_{3,1,1}^{(1)}$ cancel each other, and the value of $q_{1,1}^{(1)}$ of the full magnet of Fig. 8.4.1 is

$$q_{1,1}^{(1)} = -4J_0 \left[\Omega_1 - 2^{-1/2} \Omega_2 + (1 - 2^{-1/2}) \Omega_5 \right], \quad (8.4.30)$$

and term $l = 1$ of expansion 8.4.16 is

$$\Phi_{l=1} = -\frac{J_0}{\pi\mu_0} \left[\Omega_1 - 2^{-1/2} \Omega_2 + (1 - 2^{-1/2}) \Omega_5 \right] y. \quad (8.4.31)$$

In the limit $z_0 = \infty$ one has

$$\lim_{z_0 \to \infty} \Omega_1 = \lim_{z_0 \to \infty} \Omega_2 = \lim_{z_0 \to \infty} \Omega_5 = \frac{\pi}{2}, \quad (8.4.32)$$

and

$$\lim_{z_0 \to \infty} \Phi_{l=1} = -\frac{J_0}{\mu_0} \left[1 - 2^{-1/2} \right] y, \quad (8.4.33)$$

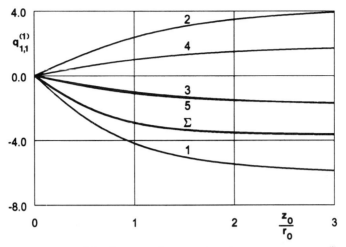

Fig. 8.4.3. Contribution of the charges on the magnet interfaces to the value of $q_{1,1}^{(1)}$ versus z_0.

which is the scalar potential inside the magnet of infinite length. Figure 8.4.3 shows the values of $q_{i,1,1}^{(1)}$ and the total value $q_{1,1}^{(1)}$ versus the length of the magnet.

Let us consider now the $l = 3$ terms of Eq. 8.4.16. Both coefficients $q_{1,1}^{(3)}$ and $q_{3,1}^{(3)}$ contribute to the $l = 3$ component of Φ. Equation 8.4.18 yields

$$q_{i,1,1}^{(3)} = \frac{1}{4} \sigma_i \int_{S_i} \frac{1}{\rho^4} (5\cos^2\theta - 1) \sin\theta \cos\psi \, dS_i \qquad (8.4.34)$$

and

$$q_{i,3,1}^{(3)} = \frac{\sigma_i}{24} \int_{S_i} \frac{1}{\rho^4} \sin^3\theta \cos 3\psi \, dS_i \ . \qquad (8.4.35)$$

In the first quadrant, the contributions $q_{i,1,1}^{(3)}$ of σ_i are

$$q_{1,1,1}^{(3)} = -\frac{\sigma_1}{6} z_0 r_0^2 \frac{5r_0^2 + 3z_0^2}{(r_0^2 + z_0^2)^2 (2r_0^2 + z_0^2)^{3/2}} \qquad (8.4.36)$$

$$q_{2,1,1}^{(3)} = -\frac{\sigma_2}{3} z_0 r_0^2 \frac{3z_0^2 + 10r_0^2}{(z_0^2 + 2r_0^2)^2 (z_0^2 + 4r_0^2)^{3/2}} \qquad (8.4.37)$$

$$q_{3,1,1}^{(3)} = \frac{\sigma_3}{2^{1/2} 3!} z_0 \left[\frac{1}{(z_0^2 + 4r_0^2)^{3/2}} - \frac{1}{(z_0^2 + 2r_0^2)^{3/2}} \right] \qquad (8.4.38)$$

306 OPEN MAGNETIC STRUCTURES

$$q_{4,1,1}^{(3)} = 2^{1/2} \frac{\sigma_4}{\sigma_3} q_{3,1,1}^{(3)} \tag{8.4.39}$$

$$q_{5,1,1}^{(3)} = -\frac{\sigma_5}{2^{1/2} 3} z_0 r_0^2 \sin\frac{\pi}{8} \frac{3z_0^2 + 2r_0^2 \left[2 + \cos^2\frac{\pi}{8}\right]}{\left[z_0^2 + 2r_0^2 \cos^2\frac{\pi}{8}\right]^2 (z_0^2 + 2r_0^2)^{3/2}} . \tag{8.4.40}$$

Again the contributions of σ_3 and σ_4 cancel each other.

Also in the first quadrant, the contributions $q_{i,3,1}^{(3)}$ to the $l = j = 3$ term of Eq. 8.4.17 are

$$q_{1,3,1}^{(3)} = -\frac{J_0}{90} \frac{z_0}{r_0^2} \frac{1}{(z_0^2 + 2r_0^2)^{3/2}} \left[z_0^2 + 3r_0^2 + \frac{1}{2} r_0^4 \frac{3z_0^2 + 5r_0^2}{(z_0^2 + r_0^2)^2}\right] \tag{8.4.41}$$

$$q_{2,3,1}^{(3)} = \frac{2^{-1/2} J_0}{180} \frac{z_0}{r_0^2} \frac{1}{(z_0^2 + 4r_0^2)^{3/2}} \left[z_0^2 + 6r_0^2 + 2r_0^4 \frac{3z_0^2 + 10r_0^2}{(z_0^2 + 2r_0^2)^2}\right] \tag{8.4.42}$$

$$q_{3,3,1}^{(3)} + q_{4,3,1}^{(3)} = -\frac{2^{-1/2} J_0}{180} \frac{z_0}{r_0^2} \left[\frac{1}{(z_0^2 + 4r_0^2)^{1/2}} + \frac{2r_0^2}{(z_0^2 + 2r_0^2)^{3/2}}\right] \tag{8.4.43}$$

$$q_{5,3,1}^{(3)} = \frac{J_0 z_0}{90 r_0^2 (z_0^2 + 2r_0^2)^{3/2}} \left[z_0^2 + 3r_0^2 + r_0^4 \cos\frac{\pi}{8} \frac{3z_0^2 + 10r_0^2 \cos^2\frac{\pi}{8}}{\left[z_0^2 + 2r_0^2 \cos^2\frac{\pi}{8}\right]^2}\right] . \tag{8.4.44}$$

Terms $q_{3,3,1}^{(3)}$ and $q_{4,3,1}^{(3)}$ do not cancel each other. The values of $q_{i,1,1}^{(3)}$ and the total value $q_{1,1}^{(3)}$ are shown in Fig. 8.4.4. The values of $q_{i,3,1}^{(3)}$ and the total value $q_{3,1}^{(3)}$ are shown in Fig. 8.4.5. Both coefficients $q_{1,1}^{(3)}$ and $q_{3,1}^{(3)}$ vanish at $z_0 \to \infty$. Figure 8.4.5 shows that the components $q_{i,3,1}^{(3)}$ have asymptotic nonzero values. The same property is exhibited by some of the components of all coefficients $q_{l,1}^{(l)}$, as shown by the graphs of $q_{j,1}^{(5)}$, $q_{j,1}^{(7)}$, $q_{j,1}^{(9)}$ presented in App. IV.

The magnitude of $q_{3,1}^{(3)}$ is small compared to $q_{1,1}^{(3)}$. However, as z_0 increases and becomes large compared to r_0, the individual components $q_{i,3,1}^{(3)}$ become large compared to the total value of $q_{1,1}^{(3)}$. This property of the structure of Fig. 8.4.1 has an important consequence in a practical magnet where fabrication and magnetization tolerances may preclude the cancellation of the individual components of $q_{3,1}^{(3)}$. As a result, in a practical situation the harmonic $l = j = 3$ may be the dominant term of the field distortion within the cavity of the magnetic structure.

Section 3.5 has shown that the geometry of a yokeless two-dimensional magnet is

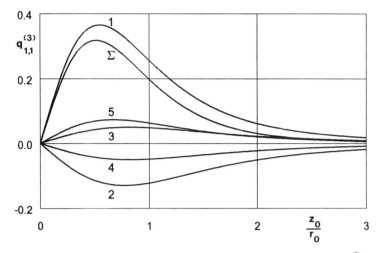

Fig. 8.4.4. Contribution of the charges on the magnet interfaces to the value of $q_{1,1}^{(3)}$ versus z_0.

independent of the orientation of the intensity $\vec{H}_0$ of the magnetic field inside the magnet cavity. However, the orientation of the remanences of the magnet components depends upon the orientation of $\vec{H}_0$, and as a consequence, in general, one may expect the scalar potential generated by a yokeless magnet of finite length to be a function of the orientation of $\vec{H}_0$.

Assume an orientation of $\vec{H}_0$ parallel to a diagonal of the cavity of Fig. 8.4.1. The distribution of the remanence in the components of the two-dimensional magnet has been computed with the vector diagram of Fig. 4.2.7 and shown in Fig. 4.2.8. The dis-

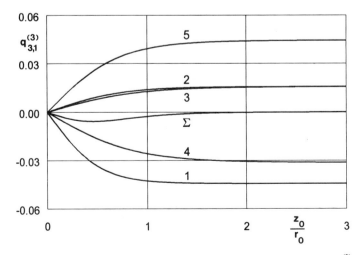

Fig. 8.4.5. Contribution of the charges on the magnet interfaces to the value of $q_{3,1}^{(3)}$ versus z_0.

308 OPEN MAGNETIC STRUCTURES

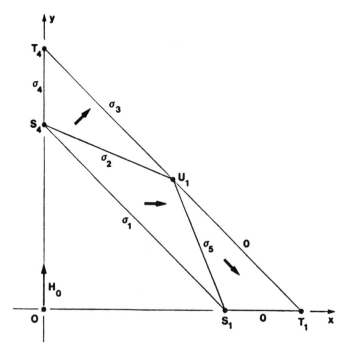

Fig. 8.4.6. Remanences and surface charge densities for $\vec{H}_0$ parallel to a diagonal of the cavity.

tribution of surface charge densities σ_i is shown in Fig. 8.4.6 and their values are

$$\sigma_1 = \sigma_4 = -\frac{J_0}{\sqrt{2}}, \qquad \sigma_2 = -(1 - 2^{-1/2})^{1/2} J_0$$
$$\sigma_3 = J_0, \qquad \sigma_5 = -\sigma_2. \qquad (8.4.45)$$

One should bear in mind that the value of σ_4 given by Eqs. 8.4.45 is half the value of the surface charge density induced by the remanences of the prisms that have interface $(S_4 T_4)$ in common with each other.

The value of coefficients 8.4.18 corresponding to the new orientation of $\vec{H}_0$ can be readily derived from the values obtained in the case of $\vec{H}_0$ oriented along a side of the square, by taking into account that the intensity $\vec{H}_0$ along the diagonal may be considered as the superposition of two intensities $\vec{H}_0/\sqrt{2}$ oriented along the x and y axes of Fig. 8.4.1. The scalar potential within a sphere of radius $\rho < r_0$ and center at O becomes

$$\Phi = \frac{1}{4\pi\mu_0} \sum_{j,l} [q_{j,1}^{(l)}]' \rho^l \cos j\phi \, P_l^j(\xi), \qquad (8.4.46)$$

where

$$[q_{j,1}^{(l)}]' = \sqrt{2} \cos(j\frac{\pi}{4}) q_{j,1}^{(l)} \qquad (8.4.47)$$

and

$$\phi = \psi - \frac{\pi}{4} ; \qquad (8.4.48)$$

ϕ is the angular coordinate relative to the axis y in Fig. 8.4.6., and $q_{j,1}^{(l)}$ is the same coefficient defined by Eq. 8.4.17.

As shown by Eq. 8.4.46 the field configuration is symmetric with respect to the $\psi = \pi/4$ plane. Coefficients $q_{j,1}^{(l)}$ are identically zero for even numbers j, l, and as a consequence, by virtue of Eq. 8.4.47, the absolute values of $[q_{j,1}^{(l)}]'$ and $q_{j,1}^{(l)}$ are identical to each other.

Coefficients with the values

$$j = 1, 7, 9, \ldots \qquad (8.4.49)$$

have the same sign, and coefficients with values

$$j = 3, 5, 11, 13, \ldots \qquad (8.4.50)$$

have opposite signs.

8.5 IMAGE METHOD OF FIELD COMPUTATION IN YOKED MAGNETS

The presence of ferromagnetic materials in an open magnet presents a much more difficult problem compared to the yokeless structures analyzed in the preceding sections. In general, a closed form solution cannot be found even under ideal conditions of linear characteristics of magnetic media and ferromagnetic materials. Thus a yoked open magnet or a magnetic structure containing high magnetic permeability components, like the one described in Chapter 6, falls in the category of problems where only numerical solutions are possible. A special situation arises in the case of a structure of magnetic material of finite length $2z_0$ with an ideal yoke of infinite magnetic permeability whose length $2z_1$ is assumed to be large compared to $2z_0$. A schematic of such a structure is shown in Fig. 8.5.1. If $z_1 \gg z_0$, the yoked magnet of Fig. 8.5.1 can be analyzed by assuming that the magnetic material has a two-dimensional yoke, in which case an exact solution of the field equation becomes possible for some simple geometries of the yoke cross-section.

Consider, for instance, a rectangular cross-section of the $\mu = \infty$ yoke, and assume the hybrid magnetic structure shown in Fig. 8.5.2, which is the hybrid equivalent of the yokeless structure of Fig. 6.5.3. If both yoke and magnetic structure have infinite

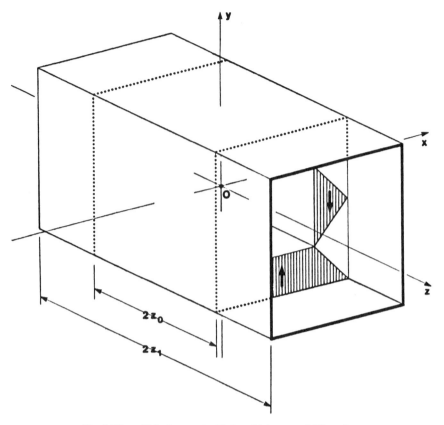

Fig. 8.5.1. Yoked magnet with long high permeability yoke.

length, the magnet of Fig. 8.5.2 corresponds to the value

$$K = \frac{1}{2} \qquad (8.5.1)$$

with the intensity $\vec{H}_0$ oriented along the axis y. In the limit $z_0 = z_1 = \infty$, the intensity is zero within the regions of nonmagnetic material that separate the yoke from the magnetic structure, and the geometry of the ideal yoke that encloses these regions has no effect on the magnet characteristics.

Assume now a finite length $2z_0$ of the magnetic structure of Fig. 8.5.2 and a two-dimensional ideal yoke ($z_1 = \infty$) of rectangular cross-section of dimensions $2x_1$, $2y_1$. The field configuration generated by the distribution of remanence of Fig. 8.5.2 can be computed by expanding the solution of Laplace's equation in a series of trigonometric harmonics in the variables x, y. Because of symmetry, the scalar potential satisfies the conditions

$$\Phi(x,y,z) = \Phi(-x,y,z) = -\Phi(x,-y,z) = \Phi(x,y,-z). \qquad (8.5.2)$$

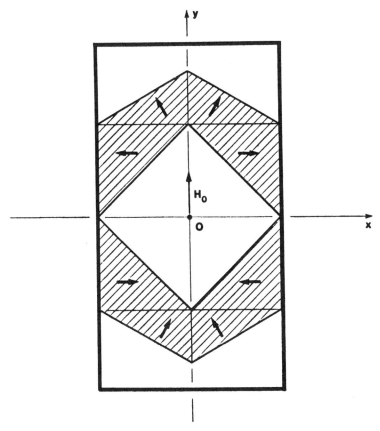

Fig. 8.5.2. Hybrid magnet with a rectangular cross-sectional yoke.

Thus the fundamental harmonic of the trigonometric series has a period λ_x along the x coordinate and a period λ_y along the y coordinate

$$\lambda_x = 4x_1, \qquad \lambda_y = 2y_1. \tag{8.5.3}$$

The periodicity of the solution of the field equations derives from the fact that the planes $x = \pm x_1$, $y = \pm y_1$ act like mirrors of the magnetic structure, and a double infinity of images results from the multiple reflections. Because of the infinite permeability of the yoke, the calculation of the field within the two-dimensional yoke of Fig. 8.5.2 can be performed by replacing the yoke with an infinite matrix of magnetic structures which correspond to the double infinity of images, as indicated in the schematic of Fig. 8.5.3.

At each point within the rectangular region $|x| < x_1$, $|y| < y_1$, the scalar potential is computed as the superposition of the potentials generated by the structure of Fig. 8.5.2 and its mirror images

$$\Phi(x,y,z) = \sum_{h,k} \Phi_{h,k}(x,y,z), \tag{8.5.4}$$

312 OPEN MAGNETIC STRUCTURES

where $\Phi_{h,k}$ is the potential generated by the magnetic structure contained within the h,k cell of the matrix. The distribution of the scalar potential is continuous throughout the matrix, and the potential is zero on the lines

$$x = (2h + 1)x_1, \quad y = (2k + 1)y_1 \tag{8.5.5}$$

for all positive and negative values of integers h,k. The region $|x| < x_1$, $|y| < y_1$ corresponds to the cell $h = k = 0$. The contribution of the h, k cell to the value of Φ given by Eq. 8.5.4 decreases as $|h|$, $|k|$ increase. Asymptotically, as the distance of the h, k cell from the magnetic structure of Fig. 8.5.3 becomes large compared to $2z_0$, the term $\Phi_{h,k}$ decreases as h^{-2} and k^{-2} as the potential of a dipole. Thus, in practice, the calculation of Eq. 8.5.4 may be limited to a finite number of reflections along both coordinates x and y depending upon the required precision.

Assume that the matrix of Fig. 8.5.3 is limited to an equal number of reflections along both coordinate axes

$$-n_0 \leq h \leq n_0, \quad -n_0 \leq k \leq n_0. \tag{8.5.6}$$

This results in a finite matrix of $(2n_0 + 1)^2$ cells. Assume the dimensions of the rectangular cross-section of the yoke

$$y_1 = 2x_1, \tag{8.5.7}$$

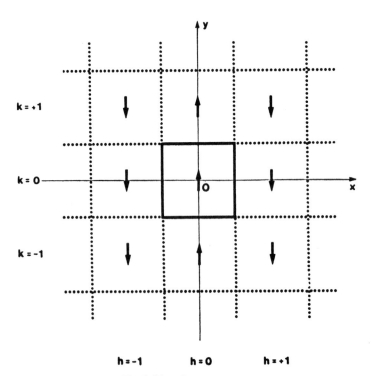

Fig. 8.5.3. Image structures.

and consider two values of the length $2z_0$ of the magnetic structure

$$z_0 = x_1, \quad z_0 = 10x_1. \tag{8.5.8}$$

The asymptotic value H_0 of the intensity at the center of the magnet for $n_0 \to \infty$ in the two cases given by Eq. 8.5.8 is

$$\lim_{n_0 \to \infty} \mu_0 \left[\frac{H_0}{J_0}\right]_{z_0=x_1} \approx 0.43, \quad \lim_{n_0 \to \infty} \mu_0 \left[\frac{H_0}{J_0}\right]_{z_0=10x_1} \approx 0.50, \tag{8.5.9}$$

i.e., the numerical result for $z_0 = 10x_1$ practically coincides with the value $K = 1/2$ of the hybrid two-dimensional magnetic structure.

The convergence of the solutions in both cases of Eq. 8.5.8 is shown in Fig. 8.5.4

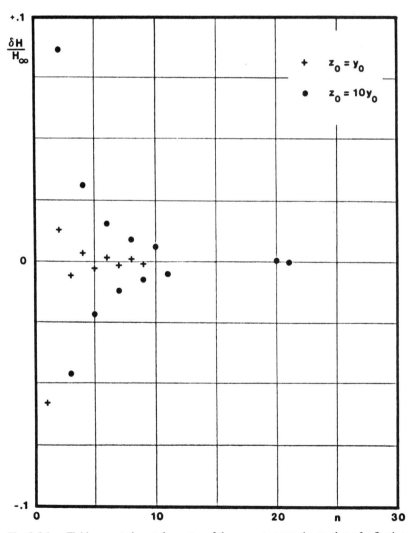

Fig. 8.5.4. Field computation at the center of the magnet versus the number of reflections.

where the value of the intensity at the center of the magnet normalized to H_0, is plotted versus n_0, i.e., versus the number of reflections included in the numerical calculation [3].

In both cases Fig. 8.5.4 shows that the results oscillate about the asymptotic values, with maxima at even values of n_0. By virtue of Eq. 8.5.3, the y components of the remanence in the columns with odd values of $|h|$ are oriented in a direction opposite to the orientation of the y components of the remanence in the columns with even values of $|h|$. In particular, the y components of the remanence in the two columns $h = \pm 1$ and the y component in the $h = 0$ column are oriented in the opposite directions with respect to each other as schematically shown in Fig. 8.5.2. Thus columns with even values of $|h|$ tend to increase the intensity generated by the $h = k = 0$ central cell at the center of the magnet, while columns with odd value of $|h|$ tend to decrease it.

Figure 8.5.4 shows that the number of reflections required to achieve a given precision is a function of the length $2z_0$ of the magnetic structure. Because the contribution of an image cell decreases rapidly as the distance of the image cell from the axis becomes large compared to the length $2z_0$, a longer magnet requires a larger number of reflections, as shown by Fig. 8.5.4.

If yoke and magnetic structure have comparable lengths, the image method is not applicable, and a numerical integration of the field equation becomes necessary.

8.6 BOUNDARY INTEGRAL EQUATIONS COMPUTATIONAL METHOD

Since the early days of the design of electric machinery, approximate methods for the field computation have been developed ranging from mechanical experimental simulations to graphic solutions of the field equations by flux plotting. The capability of handling systems of large numbers of equations with modern computers has led to the development of powerful numerical tools like the finite elements methods, where the domain of integration is divided in a large number of cells. By selecting sufficiently small cells, the variation of the field inside each cell can be reduced to any desired level. Thus the solution of Laplace's equation in each cell reduces to the dominant terms of a power series expansion, where the constants of integration are determined by the boundary conditions at the interfaces between cells. The number of cells and consequently the number of equations of the boundary conditions depend upon the required precision of calculation. Particularly in three-dimensional applications where the field within the region of interest must be determined with extremely high precision, an exceedingly large number of cells becomes the limiting factor in the use of these numerical methods. An extensive literature on finite elements methods is available to the reader interested in the formulation and application of these methods [2].

Structures composed of ideal materials with linear magnetic characteristics present a special situation where an exact formulation of the field calculation can be developed by computing the charges induced by the distribution of magnetization that satisfy the boundary conditions at the interfaces between different materials. This approach has been followed in Chapter 7 to analyze the field perturbation generated by a ferromagnetic medium. The solution of the integral equation 7.5.9 provides the charge distribution on the surface of the medium from which one computes the field as shown in Sections 5 and 6 of Chapter 7.

The boundary integral equation 7.5.9 can be generalized to a structure of several media with linear magnetic characteristics and, in particular, it can be applied to an open yoked magnet. In general the field distortion caused by the opening of a yoked structure is a far more important effect than the field perturbation caused by the finite value of permeability of the ferromagnetic materials. Hence, the effect of the magnet opening can be analyzed in the ideal limit of a $\mu = \infty$ permeability of the yoke and any other ferromagnetic component of the structure.

In the limit $\mu = \infty$, surface S_h of the schematic of Fig. 7.5.1 becomes an equipotential surface at a potential Φ_h whose value is determined by the solution of the boundary integral equation 7.5.9. At each point P of S_h, Φ_h is the sum of the potential generated by the charge distribution σ and by the external charges m_i. Thus Φ_h satisfies the equation

$$\int_{S_h} \frac{\sigma}{\rho} dS - 4\pi\mu_0 \Phi_h = -\sum_i \frac{m_i}{\rho_i}, \tag{8.6.1}$$

where σ is the solution of Eq. 7.5.9 in the limit $\mu = \infty$. Because Eq. 7.5.13 is the direct consequence of Eq. 7.5.9, the variables σ and Φ_h can be determined by solving the system of Eqs. 7.5.13 and 8.6.1.

In the integral on the left hand side of Eq. 8.6.1, the distance ρ is zero for the element of charge σdS located at the point where the potential is computed. To show that the integral does not exhibit a singularity, consider a circle on surface S_h with center at P and a small radius r. In the limit $r \to 0$, the potential due to the surface charge within the area πr^2 of the circle is

$$\frac{1}{4\pi\mu_0} \lim_{r \to 0} \int_{\pi r^2} \frac{\sigma}{r} dS = \frac{\sigma(P)}{2\mu_0} \lim_{r \to 0} \int_0^r dr = 0. \tag{8.6.2}$$

Thus, by virtue of Eqs. 7.5.13 and 8.6.1 the field in a magnetic structure of N surfaces of $\mu = \infty$ materials in a $\chi_m = 0$ medium is provided by the solution of the system of equations

$$\sum_{k=1}^{N} \int_{S_k} \frac{\sigma}{\rho_{h,k}} dS - 4\pi\mu_0 \Phi_h = -\sum_i \frac{m_i}{\rho_i}$$

$$\int_{S_h} \sigma dS = 0, \qquad h = 1, 2, \ldots, N, \tag{8.6.3}$$

where $\rho_{h,k}$ is the distance of a point of surface S_k from a point of surface S_h. The numerical solution of system 8.6.3 is obtained by dividing each surface S_k in N_k ele-

316 OPEN MAGNETIC STRUCTURES

ments and by assuming a uniform charge density $\bar{\sigma}_k$ on each element of surface. Thus system 8.6.3 transforms into a system of

$$\sum_{k=1}^{N} N_k + N \qquad (8.6.4)$$

equations with an equal number of unknown $\bar{\sigma}_k$ and Φ_h.

An example of multiple $\mu = \infty$ components is the two-dimensional structure shown in Fig. 8.6.1. The two lined rectangular areas represent the cross-section of a $\chi_m = 0$ magnetic material, uniformly magnetized in the direction of the y axis with remanence J. The heavy lines represent the cross-section of four $\mu = \infty$ components of zero thickness. Position and dimensions of the rectangular components are given by the dimensions x_0, y_0, y_1, as indicated in Fig. 8.6.1.

The right hand side of the first equation of system 8.6.3 becomes the contribution of the charges induced by the remanence $\vec{J}$. Because $\vec{J}$ is uniform, by virtue of Eq. 1.4.15, the potential generated by $\vec{J}$ reduces to the potential generated by the surface charge densities σ induced by $\vec{J}$ on the sides of the rectangles parallel to the x axis.

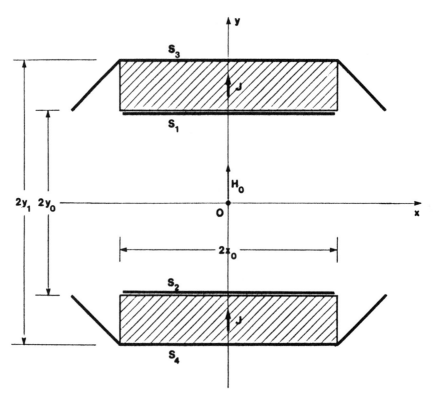

Fig. 8.6.1. Two-dimensional structure of uniformly magnetized material and zero-thickness $\mu = \infty$ plates.

The values of σ are

$$\sigma = \mp J, \quad y = \pm y_1$$
$$\sigma = \pm J, \quad y = \pm y_2. \tag{8.6.5}$$

In the limit $\mu = \infty$, surfaces S_1, S_2, S_3, S_4 are equipotential. Because of symmetry, their potentials satisfy the conditions

$$\Phi_1 = -\Phi_2, \quad \Phi_3 = -\Phi_4. \tag{8.6.6}$$

If S_3 and S_4 are assumed to be connected to each other at $z = \pm\infty$, Fig. 8.6.1 may be considered to be the ideal schematization of a yoked magnet designed to generate on the z axis a field $\vec{H}_0$ oriented in the direction of the axis y. In this case the potentials Φ_3, Φ_4 must be equal to each other, and by virtue of Eq. 8.6.6, one has

$$\Phi_3 = \Phi_4 = 0. \tag{8.6.7}$$

If, on the other hand, the $\mu = \infty$ components are not connected to each other, the four potentials $\Phi_1, \Phi_2, \Phi_3, \Phi_4$ have nonzero values that must be computed by solving the system of Eqs. 8.6.3.

It is instructive to compare the field computed under condition 8.6.7 with the field computed under the more general conditions 8.6.6. The result of the numerical solution in the case of conditions 8.6.6 is shown in Fig. 8.6.2 where the equipotential lines are plotted in the first quadrant of the structure of Fig. 8.6.1. In the limit $x_0 \to \infty$, the field is zero outside the magnetized components, and is uniform and oriented in the direction of the y axis, inside the magnetized components.

The numerical solution shown in Fig. 8.6.2 is obtained for

$$y_1 = 2y_0 = x_0, \tag{8.6.8}$$

and the numerical values of the potentials on the surfaces S_1, S_2, S_3, S_4 yield

$$\frac{\Phi_1 - \Phi_3}{\Phi_1} \approx 2.14. \tag{8.6.9}$$

Thus, with the choice of dimensions in Eq. 8.6.8, the intensity of the field within the magnetized material is more than twice the intensity within the gap between surfaces S_1 and S_2.

The concentration of the field inside the magnetic material becomes even more pronounced if the inclined sides of S_3, S_4 are disconnected, as shown in Fig. 8.6.3, and allowed to acquire a potential different from that of the segments of S_3, S_4 parallel to the x axis. The computation of the potentials of the $\mu = \infty$ components in Fig. 8.6.3

318 OPEN MAGNETIC STRUCTURES

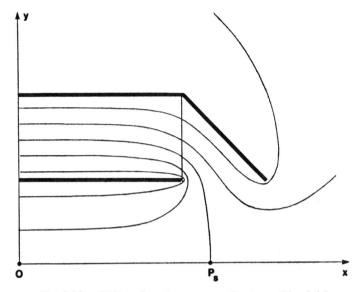

Fig. 8.6.2. Field configuration corresponding to condition 8.6.6.

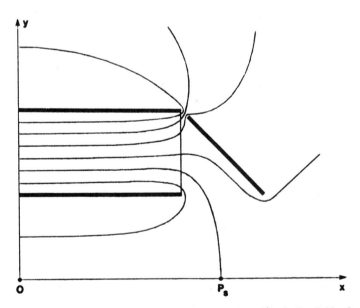

Fig. 8.6.3. Field configuration corresponding to a separation of the inclined sides from surface S_3.

yield

$$\frac{\Phi_1 - \Phi_3}{\Phi_1} \approx 3.07$$
$$\frac{\Phi_1 - \overline{\Phi}_3}{\Phi_1 - \Phi_3} \approx 0.58, \qquad (8.6.10)$$

where $\overline{\Phi}_3$ is the potential of the inclined side of the $\mu = \infty$ component.

The field in the structure of Fig. 8.6.1 changes drastically from the configurations of Figs. 8.6.2 and 8.6.3 if the potentials of surfaces S_3, S_4 are forced to satisfy condition 8.6.7. The result is shown in Figs. 8.6.4 and 8.6.5. The two figures present in a different scale the equipotential lines inside and outside the structure of Fig. 8.6.1. In Fig. 8.6.5 one of the equipotential lines that emerge from the saddle point Q_s encloses both surfaces S_1 and S_3 and the other line encloses surface S_3 only.

With the dimensions given by Eq. 8.6.8, the intensity of the field within the magnetized material and the intensity within the gap between S_1 and S_2 are essentially equal to each other. Furthermore the field within the magnetized material is essentially equal to half the value of J; i.e., the structure of Fig. 8.6.1 is the two-dimensional schematic of a yoked magnet whose magnetic material operates at the peak of the energy product curve. Condition 8.6.7 defines the role of the yoke of closing the flux of $\vec{B}$ generated by the magnetized material.

The open structure of Fig. 8.6.2 generates a field in the surrounding nonmagnetic medium. In the case of condition 8.6.7, Fig. 8.6.4 shows that this stray field is small compared to the field inside the magnet.

It is of interest to compare the properties of the field generated in the surrounding

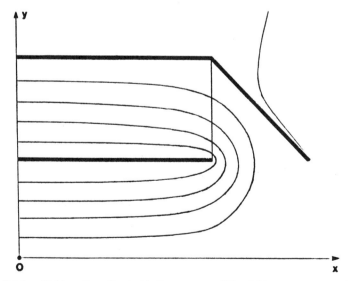

Fig. 8.6.4. Field configuration inside the structure of Fig. 8.6.1 under condition 8.6.7.

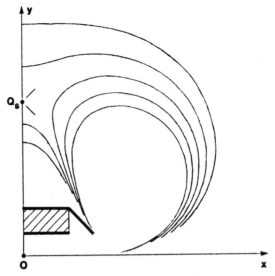

Fig. 8.6.5. Field configuration outside the structure of Fig. 8.6.1 under condition 8.6.7.

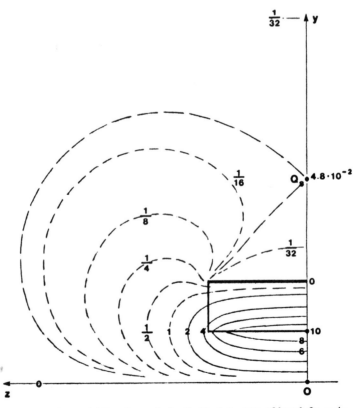

Fig. 8.6.6. Equipotential lines in a yoked cylindrical structure of length $2z_0 = 4r_1$.

medium by open yoked and yokeless magnets. The field of the cylindrical yokeless magnet of Fig. 8.1.1 has been computed in Section 8.1 and the results presented in Figs. 8.1.5 and 8.1.7 have shown that the asymptotic field outside the magnet is the field of an octupole. Consider the yoked equivalent of the structure of Fig. 8.1.1 which consists of a section of length $2z_0$ of the two-dimensional yoked magnet of Section 2.2, where the material is uniformly magnetized in the direction of the y axis. The material is confined between the two coaxial cylinders of radii r_1, $r_2 = 2r_1$. The ideal $\mu = \infty$ yoke is assumed to be a section of a cylinder of diameter $2r_2$ and length $2z_0$. The results of the numerical integration are presented in Figs. 8.6.6 and 8.6.7, which show the equipotential lines in the $x = 0$ plane for the same values $z_0 = 2r_1$ and $z_0 = 0.2r_1$ of the structures of Figs. 8.1.5 and 8.1.7. In both cases, the asymptotic equipotential lines for $\Phi \to 0$ are given by the equation

$$\frac{\sin \theta}{\rho^2} = \frac{4\pi\mu_0}{p_0} \Phi, \qquad (8.6.11)$$

where p_0 is the moment of the equivalent dipole located at the center of the magnet and oriented in the direction of the axis y as indicated in Fig. 8.6.7. In Eq. 8.6.11 ρ is the distance from the origin and θ is the angle between ρ and the $y = 0$ plane.

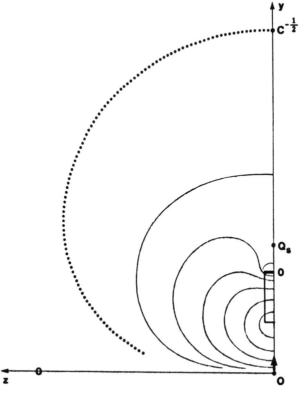

Fig. 8.6.7. Equipotential lines in a yoked cylindrical structure of length $2z_0 = 0.4r_1$.

Equation 8.6.11, which is the equation of the field of a dipole, establishes the basic differences between a yoked and a yokeless magnet where the potential decreases asymptotically as ρ^{-4} as shown in Section 1 of this chapter. The field configurations of both Figs. 8.6.6 and 8.6.7 exhibit the same property of the field in Fig. 8.6.5 characterized by the presence of a saddle point Q_s on the y axis.

REFERENCES

[1] M.G. Abele, Properties of the magnetic field in yokeless permanent magnets, Technical Report No. 18, New York University, March 1, 1988.

[2] M.G. Abele, H. Rusinek, F. Bertora, Field computation in permanent magnets, *IEEE Transactions on Magnetics.* 28(1), 1992, pp. 931-934.

[3] M.G. Abele, H. Rusinek, Field computation in permanent magnets with linear characteristics of magnetic media and ferromagnetic materials, Technical Report No. 24, New York University, August 15, 1991.

CHAPTER 9

Design Considerations

INTRODUCTION

Although design and fabrication of a magnet are outside the scope of this book, it is of importance to examine how the concepts developed in the preceding chapters are implemented in a magnet design and to assess the characteristics of some typical magnetic structures.

In a traditional design, the functions of generating and confining the field are assigned to separate components of a magnetic circuit. The design starts with the specification of an assigned field within a given region of interest and the function of shaping the field is assigned to the ferromagnetic pole pieces. The region of interest is contained within the gap between the pole pieces whose shape is usually the most important part of the overall magnet design.

The design of a magnet with the methodology developed in the preceding chapters is based on the specification of the field within an assigned closed cavity which contains the region of interest. The design starts with the computation of an ideal magnetic structure that totally encloses the magnetic cavity. Then the design proceeds to adapt the ideal structure to the conditions imposed by practical constraints and materials. This conceptual difference from the traditional approach affects all aspects of design and fabrication of a magnet.

The best way to illustrate the design procedure is to select a geometry of the closed cavity and to discuss the different types of magnetic structures capable of generating an assigned field within it.

The selection of cavity geometry and field depends on the particular application. As an example, consider the design of a magnet for nuclear magnetic resonance (NMR) clinical imaging where the minimum dimensions of the cavity are dictated by the size of the human body. The strength and the degree of uniformity of the field within the region of the body under scrutiny are selected on the basis of the diagnostic requirements. The degree of uniformity required by a medical application is of the order of tens parts per million (ppm) and such a stringent requirement is the major problem facing the magnet designer.

Two levels of field strength are selected in the following sections to illustrate designs with materials with very different energy product levels like ferrites and rare earth alloys.

9.1 COMPARISON OF DESIGNS WITH DIFFERENT MAGNETIC MATERIALS

Consider the design of two permanent magnets for two medical NMR scanners capable of imaging volumes large enough to contain the full cross-section of a human body.

324 DESIGN CONSIDERATIONS

The field $\vec{H}_0$ generated by the magnets is assumed to be oriented in a direction perpendicular to the body axis and the value of the $\vec{H}_0$ is selected at two levels,

$$\mu_0 H_0 \approx 0.2T \, , \qquad \mu_0 H_0 \approx 0.4T \, . \tag{9.1.1}$$

The patient is assumed to lie flat on his back on a horizontal support and its position is defined in a cartesian frame of reference x, y, z with the axis y oriented vertically and the axis z oriented horizontally along the axis of the body. A suitable geometry of the magnet cavity is a rectangular prism of cross-sectional dimensions

$$2x_0 = 0.9m \, , \qquad 2y_0 = 0.6m \, . \tag{9.1.2}$$

In the direction of the axis z, the prism is assumed to be open at both ends, and its length $2z_0$ is going to be determined on the basis of the design computation. The design is approached by considering the magnets as sections of two-dimensional prismatic structures which can generate a perfectly uniform field [1].

Assume that two magnetic materials with quasi-linear demagnetization curves are available to the designer: a ferrite with remanence

$$J_1 \approx 0.4T \tag{9.1.3}$$

and a rare earth material like the Nd-Fe-B alloy with remanence

$$J_2 \approx 1.2T \, . \tag{9.1.4}$$

The ferrite remanence 9.1.3 is the same as the high field level listed in Eq. 9.1.1, and as a consequence, its use for the high field level requires a multilayered structure design. Consider first the $0.2T$ low field level and examine the two-dimensional structures capable of generating a uniform field

$$\mu_0 H_0 = 0.22T \tag{9.1.5}$$

oriented in the direction of the y axis. Figure 9.1.1 shows the two single layer yokeless structures resulting from the use of the two materials 9.1.3 and 9.1.4. The ferrite magnet shown in Figure 9.1.1(a) corresponds to the value

$$K_1 = 0.550 \, , \tag{9.1.6}$$

and the rare earth magnet shown in Figure 9.1.1b corresponds to the value

$$K_2 = 0.183 \, . \tag{9.1.7}$$

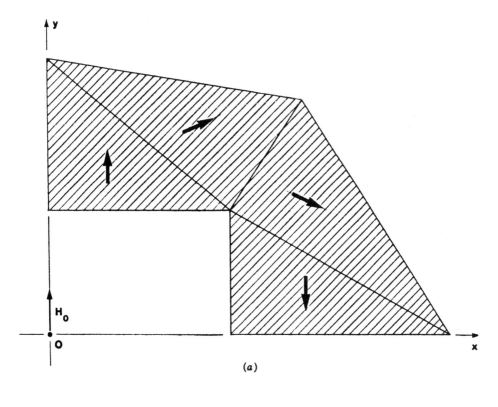

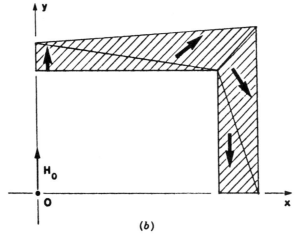

Fig. 9.1.1. Ferrite (a) and rare earth (b) yokeless magnets designed for a 0.22T field.

326 DESIGN CONSIDERATIONS

Table 9.1.1. Single layer two-dimensional magnets

	0.22T				0.44T			
	$K_1 = 0.550$ (Ferrite)		$K_2 = 0.183$ (Rare Earth)		- (Ferrite)		$K_2 = 0.367$ (Rare Earth)	
	M	W	M	W	M	W	M	W
Yokeless	0.113	6.86	0.058	2.31	-	-	0.095	5.67
Yoked	0.148	5.24	0.080	1.68	-	-	0.137	3.94
Hybrid	0.179	4.34	0.109	1.23	-	-	0.168	3.22

Table 9.1.1 lists the values of the figure of merit of the two yokeless structures of Fig. 9.1.1. The figure of merit of a yokeless single layer magnet with a rectangular cavity defined by an $x_0/y_0 = 1.5$ ratio is plotted in Fig. 9.1.2 versus K. One observes that with ferrite the value of K_1 corresponds to the optimum of M while the use of the rare earth material results in much smaller values of K_2 and M. Also shown in the table are the weights W in tons/meter for both ferrite and rare earth magnets for the cavity dimensions given by Eq. 9.1.2.

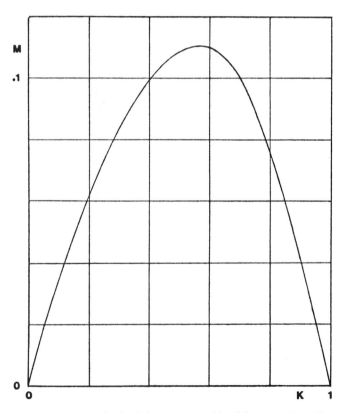

Fig. 9.1.2. Figure of merit of yokeless magnets with a 3:2 rectangular cavity versus K.

Figures 9.1.3 and 9.1.4 show the yoked and hybrid structures designed for the same values K_1 and K_2. The heavy lines in both figures represent an ideal $\mu = \infty$ external yoke. As expected, Table 9.1.1 shows that the figure of merit increases substantially for both yoked and hybrid magnets with the hybrid structures of Fig. 9.1.4 requiring the lowest amount of magnetic material [2].

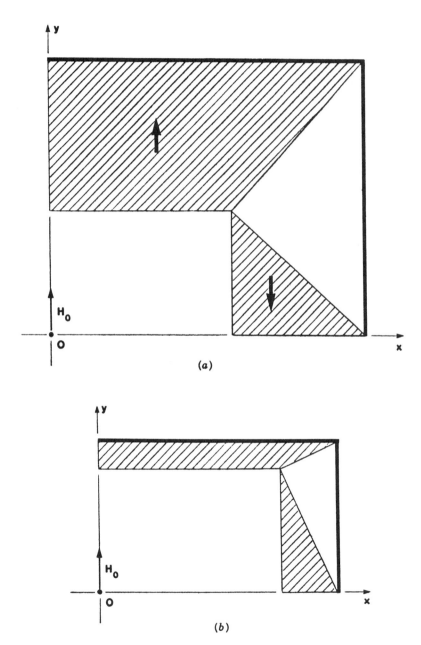

Fig. 9.1.3. Ferrite (a) and rare earth (b) yoked magnets designed for a 0.22T field.

328 DESIGN CONSIDERATIONS

In Fig. 9.1.5 the figure of merit of the hybrid structure is plotted versus K for the same $x_0/y_0 = 1.5$ ratio of the rectangular cavity. The maximum of M is found at $K = 0.5$ and it is equal to

$$M_{max} \approx 0.180 . \qquad (9.1.8)$$

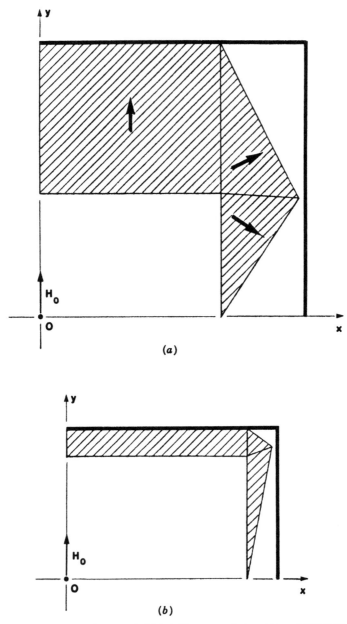

Fig. 9.1.4. Ferrite (a) and rare earth (b) hybrid magnets designed for a 0.22T field.

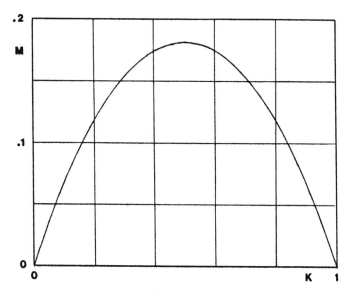

Fig. 9.1.5. Figure of merit of hybrid magnets with a 3:2 rectangular cavity versus K.

Consider now the design of two-dimensional structures capable of generating a uniform field

$$\mu_0 H_0 = 0.44 T \tag{9.1.9}$$

to be used for the $0.4T$ field level. A single layer design cannot make use of ferrite, and it must be based on the rare earth material for $K_2 \approx 0.367$. The three basic yokeless, yoked, and hybrid structures are shown in Figs. 9.1.6 and 9.1.7. The values of M and W are also listed in Table 9.1.1. Compared to the rare earth structures designed for $K_2 = 0.183$, the larger value of K_2 yields a significant increase of the figure of merit. Again the hybrid design results in the highest value of M and the lowest amount of magnetic material.

At the high field level 9.1.9, the multiple layer design approach offers several design solutions ranging from a multilayered ferrite structure to a combination of rare earth and ferrite layers. Consider, for instance, the yokeless structure of Fig. 9.1.6(a) and let A_{20} be the cross-sectional area of the magnetic material. Assume that a thin layer of rare earth material is removed on the outside of the structure and replaced by a layer of ferrite material such that the field intensity 9.1.9 within the cavity remains unchanged. Let A_r be the cross-sectional area of the removed rare earth material that is replaced by an area A_f of ferrite. If both layers are thin, the field inside the cavity remains unchanged if

$$A_r J_2 \approx A_f J_1 . \tag{9.1.10}$$

330 DESIGN CONSIDERATIONS

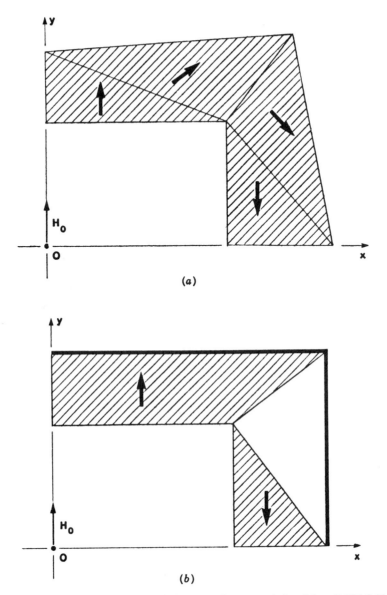

Fig. 9.1.6. Yokeless (a) and yoked (b) rare earth magnets designed for a 0.44T field.

The figure of merit of the new two-layer structure is

$$M = K_{20}^2 \frac{A_c}{(A_{20} - A_r) + A_f \dfrac{J_1^2}{J_2^2}}, \qquad (9.1.11)$$

where K_{20} is the value of K_2 for $A_r = 0$. By virtue of Eq. 9.1.10, Eq. 9.1.11 yields

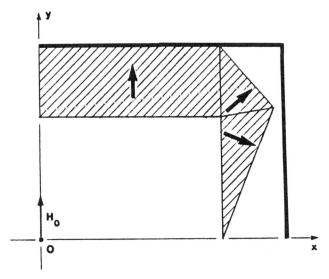

Fig. 9.1.7. Hybrid rare earth magnet designed for a 0.44T field.

$$M \approx M_{20} \frac{1}{1 - \left[1 - \dfrac{J_1}{J_2}\right] \dfrac{A_r}{A_{20}}}, \qquad (9.1.12)$$

where M_{20} is the figure of merit of the single layer, yokeless magnet designed for $K = K_{20}$,

$$M_{20} = K_{20}^2 \frac{A_c}{A_{20}}. \qquad (9.1.13)$$

Because of the thin layer assumption, $A_r \ll A_{20}$ and Eq. 9.1.12 results in

$$M > M_{20} \qquad (9.1.14)$$

as long as

$$J_1 < J_2; \qquad (9.1.15)$$

i.e., the figure of merit improves only if the thin layer of remanence J_2 is replaced by a material of lower remanence [3].

332 DESIGN CONSIDERATIONS

Assume

$$K_{20} > \frac{J_1}{J_2}, \qquad (9.1.16)$$

as in the case of the structure of Fig. 9.1.6(a). As A_r increases, a minimum value of the remaining area $A_{20} - A_r$ of the rare earth material is reached which generates a field intensity given by

$$K_2 J_2 = K_{20} J_2 - J_1. \qquad (9.1.17)$$

On the other hand the area A_f of ferrite necessary to generate a magnetic induction equal to its remanence J_1 diverges, and the figure of merit vanishes when the area of the internal rare earth material satisfies Eq. 9.1.17. Hence by virtue of Eqs. 9.1.14 and 9.1.15, as A_r increases, the figure of merit must reach a maximum before decreasing to zero as shown in Fig. 9.1.8.

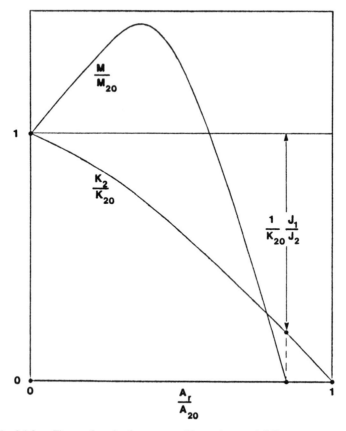

Fig. 9.1.8. Figure of merit of a magnet with two layers of different remanences.

By virtue of these considerations, the 0.44*T* field can be attained with a two-layer design that combines an inner yokeless structure of rare earth material and an outer structure of ferrite. In the schematic of Fig. 9.1.9 the inner layer is the same rare earth yokeless magnet of Fig. 9.1.1(b) designed for the value of K_2 given by Eq. 9.1.7, and the outer layer is a hybrid ferrite magnet designed around a rectangular cavity that contains the inner layer. The value of K_1 of the ferrite magnet is the same as the one given by Eq. 9.1.6. Thus in the schematic of Fig. 9.1.9 the two layers contribute equally to the field within the cavity.

The figure of merit of the double layer magnet of Fig. 9.1.9 is

$$M \approx 0.159 , \tag{9.1.18}$$

and the weights of the ferrite and rare earth layers are

$$\begin{aligned} W_{Fe} &\approx 6.46 \text{ tons}/m \\ W_{Nd} &\approx 2.31 \text{ tons}/m . \end{aligned} \tag{9.1.19}$$

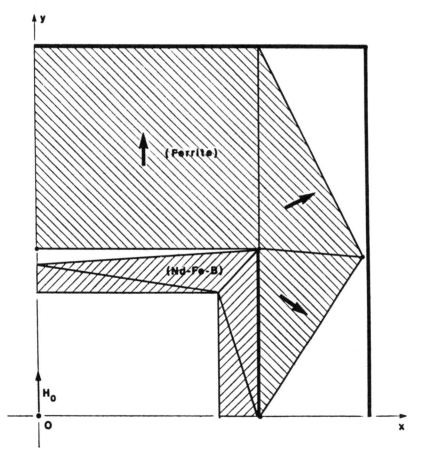

Fig. 9.1.9. Two-layer magnet with a rare earth yokeless inner layer and a ferrite hybrid outer layer.

334 DESIGN CONSIDERATIONS

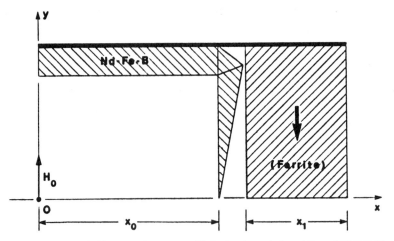

Fig. 9.1.10. Hybrid rare earth magnet with ferrite shunt designed for a 0.22 T field.

An even larger value of the figure of merit can be achieved by using the hybrid rare earth magnet of Fig. 9.1.4(b) as the inner layer of the double layered magnet. The hybrid structure of Fig. 9.1.4(b) can be made transparent to the field of the external ferrite magnet by replacing the infinite permeability yoke with the "active" yoke of magnetic material introduced in Fig. 6.5.5. Furthermore, if ferrite is used to close the flux of the rare earth magnet, the hybrid structure of Fig. 9.1.4(b) transforms into the structure of Fig. 9.1.10. The partial yoke of infinite permeability material is shunted by the rectangular ferrite component that satisfies condition 6.5.5. Thus the dimension x_1 of the ferrite shunt is

$$x_1 = K_2 \frac{J_2}{J_1} x_0 \approx 0.55 x_0 , \qquad (9.1.20)$$

where K_2 is given by Eq. 9.1.7

Figure 9.1.11 shows the hybrid two-layer structure with the external ferrite magnet designed for the same value of K_1 given by Eq. 9.1.6. The figure of merit of the structure of Fig. 9.1.11 is

$$M \approx 0.182 , \qquad (9.1.21)$$

and the weights of the ferrite and rare earth layers are

$$\begin{aligned} W_{Fe} &\approx 10.1 \text{ tons}/m \\ W_{Nd} &\approx 1.23 \text{ tons}/m . \end{aligned} \qquad (9.1.22)$$

As expected, the structure of Fig. 9.1.11 makes use of a much smaller amount of rare earth material than the structure of Fig. 9.1.9. Both structures of Figs. 9.1.9 and

9.1.11 achieve a substantial reduction of the amount of rare earth material compared to the hybrid single layer magnet designed for the same field ($K_2 = 0.367$) as listed in Table 9.1.1. The structure of Fig. 9.1.11 exhibits the highest value of the figure of merit, in spite of the much larger amount of ferrite material. In both structures the external yoke represented by the external heavy line in the schematics of Figs. 9.1.9 and 9.1.11 carries only the flux generated by the ferrite layer, because the flux generated by the rare earth layer closes within the internal layer itself.

Finally it is worthwhile pointing out that if

$$K_{20} < \frac{J_1}{J_2}, \qquad (9.1.23)$$

the high energy product rare earth material can be totally replaced with ferrite. If K_{20} is sufficiently small compared to J_1/J_2, the maximum of M in Fig. 9.1.8 is found outside the range

$$\frac{A_r}{A_{20}} \leq 1 ; \qquad (9.1.24)$$

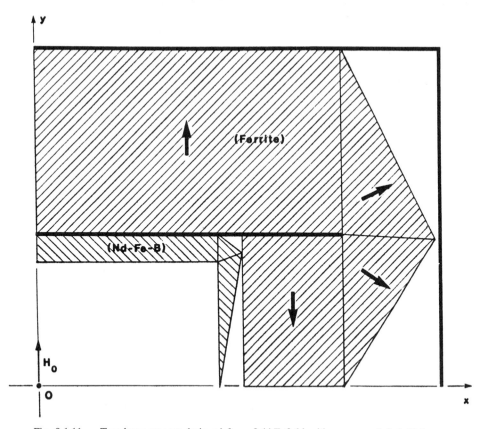

Fig. 9.1.11. Two-layer magnet designed for a $0.44\,T$ field with a rare earth hybrid inner layer and a ferrite hybrid outer layer.

336 DESIGN CONSIDERATIONS

i.e., the most efficient design is achieved with ferrite rather than the high energy product material, as shown by Table 9.1.1 for the 0.2 T field level.

9.2 COMPENSATION OF FIELD DISTORTION CAUSED BY MAGNET OPENING

The field distortion in an open prismatic magnet has been discussed in Chapter 8. In particular, the field inside the cavity has been analyzed by means of the spherical harmonic expansion of the scalar potential generated by the charges induced on the interfaces between the magnet components. The coefficients of the expansion provide the quantitative information about the field distortion within the region of interest which is the central region of the cavity. In the example of a square cross-sectional yokeless magnet, Section 8.4 has shown how the field distortion depends upon the length of the magnetic structure. In principle the distortion can be reduced to any level if the magnet is made long enough. This simple design approach may be used in applications with moderate uniformity requirements. However, as previously stated, in a magnet designed for NMR imaging, a tolerable field distortion is measured in ppm, and if the dimensions of the region of interest and the dimensions of the cavity cross-section are of the same order, the length of the magnet would be exceedingly large. Thus special compensation techniques have to be introduced in the magnet design.

Consider first the yokeless rare earth magnet of Fig. 9.1.1(b) designed for the value $K_2 = 0.183$. Because of symmetry, within a sphere of radius $\rho_0 < y_0$ whose center coincides with the center of the magnet, the spherical harmonic expansion 8.4.16 reduces to the terms with odd numbers l, j. Figure 9.2.1 shows the plotting of

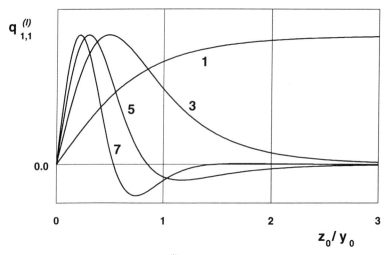

Fig. 9.2.1. Normalized coefficients $q_{1,1}^{(l)}$ versus z_0/y_0 for the yokeless magnet with a 3:2 rectangular cavity.

COMPENSATION OF FIELD DISTORTION CAUSED BY MAGNET OPENING 337

coefficients $q_{1,1}^{(l)}$ of the $l = 1,3,5,7$ terms of expansion 8.4.16 for the structure of Fig. 9.1.1(b). In the plotting of Fig. 9.2.1, the first maxima of coefficients $q_{1,1}^{(l)}$ are normalized to unity, and the length of the magnet $2z_0$ is normalized to the dimension $2y_0$ of the rectangular cavity cross-section.

As seen in Chapter 8, the $l = j = 1$ term provides the field at the center of the magnet. The finite length of the open magnet not only distorts the field but also reduces the field strength at the center of the region of interest. All $l > 1$ coefficients vanish for $z_0/y_0 \to \infty$. Thus as the length of the magnet increases, the distortion terms become smaller and smaller, and the field close to the center becomes more and more uniform. As shown by Eq. 8.4.16, for large values of l, the amplitude of the distortion terms increases rapidly as the distance ρ from the center increases. The goal of the designer is the elimination of the dominant harmonics to the extent that the desired degree of uniformity is achieved within a given distance ρ_0 from the center of the region of interest. The difficulty of eliminating the unwanted harmonics increases as the length $2z_0$ of the magnet decreases. Hence the compensation of the field distortion is a compromise between the choice of a minimum length of the magnet and a practical implementation of the compensation procedure.

The designer may follow two independent approaches: modify the distribution of the magnetized material or insert passive components of ferromagnetic material that become polarized by the field of the magnet.

The simplest way of eliminating the $q_{1,1}^{(l)}$ coefficients is a modification of the distribution of the magnetized material suggested by the observation that, for each value of $l > 1$, it is always possible to select at least two different lengths of the magnet that exhibit the same value of $q_{1,1}^{(l)}$.

Consider, for instance, coefficient $q_{1,1}^{(3)}$ that has a single maximum as shown in Fig. 9.2.1. It is possible to select two different magnet lengths $2z_0$, $2z_{0,1}$ that exhibit the same value of $q_{1,1}^{(3)}$ as indicated in Fig. 9.2.2. If z_0 is sufficiently large, so that the value of $q_{1,1}^{(3)}$ is small compared to its maximum, the value of $z_{0,1}$ is also small compared to z_0. If the two magnets of lengths $2z_0$, $2z_{0,1}$ have equal and opposite remanences, and they are superposed to each other in such a way that they have center O in common, the total value of $q_{1,1}^{(3)}$ is zero and the $l = 3$, $j = 1$ harmonic is eliminated. If $z_{0,1} \ll z_0$, the superposition of the two magnets is equivalent to a cut of width $2z_{0,1}$ in the plane $z = 0$, as indicated in the schematic of Fig. 9.2.2 [4,5].

The cancellation of the $l = 3$, $j = 1$ harmonic usually results in an increase of the amplitude of the higher order harmonics, and as a consequence, the design must include the compensation of several harmonics of increasing order. The procedure adopted in Fig. 9.2.2 can be extended, for instance, to the cancellation of the $l = 5$, $j = 1$ harmonic as long as it is done without re-introducing the $l = 3$, $j = 1$ harmonic. This can be accomplished by means of two additional cuts, in the planes where $q_{1,1}^{(3)}$ attains its maximum value, as indicated in Fig. 9.2.3. Likewise, additional cuts result in the elimination of increasingly higher order harmonics. Because $q_{1,1}^{(1)}$ increases monotonically with the length of the magnet, the elimination of each harmonic results in a decrease of the field within the region of interest, which can be compensated in the design phase by selecting either a higher value of the parameter K or a longer magnet.

Table 9.2.1 presents the values of coefficients $q_{j,1}^{(l)}$ for two lengths of the yokeless structure of Fig. 9.1.1(b) designed for $K_2 = 0.183$. The table shows that the short

338 DESIGN CONSIDERATIONS

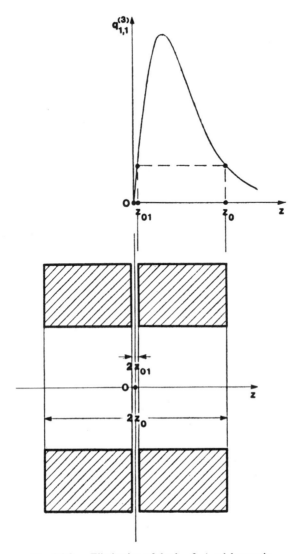

Fig. 9.2.2. Elimination of the $l = 3$, $j = 1$ harmonic.

magnet ($z_0 = 1.5 y_0$) exhibits a lower value of $q_{1,1}^{(1)}$ and significantly larger values of coefficients $q_{j,1}^{(l)}$ for $l \neq 1$. Thus wider cuts are necessary to eliminate the $l \neq 1$ coefficients, and as a consequence, the elimination of each harmonic in the short magnet results in a much larger reduction of the field within the region of interest. For instance the elimination of the $l = 3$, $j = 1$ and $l = 5$, $j = 1$ harmonics with the three cuts defined in the schematic of Fig. 9.2.3 requires the dimensions and positions of the cuts listed in Table 9.2.2 for the two values of z_0. Coordinates $z_{1,2}$, $z_{2,2}$ are the distances of the boundaries of the additional cuts in Fig. 9.2.3 from the plane $z = 0$. Also shown in Table 9.2.2 is the value $\tilde{H}$ of the field intensity at the center relative to the asymptotic value of the intensity for $z_0 = \infty$. $\tilde{H}$ in Table 9.2.3 is defined as

Table 9.2.1. Coefficients $q_{j,1}^{(l)}$ of the yokeless structure of Fig. 9.1.1(b) for $K_2 = 0.183$ and two values of length z_0 ($J_2 = 1$).

$\dfrac{z_0}{y_0}$	1.5	2.0
$q_{1,1}^{(1)}$	-1.6667 E-1	-1.7535 E-1
$y_0^2\, q_{1,1}^{(3)}$	3.8510 E-3	1.5001 E-3
$y_0^2\, q_{3,1}^{(3)}$	-8.5783 E-5	-2.6398 E-5
$y_0^4\, q_{1,1}^{(5)}$	4.7356 E-4	1.6140 E-4
$y_0^4\, q_{3,1}^{(5)}$	-8.2523 E-6	-2.6537 E-6
$y_0^4\, q_{5,1}^{(5)}$	8.1990 E-9	3.7982 E-9
$y_0^6\, q_{1,1}^{(7)}$	4.7639 E-6	1.3167 E-5

Table 9.2.2. Position and dimensions of three cuts of the yokeless structure of Fig. 9.1.1(b) for the elimination of $l = 3, j = 1$ and $l = 5, j = 1$ harmonics. Listing of $\tilde{H}$.

$\dfrac{z_0}{y_0}$	$\dfrac{z_{0,1}}{y_0}$	$\dfrac{z_{1,2}}{y_0}$	$\dfrac{z_{2,2}}{y_0}$	$\tilde{H}$
1.5	0.06601	0.40167	0.59130	0.679
2.0	0.02547	0.45624	0.52779	0.869

Table 9.2.3. Uniformity within a sphere of radius $y_0/2$ for increasing number of cuts in the yokeless structure for $K_2 = 0.183$ and $z_0/y_0 = 2$.

No. Cuts	Uniformity (ppm)	$\tilde{H}$
0	13,835	0.956
1	7,031	0.927
3	1,067	0.869
5	177	0.521

340 DESIGN CONSIDERATIONS

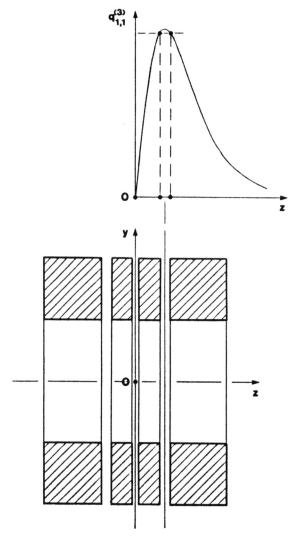

Fig. 9.2.3. Elimination of the $l = 3$, $j = 1$ and $l = 5$, $j = 1$ harmonics.

$$\tilde{H} = \frac{q_{1,1}^{(1)}}{(q_{1,1}^{(1)})_{z_0=\infty}}. \tag{9.2.1}$$

The change of the magnitude of the field within the region of interest, rather than the change of its orientation, is of importance for NMR imaging. In the case of a highly uniform field oriented in the direction of the axis y of the structures considered in Section 1, the magnitude of the field reduces to the magnitude of its y component.

In this limit the field uniformity can be defined by

$$\delta = \left| \frac{(H_y)_{\max} - (H_y)_{\min}}{(H_y)_{\mathrm{avg}}} \right|, \qquad (9.2.2)$$

where $(H_y)_{\mathrm{avg}}$ is the average value of H_y within the region of interest, and $(H_y)_{\max}$, $(H_y)_{\min}$ are the maximum and minimum values of H_y within the same region. δ is usually measured in ppm.

Define, for instance, the region of interest as the volume of a sphere concentric with the center of the magnet and radius

$$\rho_i = \frac{1}{2} y_0. \qquad (9.2.3)$$

The values of δ within the sphere of radius 9.2.3 are listed in Table 9.2.3 for increasing numbers of cuts.

Consider now the field within the cavity of the ferrite hybrid magnet of Fig. 9.1.4(a) designed for $K_1 = 0.550$. The hybrid magnet has a length $2z_0$ and is assumed to be open at both ends. Again consider the field within a sphere of radius $\rho_i < y_0$ whose center coincides with the center of the magnet. Figure 9.2.4 shows the plotting of coefficients $q_{1,1}^{(l)}$ of the $l = 1,3,5,7$ terms of the expansion of the scalar potential. The two plottings of Fig. 9.2.1 and 9.2.4 are similar to each other. A significant difference, however, is found in the normalized value of $q_{1,1}^{(1)}$. One observes that $q_{1,1}^{(1)}$ in Fig. 9.2.1 approaches its asymptotic value faster than in Fig. 9.2.4. This is due to the larger transversal dimensions of the magnetic structure of the hybrid magnet

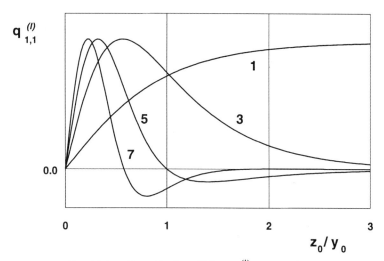

Fig. 9.2.4. Normalized coefficients $q_{1,1}^{(l)}$ versus z_0/y_0.

Table 9.2.4. Coefficients $q_{j,1}^{(1)}$ of the hybrid structure for two values of length z_0 and $K_1 = 0.55$.

$\dfrac{z_0}{y_0}$	1.5	2.0
$q_{1,1}^{(1)}$	-4.8130 E-1	-5.1864 E-1
$y_0^2\, q_{1,1}^{(3)}$	1.3324 E-2	6.3059 E-3
$y_0^2\, q_{3,1}^{(3)}$	-2.3988 E-4	-1.0353 E-4
$y_0^4\, q_{1,1}^{(5)}$	8.0638 E-4	4.7559 E-4
$y_0^4\, q_{3,1}^{(5)}$	-1.2589 E-5	-5.3479 E-6
$y_0^4\, q_{5,1}^{(5)}$	-1.5520 E-7	-4.3454 E-8
$y_0^6\, q_{1,1}^{(7)}$	-5.0298 E-5	1.0208 E-5

designed for $K = 0.550$. The higher value of K requires a longer magnet to maintain the same relative reduction of the field at the center of the magnet.

The coefficients $q_{j,1}^{(l)}$ of the hybrid magnet are presented in Table 9.2.4 for the lowest values of l and two values of the length $2z_0$. The approach of eliminating the $l > 1$ coefficients by means of a series of cuts of the magnetic structure can be used for the hybrid magnet also. For instance, Table 9.2.5 shows the dimensions and positions of three cuts in the ferrite structure of Fig. 9.1.4(a) designed to eliminate the $l = 3, j = 1$ and $l = 5, j = 1$ harmonics. The values of δ for increasing numbers of cuts are listed in Table 9.2.6.

In the two-layer structure of Fig. 9.1.9, the dimensions $2x_0'$, $2y_0'$ of the rectangular cross-section of the cavity of the external hybrid layer are related to the dimensions $2x_0$, $2y_0$ of the magnet cavity by

$$2x_0' \approx 3.66\, y_0, \qquad 2y_0' \approx 2.7\, y_0. \qquad (9.2.4)$$

Assume that the region of interest of the two-layer structure is the same sphere defined by Eq. 9.2.3. The sphere occupies a smaller fraction of the volume of the cavity of the hybrid layer. Thus for equal values of coefficients $q_{j,1}^{(l)}$ of internal and external layers, the external layer generates a more uniform field within the region of

Table 9.2.5. Position and dimension of three cuts of the $K_1 = 0.55$ hybrid structure for the elimination of $l = 3, \underline{j} = 1$, and $l = 5, j = 1$ harmonics. Listing of H.

$\dfrac{z_0}{y_0}$	$\dfrac{z_{0,1}}{y_0}$	$\dfrac{z_{1,2}}{y_0}$	$\dfrac{z_{2,2}}{y_0}$	$\tilde{H}$
1.5	0.12822	0.40381	0.77309	0.498
2.0	0.05911	0.48764	0.65522	0.768

Table 9.2.6. Uniformity within a sphere of radius $y_0/2 = y_0'/2.7$ for increasing numbers of cuts in hybrid structure for $K_1 = 0.55$ and length $2z_0' = 4y_0'$

No. Cuts	Uniformity (ppm)	$\tilde{H}$
0	19,106	0.938
1	7,571	0.885
3	603	0.770
5	63	0.556

interest, as shown by Table 9.2.6, which lists the values of δ of the external hybrid layer.

The compensation of the field distortion by means of cuts of the magnetic structure can be generalized to the use of materials with different remanences. For instance, the insertion of high energy product components may provide a way to compensate for the field distortion in a magnet built with a lower energy product material, like the ferrite magnet of Fig. 9.1.4(a).

As an example, consider again the hybrid magnet built with material of remanence J_1 and let $(q_{1,1}^{(3)})_0$ be the value of coefficient $q_{1,1}^{(3)}$ for the magnet of length $2z_0$ as indicated in Fig. 9.2.5. Following the same rationale that has resulted in the structure of Fig. 9.2.2, it is possible to select another short magnet of length $2\bar{z}_0$ positioned at

$$z = z_{0,3} \qquad (9.2.5)$$

magnetized in such a way as to generate a coefficient equal and opposite to $(q_{1,1}^{(3)})_0$. Assume, for instance, that the remanence $\vec{J}_1'$ of the short magnet is

$$\vec{J}_1' = 2\vec{J}_1 . \qquad (9.2.6)$$

If the length $2\bar{z}_0$ is small compared to $2z_0$ and the position z_{03} is larger than the abscissa of the maximum value of $q_{1,1}^{(3)}$, the length $2\bar{z}_0$ is given by

$$2\bar{z}_0 \left[\frac{d}{dz} q_{1,1}^{(3)} \right]_{z=z_{0,3}} \approx -\frac{1}{2}(q_{1,1}^{(3)})_0 . \qquad (9.2.7)$$

The superposition of the two magnets cancels coefficient $(q_{1,1}^{(3)})_0$, and by virtue of Eq. 9.2.6, it is equivalent to replacing the material of remanences $\vec{J}_1$ with a material of remanence

$$\vec{J}_2 = 3\vec{J}_1 \qquad (9.2.8)$$

which corresponds to the Nd-Fe-B alloy.

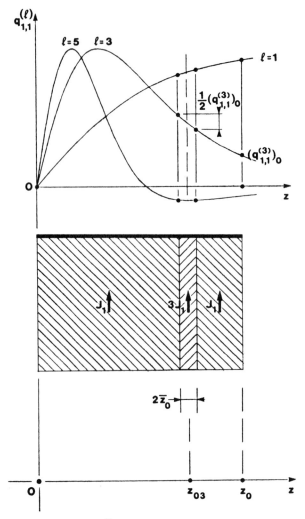

Fig. 9.2.5. Elimination of the $q_{1,1}^{(3)}$ harmonic by means of an insert of higher remanence material.

As shown in Fig. 9.2.5 the short section of remanence $\vec{J}_2$ may be located at the position of the minimum value of coefficient $q_{1,1}^{(5)}$, in which case the elimination of coefficient $q_{1,1}^{(3)}$ is achieved without affecting $q_{1,1}^{(5)}$. The procedure defined by the schematic of Fig. 9.2.5 can be extended to the elimination of increasingly higher order harmonics by replacing the J_1 remanence material at selected positions with short sections of material of remanence J_2.

Table 9.2.7 shows the dimensions and positions of the two inserts of remanence J_2 designed to eliminate the $l = 3$, $j = 1$ harmonic and the dimensions and positions of four slices designed to eliminate the $l = 3$, $j = 1$ and $l = 5$, $j = 1$ harmonics. Listed in the same table are the corresponding values of δ.

An important difference exists between the approach of Fig. 9.2.5 and the compensation approach with cuts defined in Fig. 9.2.2. The elimination of each harmonic with

Table 9.2.7. Position and dimension of Nd-Fe-B inserts in a ferrite hybrid magnet for $K_1 = 0.55$ and $z_0 = 2y_0$ to eliminate the $l = 3, j = 1$ and $l = 5, J = 1$ harmonic value of $\tilde{H}$.

No. Inserts	$\dfrac{z_1}{y_0}$	$\dfrac{z_2}{y_0}$	ppm	$\tilde{H}$
0	-	-	19,106	0.938
2	1.3261	1.4705	1,229	0.992
4	0.2858	0.2910	170	1.02
	1.7106	2.0000		

the approach of Fig. 9.2.5 results in an additional increase of $q_{1,1}^{(1)}$ regardless of the position of the inserts of higher energy product material. Table 9.2.7 shows that the two Nd-Fe-B inserts practically restore the nominal value of the field intensity at the center of the magnet.

9.3 COMPENSATION OF FIELD DISTORTION WITH FERROMAGNETIC MATERIALS

The correction of the field distortion discussed in Section 2 is an example of shimming technique based on the use of active components consisting of magnetized materials inserted in the magnetic structure. If the materials are rigid magnets transparent to the field of the structure, the designer enjoys the advantage that the shimming is not affected by the field of the magnet. A different situation is found in shimming techniques based on the use of elements of ferromagnetic materials inserted in the magnetic structure. These elements are polarized by the field of the magnet in a way that depends not only upon their geometry but also upon their position relative to the field.

Chapter 6 has shown the effect of the insertion of a thin $\mu = \infty$ layer in a magnetic structure. If the layer follows an equipotential surface of the field generated by a closed magnet, it has no effect on the field. However, any field where the layer is not an equipotential surface is affected by the presence of the layer. Thus, structures of high permeability layers may be designed to improve the uniformity of the field by filtering out the unwanted harmonics generated by the opening of a magnet.

Consider, again, for instance, the yokeless rare earth magnet of Fig. 9.1.1(b) designed for the value $K_2 = 0.183$ and assume a length $2z_0 = 4y_0$. The values of coefficients $q_{j,1}^{(l)}$ for the lowest order harmonics have been computed in Table 9.2.1. Figures 9.3.1 and 9.3.2 show the equipotential lines in the planes $x = 0$ and $z = 0$, respectively. The configuration of equipotential lines reflect the fact that within the cavity of the yokeless magnet of Fig. 9.1.1(b) the field intensity decreases along the axis z, and it increases along both axes x and y as the distances from the center of the magnet increase, as shown by Fig. 9.3.3 where the field intensity is plotted on the three axes. Figures 9.3.1, 9.3.2, as well as 9.3.3, show that the 3:2 ratio of the x and y

346 DESIGN CONSIDERATIONS

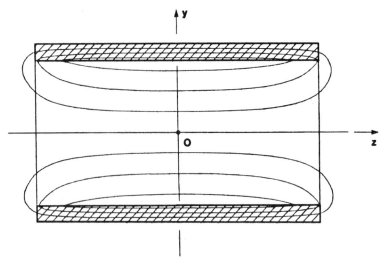

Fig. 9.3.1. Equipotential lines in the $x = 0$ plane of the $K_2 = 0.183$ yokeless magnet.

dimensions of the cavity results in a better uniformity of the field in the $z = 0$ plane compared to the field configuration in the $x = 0$ plane.

An improvement of the field uniformity can be expected by inserting two high permeability plates at the interfaces between magnetic material and cavity on the $y = \pm y_0$ planes. Ideally, assume that two $\mu = \infty$ rectangular plates S_1, S_2 of zero thickness are located on the $y = \pm y_0$ planes as indicated in the schematic of Fig. 9.3.4. The effect of the plates can be analyzed by applying the boundary integral equation method developed in Section 8.6 to the three-dimensional structure of the yokeless magnet of length $2z_0$. The surface charges σ on the two plates S_1, S_2 as well as the potentials of

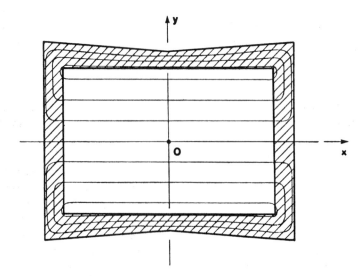

Fig. 9.3.2. Equipotential lines in the $z = 0$ plane of the $K_2 = 0.183$ yokeless magnet.

COMPENSATION OF FIELD DISTORTION WITH FERROMAGNETIC MATERIALS 347

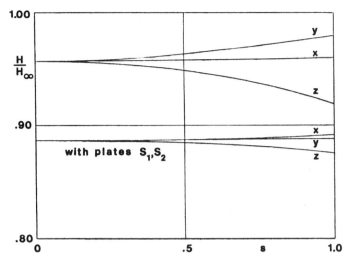

Fig. 9.3.3. Plotting of the field intensity on the three axes. The distances s from the center O are normalized to the dimension y_0. The intensity is normalized to the value of H for $z_0 = \infty$.

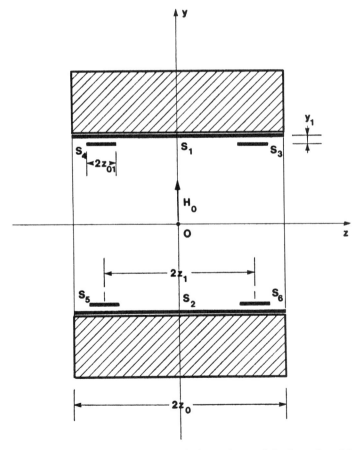

Fig. 9.3.4. Insertion of $\mu = \infty$ plates perpendicular to the y axis in the cavity of the magnet.

348 DESIGN CONSIDERATIONS

S_1, S_2 are provided by the numerical solution of the system of Eqs. 8.6.3. Then the values of σ on all the interfaces of the three-dimensional structure are used to compute the scalar potential Φ by numerical integration of the equation

$$\Phi = \frac{1}{4\pi\mu_0} \int \frac{\sigma}{\rho} dS . \qquad (9.3.1)$$

Furthermore, the values of σ are used in the numerical computation of coefficients $q_{j,1}^{(l)}$ of the spherical harmonic expansion of Φ for the quantitative analysis of the field properties within a spherical region of the magnet cavity with center at O [6].

The result of the numerical computation based on the system of Eqs. 8.6.3 and 9.3.1 is presented in Figs. 9.3.5 and 9.3.6, which show the equipotential lines in the planes $x = 0$ and $z = 0$ in the presence of the two plates S_1, S_2. The field intensity on the three axes is plotted in Fig. 9.3.3, and Table 9.3.1 shows that the insertion of the two plates results in a partial compensation of the field distortion at the expense of a reduction of the field intensity at the center of the magnet.

The significant reduction of the value of coefficient $q_{3,1}^{(3)}$ in the presence of the two plates S_1, S_2 implies an improvement of the field uniformity in the x, y plane, as apparent by comparing the plottings of H_y on the x and y axes in Fig. 9.3.3 with and without the plates. In the particular example of Fig. 9.1.1(b), the presence of the two plates results in a field configuration within the region of interest that is very close to that generated by a two-dimensional structure of infinite length along the axis x. The results of Fig. 9.3.3 and Table 9.3.1 suggest that further improvement of the field uniformity in the $x = 0$ plane can be achieved by modifying the geometry of the high permeability components in a direction parallel to the axis z. Consider, for instance, the

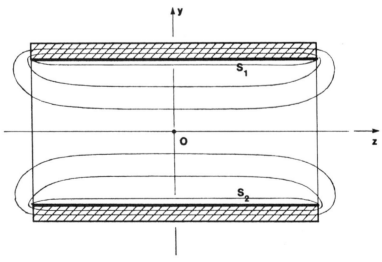

Fig. 9.3.5. Equipotential lines in the $x = 0$ plane of the $K_2 = 0.183$ magnet in the presence of plates S_1, S_2.

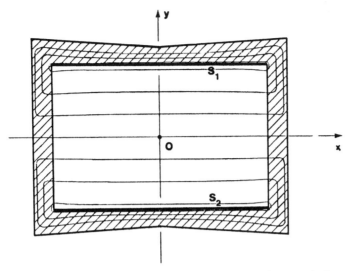

Fig. 9.3.6. Equipotential lines in the $z = 0$ plane of the $K_2 = 0.183$ magnet in the presence of plates S_1, S_2.

insertion within the cavity of a second $\mu = \infty$ layer of zero thickness in close proximity to the rectangular plates S_1, S_2. In the schematic of Fig. 9.3.4 the second layer consists of four rectangular plates S_3, S_4, S_5, S_6 symmetrically arranged with respect to the center of the magnet. The plates are assumed to have identical rectangular geometries with dimensions $2x_0$, $2z_{0,1}$, and they are positioned at $z = \pm z_1$ at a distance y_1 from plates S_1 and S_2, as shown in Fig. 9.3.4.

Figure 9.3.7 shows the value of coefficient $q_{1,1}^{(3)}$ versus the position z_1 of the four rectangular plates whose dimension $2z_{0,1}$ and distance y_1 are assumed to be

$$2z_{0,1} = 0.4y_0, \quad y_1 = 0.1y_0. \tag{9.3.2}$$

Table 9.3.1. Coefficients $q_{j,1}^{(l)}$ with and without plates S_1, S_2.

	No Plate	Plates S_1, S_2
$q_{1,1}^{(1)}$	-1.7535 E-01	-1.6248 E-01
$y_0^2 \, q_{1,1}^{(3)}$	1.5001 E-03	1.8313 E-04
$y_0^2 \, q_{3,1}^{(3)}$	-2.6398 E-05	1.0323 E-05
$y_0^4 \, q_{1,1}^{(5)}$	1.8140 E-04	1.0711 E-04
$y_0^4 \, q_{3,1}^{(5)}$	-2.6537 E-06	-1.8750 E-06
$y_0^4 \, q_{5,1}^{(5)}$	3.7982 E-09	-1.3780 E-07
$y_0^6 \, q_{1,1}^{(7)}$	1.3167 E-05	1.7431 E-05

350 DESIGN CONSIDERATIONS

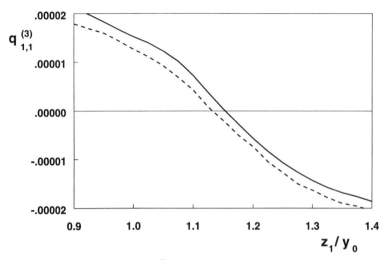

Fig. 9.3.7. Coefficient $q_{1,1}^{(3)}$ versus position of plates S_3, S_4, S_5, S_6.

The result of Fig. 9.3.7 is obtained by imposing the condition

$$\Phi_3 = \Phi_4 = \Phi_1, \qquad \Phi_5 = \Phi_6 = \Phi_2 = -\Phi_1 ; \qquad (9.3.3)$$

i.e., by assuming that plates S_3, S_4 are connected to S_1 and plates S_5, S_6 are connected to S_2. Then in Eqs. 8.6.3 the charge distribution satisfies the condition

$$\int_{S_1 + S_3 + S_4} \sigma \, dS = \int_{S_2 + S_5 + S_6} \sigma \, dS = 0 . \qquad (9.3.4)$$

The cancellation of coefficient $q_{1,1}^{(3)}$ occurs at the position z_1 of the plates of the second layer:

$$z_1 \approx 1.15 y_0 . \qquad (9.3.5)$$

A similar result is obtained with the arrangement of Fig. 9.3.8 where the surfaces S_3, S_4, S_5, S_6 are perpendicular to surfaces S_1 and S_2. Again the two sets of surfaces are assumed to satisfy solution 9.3.3. The plotting of coefficient $q_{1,1}^{(3)}$ versus the position of surfaces S_3, S_4, S_5, S_6 is also shown by the dashed line in Fig. 9.3.7 for a dimension y_1,

$$y_1 = 0.1 y_0 , \qquad (9.3.6)$$

and the cancellation of coefficients $q_{1,1}^{(3)}$ occurs at the position

$$z_1 \approx 1.13 y_0 . \qquad (9.3.7)$$

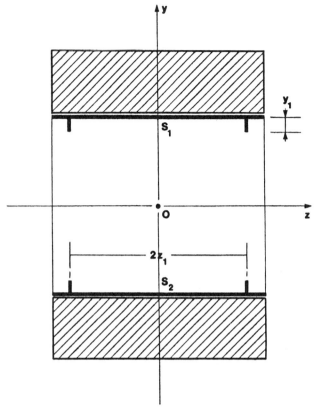

Fig. 9.3.8. Second set of surfaces S_3, S_4, S_5, S_6 perpendicular to S_1 and S_2.

9.4 COMPENSATION OF FIELD DISTORTION CAUSED BY NONZERO MAGNETIC SUSCEPTIBILITY

The compensation techniques discussed in the previous sections are based on the assumption that the magnet opening is the dominant cause of the field distortion within the region of interest. This is certainly true in the majority of applications. However, as the uniformity requirements become more and more stringent, the elimination of the field harmonics generated by the opening gets to a point where the residual distortion of the field is due primarily to the real magnetic characteristics of the materials and, in particular, to the demagnetization characteristics [7]. The analysis of the effect of a small nonzero value of the magnetic susceptibility χ_m has been developed in Chapter 7, and an example of field distortion caused by χ_m has been shown in Section 7.4.

To assess the magnitude of the effect of χ_m on the type of structures discussed in Section 1 of this chapter, consider again a section of length $2z_0$ of the yokeless rare earth magnet of Fig. 9.1.1(b), and assume that the field distortion due to the opening

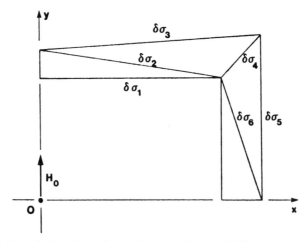

Fig. 9.4.1. Surface charges induced by χ_m in the $K = 0.183$ yokeless magnet.

has already been reduced to a small level by means of the compensating techniques described in Section 9.2. Then, as seen in Chapter 7, in a first order approximation, if $\chi_m \ll 1$, the effect of χ_m can be assessed by assuming that the magnetic material is polarized by the field computed in the perfect two-dimensional magnet of infinite length.

Figure 9.4.1 shows the first quadrant of the magnet cross-section with the surface charge densities $\delta\sigma_k$ induced by the field. For $\chi_m \ll 1$, coefficients $q_{j,1}^{(l)}$ of the spherical harmonics expansion are

$$q_{j,1}^{(l)} \approx \left[q_{j,1}^{(l)}\right]_{\chi_m = 0} + \delta q_{j,1}^{(l)}, \qquad (9.4.1)$$

where $\delta q_{j,1}^{(l)}$ are derived from the surface charge densities $\delta\sigma_k$ that result from the induced polarization of the magnetic material. As shown in Section 7.1, $\delta\sigma_k$ are proportional to χ_m, and for $K = 0.183$ their values are

$$\begin{aligned}
&\delta\sigma_1 \approx 0.707\,\chi_m J_2, \quad \delta\sigma_2 \approx -0.160\,\chi_m J_2, \quad \delta\sigma_3 \approx -0.555\,\chi_m J_2, \\
&\delta\sigma_4 \approx 0.779\,\chi_m J_2, \quad \delta\sigma_5 = \delta\sigma_3, \quad \delta\sigma_6 \approx 0.650\,\chi_m J_2.
\end{aligned} \qquad (9.4.2)$$

The normalized values of coefficients $\delta q_{j,1}^{(l)}$ for $l = 1, 3, 5$ are plotted in Fig. 9.4.2 versus the normalized value of z_0. Coefficients $l = j > 1$ reflect the most significant effect of the nonzero value of the magnetic susceptibility. As seen in Chapter 8, all terms $l > 1$ of the field computed in the ideal limit $\chi_m = 0$ vanish for $z_0 \to \infty$. This is no longer true for $\chi_m \neq 0$ as shown by the plotting of $q_{3,1}^{(3)}$ and $q_{5,1}^{(5)}$ in Fig. 9.4.2. The contributions of the individual charges $\delta\sigma_k$ do not cancel each other asymptotically, and one has

$$\begin{aligned}
\lim_{z_0 \to \infty} \delta q_{1,1}^{(1)} &\approx +1.54\,\chi_m J_2 \\
\lim_{z_0 \to \infty} \delta q_{3,1}^{(3)} &\approx -7.39 \cdot 10^{-3}\,\chi_m J_2 y_0^{-2} \\
\lim_{z_0 \to \infty} \delta q_{5,1}^{(5)} &\approx +8.33 \cdot 10^{-6}\,\chi_m J_2 y_0^{-4}.
\end{aligned} \qquad (9.4.3)$$

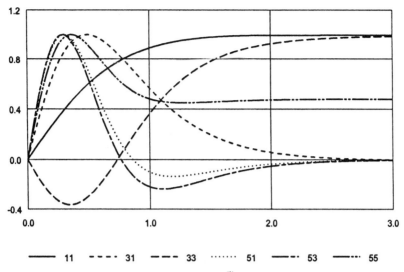

Fig. 9.4.2. Normalized coefficients $\delta q_{j,1}^{(l)}$ induced by χ_m versus z_0/y_0.

——— 11   - - - - 31   — — 33   51   —··— 53   —···— 55

The nonzero limit of $\delta q_{1,1}^{(1)}$ results in a reduction $\delta \vec{H}$ of the field intensity at the center of the magnet,

$$\lim_{z_0 \to \infty} \delta \vec{H} \approx -0.123 \chi_m \frac{J_2}{\mu_0} \vec{y}, \quad (9.4.4)$$

whose magnitude is of the same order as the results obtained in Section 7.4 for the yokeless square cross-sectional magnet.

The asymptotic value of the $l = 3$ term of the perturbation $\delta \Phi$ of the scalar potential is

$$\lim_{z_0 \to \infty} \delta \Phi_3 \approx -8.82 \cdot 10^{-3} \frac{\chi_m}{\mu_0 y_0^2} J_2 (y^2 - 3x^2) y, \quad (9.4.5)$$

which corresponds to a perturbation δH_y of the y component of the field intensity

$$(\delta H_y)_{l=j=3} \approx 2.65 \cdot 10^{-2} \frac{\chi_m}{\mu_0 y_0^2} J_2 (y^2 - x^2). \quad (9.4.6)$$

Term 9.4.5 is independent of the z coordinate. Thus its compensation can be accomplished by means of a z independent distribution of sources located outside the region of interest. Assume, for instance, that two distributions of magnetic dipoles are added to the structure of Fig. 9.1.1(b) in the plane $y = 0$ on the lines $x = \pm x_d$ as indicated in Fig. 9.4.3. The dipole distributions have a uniform dipole moment per unit length $\vec{p}_3$

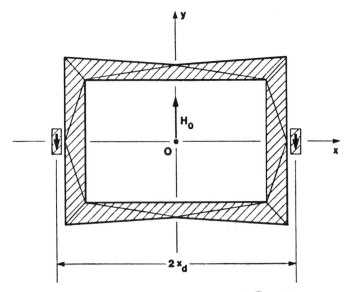

Fig. 9.4.3. Dipole distribution to cancel the $\delta q_{3,1}^{(3)}$ coefficient.

oriented in the direction of the axis y. The potential Φ_3 generated by the two dipole distributions is given by

$$\Phi_3 = \frac{p_3 y}{2\pi\mu_0} \left\{ \left[(x_d - x)^2 - y^2\right]^{-2} + \left[(x_d + x)^2 - y^2\right]^{-2} \right\}. \tag{9.4.7}$$

In a region close to the axis z such that

$$r = (x^2 + y^2)^{1/2} \ll x_d \tag{9.4.8}$$

Eq. 9.4.7 reduces to

$$\Phi_3 \approx \frac{p_3 y}{\pi\mu_0 x_d^2} \left[1 + \frac{1}{x_d^2}(3x^2 - y^2) \right]. \tag{9.4.9}$$

Thus close to the axis of the magnet, the field perturbation 9.4.6 is canceled if the dipole moment per unit length $\vec{p}_3$ is

$$\vec{p}_3 = -8.82 \cdot 10^{-3} \pi \chi_m \frac{x_d^4}{y_0^2} J_2 \vec{y}, \tag{9.4.10}$$

where $\vec{y}$ is a unit vector oriented in the direction of the axis y.

In the schematic of Fig. 9.4.3 the two line distributions of magnetic dipoles are added outside the structure of the yokeless magnet, where the field intensity is expected to be small compared to the intensity within the cavity. The dipole distribution can be implemented by means of two rods of a magnetic material of remanence $\vec{J}_3$ oriented parallel to the axis y, as indicated in Fig. 9.4.3. The area A_3 of the cross-section of the rod is given by

$$A_3 \vec{J}_3 \approx \vec{p}_3. \qquad (9.4.11)$$

Thus for $\chi_m \ll 1$ and $\vec{J}_3$ of the same order of $\vec{J}_2$, Eqs. 9.4.10 and 9.4.11 show that the rods have a cross-sectional area small compared to the area of the magnetic structure.

9.5 COMPENSATION OF MAGNETIZATION TOLERANCES

The partial compensation of the field distortion caused by magnet opening and magnetic characteristics of the materials is a design problem that is solved by means of the approaches described in the preceding sections.

The fabrication tolerances, on the other hand, result in random patterns of the field distortion that change from magnet to magnet. The most important cause of the field distortion is due to the magnetization tolerances of the magnetized materials, whose remanence may change from point to point within the same material both in magnitude and in orientation as well [8].

Consider, for instance, the fabrication of the structures presented in Section 1 of this chapter. In practice, each component of these structures is fabricated by machining and assembling a number of individual blocks of magnetic material. In a large magnet, such as those used in the examples of Section 1, the dimensions of the magnetic structure are large compared to the size of the individual blocks. Thus the fabrication of these large magnets may require a number of blocks in the order of hundreds or even thousands. Each of the blocks may exhibit a slightly different value and orientation of its remanence. Variations of a few percent are normally found in the magnitude of the remanence, whose orientation may also differ by a few degrees relative to the nominal easy axis.

The effect of the magnetization tolerances on the field inside the cavity may be greatly reduced because of statistical averaging of the random fluctuations of $\vec{J}_0$ over a large number of blocks. Moreover, a significant reduction of the field distortion may result from the insertion of high permeability components in the magnetic structure. Chapter 6 has shown that a high permeability plate acts as a shield or spatial filter of the field distortion caused by a nonuniform distribution of the remanence of the material. Thus the high permeability plates used in the schematic of Fig. 9.3.4 to offset the effect of the magnet opening, have the additional beneficial effect of reducing the effect of magnetization tolerances, particularly if their dimensions are larger compared to the size of the individual blocks of magnetic material.

The filtering effect coupled with statistical averaging may be enough to satisfy the uniformity requirements in the majority of industrial applications. In NMR imaging,

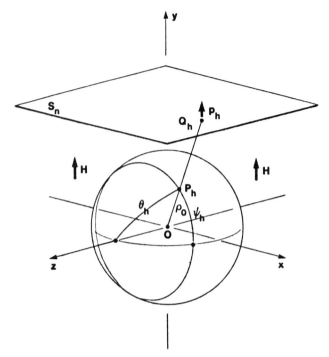

Fig. 9.5.1. Distribution of shimming dipoles.

however, magnet shimming for compensation of fabrication tolerances is usually necessary. Consider a fully assembled magnet that incorporates the compensation of the major causes of field distortion discussed in the preceding sections, and assume that a measurement of the field is performed over a sphere that includes the whole region of interest. As shown in Fig. 9.5.1, ρ_0 is the radius of the sphere, whose center can be chosen, for instance, to coincide with the nominal center F of the field configuration. In practice, the mapping of the field on the sphere is done by measuring the components of the field intensity $\vec{H}$ rather than the scalar potential Φ. In the case of a highly uniform field the dominant component of $\vec{H}$ within the sphere is oriented along the axis y, as indicated by the heavy arrows of Fig. 9.5.1. Then the mapping of the magnitude of $\vec{H}$ can be obtained by the measurements of the H_y component only, without any appreciable error.

Assume that H_y is measured over a large number of points P_h of the sphere whose position is determined by their angular coordinates θ_h, ψ_h in the spherical frame of reference ρ, θ, ψ used for the field mapping as indicated in the schematic of Fig. 9.5.1. Let $(H_y)_{\text{avg}}$ be the average of the measured values over the sphere. At each point P one can define the y component of the field perturbation as

$$\delta H_h = (H_y)_h - (H_y)_{\text{avg}}. \qquad (9.5.1)$$

Assume that points P_h are projected from center O on a surface S that encloses the sphere. In the ideal case of a closed cavity, S could be the same polyhedral surface of the cavity. As indicated in the schematic of Fig. 9.5.1 let Q_h be the projection of P_h on face S_n of the polyhedron. The basic approach for the compensation of the magnetization tolerances is to locate at each point Q_k a dipole moment $\vec{p}_k$, such that the field generated by the distribution of dipoles on surface S cancels the perturbation δH_h of the field of the magnet at each point P_h.

Assume at each point Q_k of surface S a dipole oriented in the direction of the axis y with a dipole moment equal to unity. The potential generated by the unit dipole at a point P_h of the sphere of radius ρ_0 is

$$\Phi_{h,k}^{(1)} = -\frac{1}{4\pi\mu_0} \vec{y} \cdot \nabla_h \left[\frac{1}{\rho_{h,k}}\right], \qquad (9.5.2)$$

where $\vec{y}$ is a unit vector oriented in the direction of the axis y, $\rho_{h,k}$ is the distance between points Q_k and P_h, and the gradient of $\rho_{h,k}^{-1}$ is computed at point P_h. Assume that points P_h coincide with the measuring points. The field perturbation δH_h in Eq. 9.5.1 can be canceled at each point P_h if moments $\vec{p}_k$ of the dipoles oriented in the direction of the axis y satisfy the system of equations

$$\sum_{k=1}^{n_0} H_{h,k}^{(1)} p_k = -\delta H_h \qquad (h = 1, 2, \cdots, n_0), \qquad (9.5.3)$$

where p_k is the nondimensional value of the moment of the dipole $\vec{p}_k$ and coefficients $H_{h,k}^{(1)}$ are the y components of the intensities generated by the unit dipoles at points P_h,

$$H_{h,k}^{(1)} = -\frac{\partial}{\partial y} \Phi_{h,k}^{(1)}. \qquad (9.5.4)$$

The selection of the number n_0 of measuring points on the sphere of radius ρ_0 is dictated by the condition that the difference between the values of the field perturbation at two adjacent points is small compared to the value of the perturbation at either point. The selection of the position of the correcting dipole relative to the measuring points on the sphere of radius ρ_0 is dictated by the condition that the correcting dipoles cancel the field perturbation at the measuring points without introducing an equally large or even larger perturbation at any other point of the sphere. For instance, if point Q_k is infinitely close to P_h, a dipole located at Q_k that cancels the field perturbation at P_h not only may be totally ineffective every place else but it may introduce higher order harmonics within the sphere of radius ρ_0. On the other hand, the larger the distance of the correcting dipoles from the measuring sphere, the larger the dipole moment of each dipole is going to be in order to satisfy the system of Eqs. 9.5.3. As the distance of the correcting dipoles from the measuring sphere increases, dipole moment p_k must increase approximately as the cube of $\rho_{h,k}$, in order to generate the same field at point P_h.

Each correcting dipole can be implemented with a small block of magnetized material. The order of magnitude of the volume V_k of each dipole is

$$O(V_k) \approx \frac{p_k}{J_s}, \qquad (9.5.5)$$

where J_s is the remanence of the dipole material. Hence the volume V_k also increase with the cube of distance $\rho_{h,k}$. The optimum position of the dipole is finally dictated by the available space within and without the magnet cavity. In particular, because of the transparency of the magnetic structure, the shimming of a yokeless magnet can be performed with correcting dipoles outside the magnets.

The final shimming of a magnet may involve a sequence of converging steps. After installing the correcting dipoles computed with the system of Eqs. 9.5.3, a new measurement of the field is performed to determine the new value of the field perturbation $\delta' H_h$ at the sampling points P_h. Then, by means of the same system of Eqs. 9.5.3, the values $\delta' H_h$ are used to determine the additional correction of the dipole distribution. This second correction may be implemented in several ways by adding additional blocks of magnetized material and/or changing the orientation of dipoles $\vec{p}_k$ wherever possible.

It is of importance to point out that a total cancellation of the field perturbation can never be achieved, regardless of the distribution of the correcting dipoles. At best the shimming procedure may yield a degree of field uniformity equal to or better than what is required for a specific application. The limitation of the shimming procedure is due to the fact that as the magnet is open to the surrounding medium, the correcting dipoles are also distributed on a surface S that cannot be closed and the opening of S determines the limit of the correction that can be achieved.

In the schematic of Fig. 9.5.2, A_0 denotes the area of the opening of the magnet of length $2z_0$. The part of the magnetic structure missing within the solid angle Ω_0, sub-

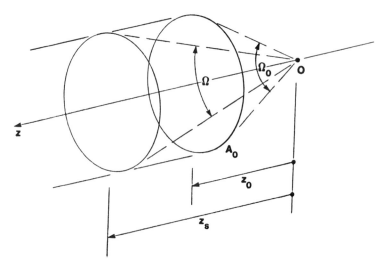

Fig. 9.5.2. Location of shimming dipoles outside the magnetic structure.

tended by area A_0 from center O of the magnet, determines the spectrum of the high order spatial harmonics of the field distortion caused by the magnet opening.

In the case of a yokeless magnet, the correcting dipoles can be located outside the magnetic structure, and as a consequence, they can be distributed over a length $2z_s > 2z_0$ of the cylinder of cross-section A_0, as schematically indicated in Fig. 9.5.2. The effect of this extended distribution of the dipoles is equivalent to a lengthening of the magnet, with the result of improving the field uniformity by shifting the spectrum of the field distortion towards higher order harmonics.

9.6 CONCLUDING REMARKS

This chapter has shown examples of the techniques available to the magnet designer to compensate for the field distortion caused by actual characteristics of magnetic materials and changes of the magnet geometry dictated by the requirements of individual applications. The types of compensating techniques described in this chapter are applicable as long as they are intended to correct relatively small distortions of the field within the region of interest. This is the case when the magnetic materials exhibit quasi-linear characteristics and the departure of the magnet geometry from an ideal closed structure does not affect in a substantial way the field within the region of interest. If, on the other hand, a major correction of the field is required, a large compensating structure would have to be added to the magnet, which is equivalent to saying that the magnet should be redesigned to meet the design requirements. For instance, if a magnet is designed on the basis of an open section of finite length of a two-dimensional structure, one cannot expect to start the design by choosing a length of the magnet that is short compared to the dimensions of the region of interest and then assume to be able to regenerate the field uniformity by means of techniques such as the one described in Sections 2 and 3 of this chapter. Such a short magnet would lose completely the properties that characterize the structures developed in this book. As previously stated, the size of a magnet opening is the most important factor to be considered in the initial selection of the magnet geometry, and its effect could be reduced to any desired level as long as the magnetic structure is large enough. In practice, the initial selection of the magnet dimensions will result from a trade-off of several factors, including cost and complexity of the compensating techniques. Magnet dimensions like the example considered in Section 3 of this chapter are well within practical limits, and as indicated by Tables 9.2.6 and 9.3.1 an initial distortion of the field within the region of interest of the order of 1% can be reduced by one or two orders of magnitude to a level of the order of 10^2 ppm by means of the techniques described in Sections 2 and 3 of this chapter. Typically, one can expect such a level of field uniformity in the random pattern of the field caused by the magnetization tolerances in a structure composed of a large number of individual blocks of magnetic material.

As a consequence, a uniformity of the order of 10^2 ppm may be considered as the limit of the field correction that the magnet designer may want to achieve in the design phase. Any improvement beyond this point should be left to a final shimming of the magnet following, for instance, the logic outlined in Section 5 of this chapter.

REFERENCES

[1] F. Bertora, M. G. Abele, H. Rusinek, Permanent magnet design for magnetic resonance imaging. Twelfth International Workshop on Rare-Earth Magnets and their Applications, p. 618-629, Jul 12-15, 1992, Canberra, Australia.

[2] M. G. Abele, Optimum design of two-dimensional permanent magnets, Technical Report No. 21, New York University, Oct 15, 1989.

[3] M. G. Abele, High field permanent magnets. Twelfth International Workshop on Rare-Earth Magnets and their Applications, p. 67-78, Jul 12-15, 1992, Canberra, Australia.

[4] M. G. Abele, H. Rusinek, Generation of a uniform field in a yokeless permanent magnet for NMR clinical applications. Technical Report No. 19, New York University, July 1, 1988.

[5] M. G. Abele, H. Rusinek, Yokeless permanent magnet for NMR application, Proceedings of 8th Meeting, Society of Magnetic Resonance in Medicine, Amsterdam, The Netherlands, p. 870, Aug 12-18, 1989.

[6] M. G. Abele, H. Rusinek, F. Bertora, Field computation in permanent magnets. *IEEE Transactions on Magnetics*, 28(1), p. 931-934, 1992.

[7] M. G. Abele, H. Rusinek, Effects of demagnetization characteristics in two-dimensional permanent magnets, Technical Report No. 22, New York University, June 1, 1990.

[8] M. G. Abele, R. Chandra, H. Rusinek, H.A. Leupold, E. Potenziani, Compensation of non-uniform magnetic properties of components of a yokeless permanent magnet, *IEEE Transactions on Magnetics* 25, p. 3904-3908, 1989.

APPENDIX I

Units

In the rationalized MKS system of units length (l) is measured in meters, mass (m) is measured in kilograms, time (t) is measured in seconds, and electric charge (Q) is measured in coulombs.

Table A.I.1 MKS electromagnetic units.

Quantity	Symbol	Dimensions	MKS Units
Current	I	$t^{-1}Q$	Ampere
Energy	W	ml^2t^{-2}	Joule
Magnetic flux	Ψ	$ml^2t^{-1}Q^{-1}$	Weber
Magnetic induction	B	$mt^{-1}Q^{-1}$	Tesla
Magnetic field intensity	H	$l^{-1}t^{-1}Q$	Ampere-turn/meter
Magnetic moment	p	$ml^3t^{-1}Q^{-1}$	Weber-meter
Magnetic potential (magnetomotive force)	Φ	$t^{-1}Q$	Ampere-turn
Permeability	μ	mlQ^{-2}	Henry/meter
Remanence	J	$mt^{-1}Q^{-1}$	Tesla
Susceptibility	χ_m	no dim.	

Table A.I.2 Conversion Table

Multiply MKS Units	by	cgs electromagnetic units (nonrationalized)
Ampere-turns	$4\pi 10^{-1}$	Gilbert
Tesla	10^4	Gauss
Ampere-turn/meter	$4\pi 10^{-3}$	Oersted
Joule	10^7	Erg
Weber	10^8	Maxwell

APPENDIX II

Legendre Polynomials of the First Kind

The Legendre polynomial of the first kind of order l is defined by the equation

$$P_l(\xi) = \frac{1}{2^l l!} \frac{d^l}{d\xi^l} (\xi^2 - 1)^l .$$

The polynomials can be derived from the recurrence equation

$$l\, P_l(\xi) - (2l - 1)\xi P_{l-1}(\xi) + (l - 1) P_{l-2}(\xi) = 0 .$$

The first nine polynomials are

$$P_0(\xi) = 1$$

$$P_1(\xi) = \xi$$

$$P_2(\xi) = \frac{1}{2}(3\xi^2 - 1)$$

$$P_3(\xi) = \frac{1}{2}(5\xi^3 - 3\xi)$$

$$P_4(\xi) = \frac{1}{8}(35\xi^4 - 30\xi^2 + 3)$$

$$P_5(\xi) = \frac{1}{8}(63\xi^5 - 70\xi^3 + 15\xi)$$

$$P_6(\xi) = \frac{1}{16}(231\xi^6 - 315\xi^4 + 105\xi^2 - 5)$$

$$P_7(\xi) = \frac{1}{16}(429\xi^7 - 693\xi^5 + 315\xi^3 - 35\xi)$$

$$P_8(\xi) = \frac{1}{128}(6435\xi^8 - 12012\xi^6 + 6930\xi^4 - 1260\xi^2 + 35)$$

$$P_9(\xi) = \frac{1}{128}(12155\xi^9 - 25740\xi^7 + 18018\xi^5 - 4620\xi^3 + 315\xi) .$$

APPENDIX III

Complete Elliptic Integrals

Complete elliptic integral of the first kind is defined as

$$F(k) = \int_0^{\pi/2} \frac{d\gamma}{(1 - k^2 \sin^2\gamma)^{1/2}} \ .$$

Complete elliptic integral of the second kind is defined as

$$E(k) = \int_0^{\pi/2} (1 - k^2 \sin^2\gamma)^{1/2} \, d\gamma \ .$$

Functions D and G are defined in terms of the elliptic integrals as follows

$$D(k) = \frac{1}{k^2}[F(k) - E(k)] \ , \qquad G(k) = E(k) - (2-k^2)D(k) \ .$$

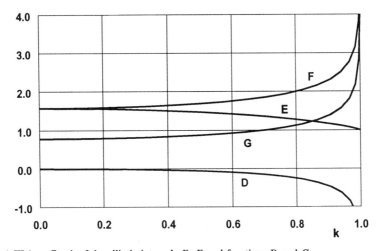

Fig. A.III.1. Graph of the elliptic integrals F, E and functions D and G.

APPENDIX IV

Coefficients of the Spherical Harmonics for the Yokeless Square Cross-Sectional Magnet

The coefficients $q_{i,j,1}^{(l)}$ of order $l = 5, 7$ and 9 of the square cross-section yokeless magnet designed for the value $K = 1 - 2^{-1/2}$ are plotted versus the normalized length z_0/r_0 for the five interfaces shown in Figs. 8.4.2 and 8.4.3. Solid lines indicate the total value $q_{j,1}^{(l)}$, dashed lines indicate interface $i = 1$, long dashes interface 2, dots interface 3, dash-dots interface 4 and dash-dot-dot interface 5.

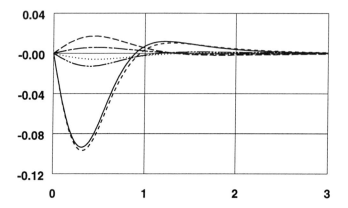

Fig. A.IV.1.   Coefficients of the spherical harmonics of order $l = 5, j = 1$.

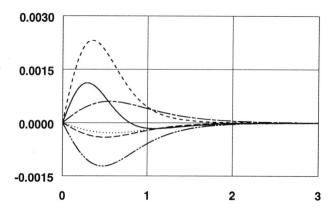

Fig. A.IV.2.   Coefficients of the spherical harmonics of order $l = 5, j = 3$.

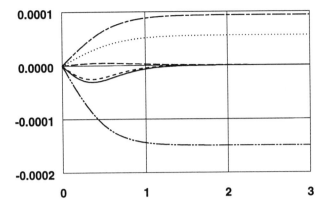

Fig. A.IV.3. Coefficients of the spherical harmonics of order $l = 5$, $j = 5$.

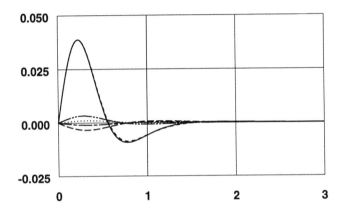

Fig. A.IV.4. Coefficients of the spherical harmonics of order $l = 7$, $j = 1$.

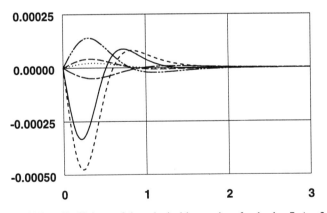

Fig. A.IV.5. Coefficients of the spherical harmonics of order $l = 7$, $j = 3$.

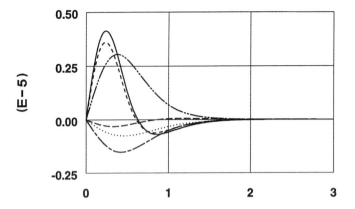

Fig. A.IV.6. Coefficients of the spherical harmonics of order $l = 7$, $j = 5$.

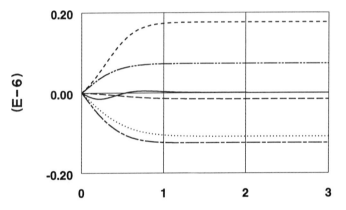

Fig. A.IV.7. Coefficients of the spherical harmonics of order $l = 7$, $j = 7$.

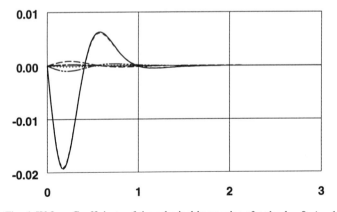

Fig. A.IV.8. Coefficients of the spherical harmonics of order $l = 9$, $j = 1$.

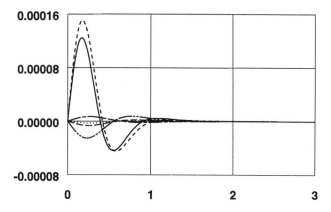

Fig. A.IV.9. Coefficients of the spherical harmonics of order $l = 9$, $j = 3$.

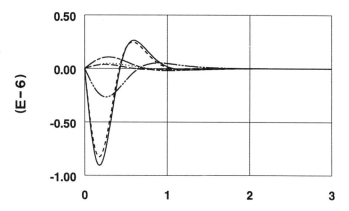

Fig. A.IV.10. Coefficients of the spherical harmonics of order $l = 9$, $j = 5$.

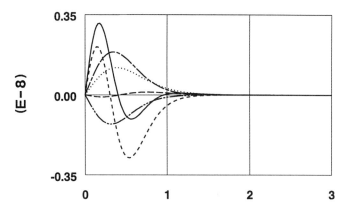

Fig. A.IV.11. Coefficients of the spherical harmonics of order $l = 9$, $j = 7$.

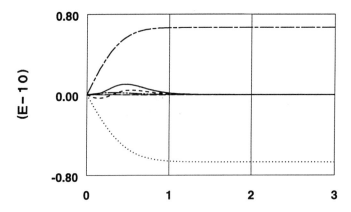

Fig. A.IV.12. Coefficients of the spherical harmonics of order $l = 9$, $j = 9$.

Index

Active yoke of magnetic material, 227

Biot-Savart law, 22
Boundary conditions, 5–7

Characteristic point F, 105
Clad magnet, 228
Closed cavity in a traditional magnet, 235
Coercive force, 34
   intrinsic, 35
Compensation of finite length effects, 338, 340, 343
Compensation of finite susceptibility in cylindrical magnet, 63
Compensation of magnetization tolerances, 355
Complete elliptic integrals, 277, 381
Cubic cavity:
   yoked magnet, 183
   yokeless magnet, 139
Curl in curvilinear coordinates, 9
Current in a closed loop, 22
Current density, 2
   surface density, 5
Currents, magnetic field generated by, 21
Curvilinear cavity boundary, 171
Curvilinear coordinates, 7–11
Cylindrical magnet:
   compensation of finite susceptibility, 63
   field confinement, 53, 55
   figure of merit, 126
   finite length, 276
   hollow cylinder, general solution, 49, 51
   multipole moments, 280
   nonzero susceptibility, 62–66
   octupole field, 285, 286
   yoked, 42–46
      multipole moments, 320–321

   yokeless, multipole moments, 285, 286, 294
Cylindrical polar coordinates, 10

Demagnetization characteristics, 35
Dipole, 24
Dipole moment, 28
Divergence in curvilinear coordinates, 9
Double layer magnet:
   hybrid magnet, 333, 335
   yoked magnet, 152
   yokeless magnet, 138

Electric current, field generated by, 21
Elliptic integrals, 277, 381
Energy of a magnetic field, 32
Energy product, 34
Equation of continuity, 2
Equivalence theorem, 25

Ferromagnetic layers, 208
   filter properties, 213
Ferromagnetic materials, field inside, 269
Field confinement, cylindrical magnet, 53, 55
Field distortion due to nonzero magnetic susceptibility:
   in yoked magnet, 254–257
   in yokeless magnet, 244–249
Field inside ferromagnetic materials, 269
Field in open yoked magnet, 318–320
Field in yokeless magnet, 346
Field in yokeless square cross-sectional magnet, 305, 306, 364–368
Figure of merit:
   cylindrical magnet, 126
   hybrid magnet, 200
   multilayered prismatic structures with different materials, 333, 334

Figure of merit (*Continued*)
  optimization, 331
  and point *F*, yokeless magnet, 131
  traditional magnet, 234, 235
  yoked magnet, polygonal cavity, 157, 158
  yokeless magnet, polygonal cavity, 127
Filter properties of ferromagnetic layers, 213
Finite length effects, 338, 340
Finite susceptibility in cylindrical magnet, 63
Flux lines:
  hexagonal cavity yoked magnet, 195
  square cavity yoked magnet, 156, 163
Flux of $\vec{B}$, 177
Flux plot:
  yoked magnet, 91–93
  yokeless magnet, 104, 105

Gradient in curvilinear coordinates, 8
Green's first identity, 18
Green's second identity, 19

Harmonic functions, 3
Hexagonal cavity:
  hybrid magnet, 195, 197, 198
  yoked magnet, 148, 194
  yoked magnet, flux lines, 195
  yokeless magnet, 112
Hollow cylinder, general solution, 49, 51
Hybrid magnet:
  computation of two-dimensional geometry, 200
  double layer, 333, 335
  figure of merit, 200
  hexagonal cavity, 195, 197, 198
  multilayered prismatic structures, 333, 335
  rectangular cavity, 328, 331
  rhombic cavity, 206
  unidirectional flux of $\vec{B}$, 212
  and yoked magnet, open 309, 316–320
Hysteresis cycle, 34

Image computational method, 309
Induction, magnetic, 1
Insertion of layers of different magnetic materials, 343
Insertion of layers of high permeability materials, 345, 357
Intensity of the magnetic field, 1

Interface between cavity and magnetic material, 88
Intrinsic coercive force, 35
Invariance theorem, 102

K, definition, 41

Laplace's equation, 3
Laplacian in curvilinear coordinates, 9
Layers of different magnetic materials, 343
Layers of high permeability materials, 345, 357
Legendre's functions, 17
Legendre's polynomials, 18, 380

Magnetic charge, 21, 24
  surface charge density, 25
  volume charge density, 3
Magnetic field generated by electric currents, 21
Magnetic field intensity, 1
Magnetic induction, 1
Magnetic susceptibility, 2
Magnetized sphere, 56
Multilayered magnet:
  nonzero magnetic susceptibility, 249
  prismatic structures with different materials, figure of merit, 333, 334
  prismatic structures of hybrid magnet, 333, 335
  yokeless magnet, 122
Multipole moments:
  cylindrical magnet, 280
  prismatic magnet, 294
  spherical magnet, 290, 291
  square cross-sectional magnet, 300–302
  yoked cylindrical magnet, 320–321
  yokeless cylindrical magnet, 294, 285, 286

Nonzero magnetic susceptibility:
  computation, 354
  equivalent structures, 258, 263
  field distortion in yoked magnet, 254–257
  field distortion in yokeless magnet, 244–249
  field in square cross-sectional magnet, 252, 257–265
  multilayered magnet, 249
  perturbation theory, 242

Octahedral cavity magnet, 138
Octupole field, cylindrical magnet, 285, 286

INDEX

Octupole moment, 29
One-dimensional magnet, 40
Open yoked and hybrid magnet, 309, 316–320
Open yoked magnet, magnetic field, 318–320
Operators in curvilinear coordinates, 7–11
 curl, 9
 divergence, 9
 gradient, 8
 Laplacian, 9
Optimization of the figure of merit, 331

Permeability, 2
Perturbation theory, nonzero magnetic susceptibility, 242
Poisson's equation, 3
Polarization, 1
Pole pieces of traditional magnet, 231, 234, 235, 237, 239
Position of point F and figure of merit, 131
Position of $\Phi = 0$ line, yoked magnet, 168

Quadrangular cavity yokeless magnet, 129
Quadrupole moment, 29

Rectangular cavity:
  hybrid magnet, 328, 331
  yoked magnet, 327, 330
  yokeless magnet, 228, 325, 330
Remanence, 1
Rhombic cavity:
  hybrid magnet, 206
  yoked magnet, 202
  yokeless magnet, 203
Rigid magnet, 36

Saddle point, 13
Scalar potential, 3
Shunt of magnetic material, 334
Similar geometrical boundaries of yokeless magnets, 116, 123
Sphere, magnetized, 56
Spherical harmonics expansion, 26, 27
Spherical magnet, multipole moments, 290, 291
Spherical polar coordinates, 11
Square cavity yoked magnet, 154
Square cavity yoked magnet, flux lines, 156, 163
Square cross-sectional magnet:
  multipole moments, 300–302
  nonzero susceptibility, 252, 257–265

yokeless, 120, 305, 306, 364–368
Surface current density, 5
  magnetic charge, 25
Surface integral method, 265, 314
Susceptibility, 2
  computation, 354
  cylindrical magnet, 62–66
  equivalent structures, 258, 263
  field distortion in square cross-sectional magnet, 252, 257–265
  field distortion in yoked magnet, 254–257
  field distortion in yokeless magnet, 244–249
  multilayered magnet, 249
  perturbation theory, 242
Tolerances, compensation of, 355
Traditional magnet:
  closed cavity, 235
  figure of merit, 234, 235
  pole pieces, 231, 234, 235, 237, 239
  traditional magnet design, 233
  traditional yoked magnet, 231
Transition wedges, 79–88
  yoked magnet, 89–95
Trapezoidal cavity yoked magnet, 166
Triangular cavity:
  yoked magnet, 169
  yokeless magnet, 118
Two-dimensional geometry:
  hybrid magnet, 200
  yokeless magnet, 108, 111

Unidirectional flux of $\vec{B}$, 177
  hybrid magnet, 212
Uniform field, theorem of existence, 72

Vector diagram, 88, 89, 97
  yoked magnet, 147
  yokeless magnet, 109
Vector potential, 2
Volume charge density, magnetic charge, 3

Wedges, transition, 79–88

Yoked magnet:
  computation of two-dimensional geometry, 146, 148
  cubic cavity, 183
  curvilinear cavity boundary, 171
  cylindrical, 42–46
  cylindrical magnet, multipole moments, 320–321

Yoked magnet (*Continued*)
  double layer, 152
  field distortion due to nonzero magnetic susceptibility, 254–257
  figure of merit of magnet with polygonal cavity, 157, 158
  flux plot, 91–93
  hexagonal cavity, 148, 194
    flux lines, 195
  and hybrid magnet, open, 309, 316–320
  open, 318–320
  position of $\Phi = 0$ line, 168
  rectangular cavity, 327, 330
  rhombic cavity, 202
  square cavity, 154
    flux lines, 156, 163
  transition wedges, 89–95
  trapezoidal cavity, 166
  triangular cavity, 169
  vector diagram, 147
Yoke of magnetic material, 227
Yokeless magnet:
  characteristic point F, 105
  computation of two-dimensional geometry, 108, 111
  cubic cavity, 139
  cylindrical magnet, multipole moments, 285, 286, 294
  double layer, 138
  field distortion due to nonzero magnetic susceptibility, 244–249
  figure of merit of magnet with polygonal cavity, 127
  figure of merit vs. position of point F, 131
  flux plot, 104, 105
  geometric invariance theorem, 102
  hexagonal cavity, 112
  magnetic field, 346
  multilayered, 122
  quadrangular cavity, 129
  rectangular cavity, 228, 325, 330
  rhombic cavity, 203
  similar geometrical boundaries, 116, 123
  square cross-section, 305, 306, 364–368
  square cross-sectional cavity, 120
  triangular cavity, 118
  vector diagram, 109